a pictorial history of

sea monsters

and other dangerous marine life

Also by the Author

A PICTORIAL HISTORY OF OCEANOGRAPHIC SUBMERSIBLES

a pictorial history of
sea monsters
and other dangerous marine life

by JAMES B. SWEENEY

BONANZA BOOKS · NEW YORK

Library of Congress Catalog Card Number: 76-185087

This edition is published by Bonanza Books
a division of Crown Publishers, Inc.
b c d e f g h
Manufactured in the United States of America

CONTENTS

PART ONE SEA MONSTERS

PART TWO OTHER DANGEROUS MARINE LIFE

SECTION ONE

Killers and Their Close Relatives

SECTION TWO
Dangerous Through Size

SECTION THREE
Stingers and Shockers of the Sea

SECTION FOUR
Poisonous to Eat

PART ONE

sea monsters

Giant sea devils of the genus Endoceras, *each about 20 feet in length, shown stranded on beach of about 400 million years ago.* Photo, Field Museum of Natural History, Chicago, Illinois.

I

Setting the Stage

The history of the Stone Age is replete with dinosaurs and flying reptiles having wingspreads of up to 25 or 30 feet. Although civilized man has never seen such creatures, today, in the light of geological findings, we accept their past existence as historical fact. But mention sea monsters and watch the look of incredulity with which the subject is received. In the first instance, there is enough fossil evidence to substantiate their existence; but in the case of sea monsters, the evidence has been mostly in the form of actual sightings and thus is considered doubtful.

Let us look at two historical statements and see how each has been accepted as true or false. The first concerns a sea serpent no less than 45 feet in length. "It had black eyes, and a long white mane, that hung down from the neck to the surface of the water. Besides the head and neck, we saw seven or eight folds, or coils, of this snake, which were very thick, and as far as we could guess there was a fathom (6-foot) distance between each fold." This was written in 1746 by Captain Lawrence de Ferry, who was not only a high-ranking officer in the Norwegian navy but also a citizen of considerable reputation.

The second statement concerns a man of little

or no worldly consequence prior to the time of his achievements. Nevertheless, he begged enough money to launch an expedition and set sail in a vessel hardly larger than a small offshore fishing boat. "Columbus, in 1492, crossed the Atlantic and discovered America."

Do we give equal credence to these two statements? No, for in the case of Columbus we have been drilled into accepting his voyage as historical fact. But the statement of Captain de Ferry evokes nothing more than skepticism.

On March 25, 1941 the British merchant steamer *Britannia*, out of her home port of Glasgow, was sunk in the South Atlantic by the German raider *Thor*. A sizable number of her crew managed to grab hold of a single piece of wreckage and either clamber aboard or cling to its side. This ordeal lasted for several days, during which, one night, a marine animal of some sort crawled aboard the makeshift raft, selected a screaming, fighting victim, and slid back into the sea with him. Shortly thereafter the monster reappeared, grabbed Lieutenant R. E. G. Cox, a member of the armed gun crew, and tried to make off with him. The valiant officer put up such a struggle that the beast let loose and returned to the depths. Later, after a rescue had been completed and the men hospitalized, doctors examined Cox's wounds and concluded that the beast had been a giant squid 23 feet long. As squids go, this is small, for they have been recorded at lengths of up to 80 feet. Now let us compare this with a report of 108 years earlier.

On May 15, 1833, a group of British naval officers, headed by Captain T. R. Sullivan, while aboard a small boat, "saw at a distance of a hundred and fifty to two hundred yards, the head and neck of some denizen of the deep, precisely like those of a common snake in the act of swimming, and the head so far elevated and thrown forward under and beyond it. The creature rapidly passed, leaving a regular wake, from the commencement of which to the fore part, which was not out of water, they judged its length to be about *eighty feet*. They set down the head at about six feet in length (considerably larger than that of a horse), and that portion of the neck which they saw at the time. There could be no mistake—no delusion and we were all perfectly satisfied that we had been favoured with a view of the true and veritable sea serpent." This account was published in the London *Zoologist* in 1847, and at that date all the officers involved were still living.

Are we less inclined to believe Captain Sullivan's statement than Lieutenant Cox's simply because one is more recent than the other? Is it less true that plesiosaurs or dinosaurs existed than that giant whales still exist simply because one is present and the other is not? The truth of a historical event cannot be judged by its nearness to the present. It can only be evaluated on the basis of facts. And to find these facts about sea monsters, we must go back into the musty caverns of time. Once we have explored the facts about sea monster sightings, we must come to the inevitable conclusion that *sea monsters did and do exist.*

We live today in a world of scientific sophistication, in an era of miracle drugs, transistors, and computers. Our society has become case-hardened against itself. Consequently, in entering the black depths of sea monster logic, we must show convincing truths if we are to be believed. For one recent example, let us go aboard a South African trawler at sea. The massive gill nets have just been pulled aboard, and an odd-looking fish has been spotted. Professional fishermen are a unique lot. Although many of them have had no formal education, they are generally guided, in making their judgments, by a true love of the sea. Consequently, even amongst the rough deckhands who pour millions of pounds of fish into the holds of ships every month, a peculiar fish is an item worthy of serious note. Fortunately, the captain of the trawler was of such a mind. He packed this strange "thing" in ice and brought it ashore for ichthyologists to study. This was in 1938, and the world of sea monsters shuddered from the impact. It turned out to be a long-extinct coelacanth, or lobe-finned fish, which lived about 300 million years ago, but which science had declared extinct. Therefore, it was believed that it could no longer be in existence. They stated that the last coelacanth had expired over 50 million years before our time. Yet here was one just caught—for all the world to see.

Let us look at other contemporaries of the coelacanth. The *Titanichthys* serves as a good example. Here was a massive fish thought to have gen-

Sea monsters are envisioned by many people in many different forms. This print was taken from a French book written in the early 1600s by Jacques Damase. Photo, Boston Public Library, Boston, Massachusetts.

erally attained a length of 20 to 50 feet. It had a protective layer of armored plate across its skull and a cavernous mouth armed with cutting edges formed in the upper and lower jawbones themselves. It resembled a crocodile, in that the upper part of the head moved, rather than just the lower jaw. Because of the fact that such a monster's skull would measure from 3 to 5 feet across, this action was akin to the cutting edge of a large pair of mechanical shears.

Science tells us that the *Titanichthys* just faded away. Yet the less-powerful coelacanth, equipped with nothing more favorable than fins, which are indeed the forerunners of those appendages later to become arms and legs, continued to exist despite what science has been telling us. There is an old axiom that only the fittest shall survive. Who then is to say, with positive authority, that the *Titanichthys* has proved itself less fit than the coelacanth?

Perhaps the truth of this matter can best be found in man himself. We fish the deep with hooks and nets that, in fact, never reach the true depths. Some commercial trawlers drag their nets at about 10 fathoms, or approximately 60 feet. Such techniques are hardly likely to catch a monster that probably lives in an environment 20,000 to 30,000 feet down. If we are to be conclusive in this matter, we must first learn to fish the ocean bottoms and look into the caverns, hills, fractures, and mountains that have never been explored in all the millions of years of man's history. Obviously, some scientists have probed the depths, but certainly the odds are very great against their catching a *Titanichthys*. Fish are not stupid. Ask any Izaak Walton. He will tell you of the wiles of bass, cod, tuna, or big game fish such as broadbills. Should we not attribute to *Titanichthys* the same instinct to survive? The whale made it; why not this armored version of a giant shark?

In geologic time, we find the first traces of marine animal life before there were such things as fish or amphibians or dinosaurs. Buried under immense eons of silt are the unmistakable remains of wormlike beasts that are now referred to as "little sea monsters," for that, in reality, is what they looked like.

There were no other animals in the sea, in the air, or on the ground, just these heavily armored sea worms, some of which attained a length of 2 feet. In truth, the protective scale they wore was an outside skeleton, for they possessed no backbone. Some were able to swim, others could only crawl, and the less highly developed seemed con-

Around 1600, one Hungarian artist thought a sea serpent looked like this. Photo, Library of Kiskunhalas, Hungary.

tent to burrow into bottom ooze and spend their entire existence without ever moving. This period is known as the Paleozoic era.

Most of these wormlike creatures are now called *trilobites*, meaning "three-lobed." All they could do if the need arose was to curl themselves into a tight ball to protect their soft underbellies. However, as the millenniums passed, this curl-shaped configuration became permanent and developed into what are today called *mollusks*. They had curved shells, out of which claws could be projected in order to feed upon less well developed trilobites. The world of the little monsters was beginning to wane. New and more formidable forms of life were developing. Starfish crawled over coral bottoms; jellyfish commenced to float through the water that covered the world. There were massive sharks (far larger than present-day whales), giant sea scorpions, huge turtles, horseshoe crabs, and sea serpents of incredible length.

On land during the Paleozoic era, there were big centipedes, giant cockroaches, and ungainly amphibians. Enormity seems to have been the key to survival. The bigger the species, the better it prospered. And out of such enormity evolved a ferocious creature of the sea, capable of breathing underwater, which possessed jaws so cavernous it could swallow two dinosaurs. About 100 million years later, the descendants of this sea monstrosity were to be known as *sharks*. They ruled the seas, devouring whatever living matter came within their reach. But competition was developing. Massive skates and rays got into the act; lungfish started to appear; and others came equipped with deadly protective devices for use against the world of sharks. They had armored bodies, stinging devices, and a tendency to crawl into caves and hide. Man has since looked upon the remains of these frightening oddities, for their skeletal residues are pressed into rocks that were once part of the floor of the ocean; but man has never been able to probe fully into the deep watery caverns wherein they dwelt. These caves, grottoes, blue holes, and subterranean openings are still submerged at a depth that is beyond investigative abilities.

Amphibians were also doing well for themselves. Reptiles weighing as much as eight or ten present-day elephants sampled the sea and liked what they found. They went to the oceans for their food and lugged their fat bellies onto land in order to enjoy the comforting warmth of the sun. The water planet was now abundant with monster fish and massive serpents. Nature adjusted the matter by arranging for fins to develop into paddles, and then into legs; gills evolved from underwater equipage into bladders and lungs that enabled the animals to breathe on land.

Perhaps all this started to take place about 200 million or more years ago. We cannot pinpoint the time zone closer than within 80 or 90 million years. But it is certain that the earth's environment was suitable for the rearing of amphibious monsters. Giant swamps, moist hot air, soft earth that spawned ferns and bushes 100 feet high—the entire setting was in need of brutes big enough to flatten these forests of wild growth and trample

During the Jurassic period, about 150 million years ago, these plesiosaurs (left) *and ichthyosaurs ruled the seas.* Photo, Field Museum of Natural History, Chicago, Illinois.

them into the earth, and some day man would find a need for this forgotten vegetation, which he called *coal.*

This was the age of reptiles. They swam across every ocean, migrated to every corner of the earth, and developed numerous ways of eking out an existence. Some were fish eaters; others hunted for live meat; many were vegetarians. For example, the *Plesiosaurus* was a seagoing, fish-eating reptile, with a barrellike body, an enormously long neck topped by a small head, and four diminutive flippers resembling limbs that never quite attained their intended growth. Interestingly enough, long before Columbus, which was hundreds of years before man knew of the previous history of such beasts, sailors described sea monsters resembling the *Plesiosaurus.* Proof of this may be found in the catacombs in Rome, where a sculptured coffin depicts a sea monster swallowing Jonah. The beast is obviously not a whale, as most of us have been taught to believe, but something resembling a plesiosaur. Henry Lee, who investigated this figure, published the following description during 1883 in his book *Seamonsters Unmasked:* "Amongst them also is Jonah and the 'whale.' A facsimile of this sculpture is one of the illustrations. . . . It will be seen that Jonah is being swallowed feet foremost, or possibly being ejected head first, by an enormous sea monster, having the chest and fore-legs of a horse, a long arching neck, with a mane

Coelacanth, a lobe-finned fish known to have lived over 300 million years ago. Drawing by the author.

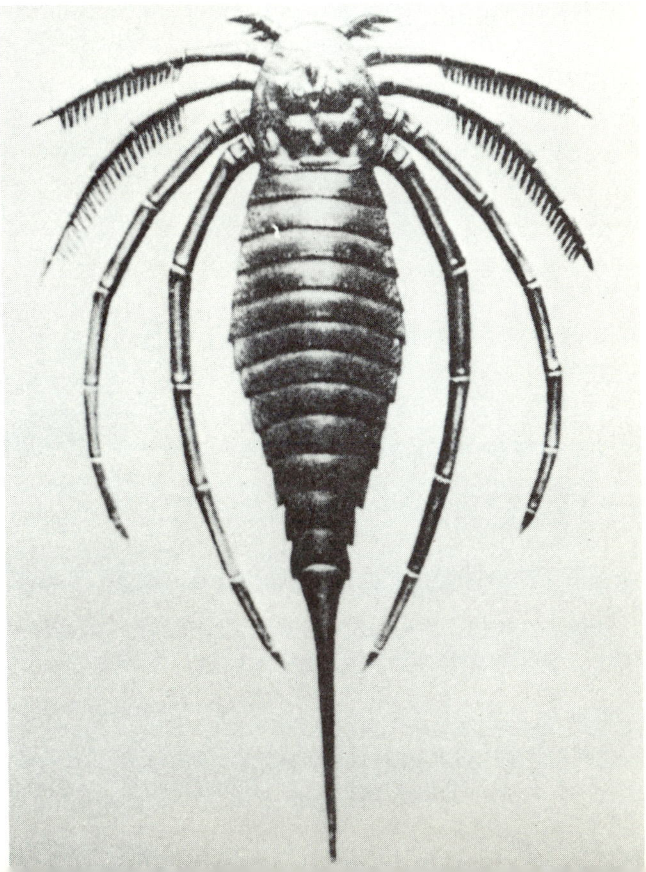

The giant sea scorpion stylonurus, belonging to a group named Eurypterus remipes, *grew to about 15 feet in length.* Photo, American Museum of Natural History, New York City.

Sea lizards, turtles, and flying reptiles were common at the close of the Mesozoic era, 100 million years ago. Photo, Field Museum of Natural History, Chicago, Illinois.

at its base, near the shoulders, a head like nothing in nature, but having hair upon and beneath the cheeks, the hinder portion of the body being of a serpent of prodigious length, undulating in several vertical curves. This sculpture appears to have been cut before the beginning and the middle of the third century, about A.D. 230, but it probably represents a tradition of far greater antiquity."

Then, too, there was the ichthyosaur, or fish-lizard. This specimen possessed a body that looked like a shark's, a tail fitted with something resembling a rudder, a short neck, and a long nose. His scaleless, 20-foot body retained what once might have been legs. Like the plesiosaur, he dominated the oceans for many millions of years—far longer than the known presence of mankind—and probed every portion of the sea during what scientists now refer to as the *Mesozoic era*.

Nothing lasts forever, not even the age of ichthyosaurs and plesiosaurs. They were followed by another seagoing monster, the mosasaur, or Mosa lizard, whose remnants were first discovered by man in the European section of the Meuse River. These 40-foot reptiles were powerful swimmers and mighty warriors of the deep. Their bones are to be found in the coal fields and chalk beds formed during the Ice Age. We know that their land-bred counterparts, such as the pterosaurs, the flying reptiles, and the brontosaurus, the thunderbeast, were killed off by time. Many, such as the elephantlike giants called *mammoths*, split into factions, some preferring the sea. We now know them as the familiar manatee, or sea cow.

As the age of the great mammals receded into history, saber-toothed tigers, mastodons, hairy elephants, and giant buffalolike creatures disappeared from the surface of the earth. But what happened deep in the sea has never been determined. Whether such monsters survived or became extinct, we do not know.

We have seen that there *were* sea monsters, very dangerous sea monsters, and that they possessed an inordinate skill at hiding from enemies. The question is: Did this ability to seek out and sneak into the depths preserve them from the fate

Ichthyosaurs (left) *and plesiosaurs* (right) *lived during the Mesozoic era.* Photo, American Museum of Natural History, New York City.

of the monstrous land mammals? Logically, it very well could have, for the seas *did* preserve a coelacanth. The same thing might have happened to a *Titanichthys* or a *plesiosaur*. Enough positive evidence remains that would allow us to assume that sea monsters could exist! For in 1930, a Dr. Anton Bruun brought up, from an ocean depth of almost 1,000 feet, the larva of an eel that was 6 feet long. The assumption that we must make is that all eel larvae grow at maturity to a size that is at least eighteen times longer than the grub form. These figures are based on the normal recorded patterns of those numerous species known as common eels. In some cases, such as that of the conger eel, the larvae have been known to extend as much as thirty times the grub length. Based on Dr. Bruun's discovery it can be assumed that a snakelike animal over 100 feet long could exist in the depths of the ocean.

Sea scorpions, members of the order Eurypterida, had powerful claws and grew up to forty times larger than present-day scorpions. Photo, New York Natural History Museum.

Oceanologists will admit that little is known about the growth patterns of creatures living at great depths, where food is scarce and there is no light. Such an environment is cold and barren. Yet that world is well populated with monstrous-looking fish, any one of which, if given a growth multiple such as that of common eel larvae, could be considered a sea monster. The popular names that have been given these miniature terrors of the deep adequately portray their appearance and viciousness. *Linophryne arborifer* is called a deep sea devil fish and looks the part; *Chauliodus sloanei* has been named viperfish. There are gulper eels, tiger fish, lionfish, hatchetfish, and hundreds of other frightening-looking fishes.

Many scientists object to the notion that a vertebrate animal can exist in the freezing cold depths indicated by Dr. Bruun. This argument can be refuted by pointing out the recent findings of the research vessel *Pillsbury*. That ship, while operating over the Puerto Rico trench, retrieved a fish of the genus *Bassogigas* from a depth of 26,132 feet. This was a new world depth record for the collection of a vertebrate, and it revealed one truly amazing factor: The fish, although existing in total blackness, still retained two small eyes.

The obvious question then presents itself: If this fish (or any living sea monster) existed in the terribly cold depths, why did its blood not freeze? This very interesting problem was studied by a group of marine biologists working in the fjords of Labrador. Fish were brought up from great depths, out of water that should have frozen even the hardiest of blood. The biologists compared the blood of these fish with the blood of surface fish,

Sea monster, as reported by an early English mariner. Photo, The Mariners Museum, Newport News, Virginia.

but they found no difference. However, one peculiarity was noted for all fish studied. They remained alive in superchilled water until coming in contact with ice crystals. When crystals were formed and did indeed come in contact with the fish, corresponding crystals formed within the fish and they froze to death. This, as we know, could never happen at any great depth because immense pressure does not allow for the formation of crystals.

Sea creatures have clearly indicated over the millenniums that they possess a great ability to survive by adaptation. One species of African lungfish (*Protopterus*) is known to be able to live out of water for as long as four years; man-eating sharks have managed to adapt themselves to living in fresh water; and salmon have learned to scale rapids in order that they might survive. Certainly, from our point of view at least, the eel, as well as other large marine animals, should be credited with this same ability to adapt and survive.

If nature can make the many wonderful adaptations that we know about, there must be many more just as spectacular about which we have no knowledge. It is in search of these superb subterfuges that we must look for our sea monsters. The question then remains: How well documented is the history of sea monster sightings? Discounting hysteria, fraud, and insanity, we find that any number of sightings were recorded by reliable people and witnessed by other responsible people. Let us give one typical example. This sighting, picked almost at random, typifies hundreds of others.

In 1847, there appeared in a London daily newspaper a long account, translated from Norse sea journals, of appearances made by a sea serpent. The statement said that the creature had recently and frequently been seen in the neighborhood of Christiansand and Molde, two cities along the Norwegian Sea.

"In the large bight of the sea at Christiansand it has been seen every year, only in the warmest of weather, and when the sea was perfectly calm, and the surface of the water unruffled. The evidence of three respectable persons was taken, namely, Nils Roe, a workman at Mr. William Knudtson's, who saw it twice there, John Johnson, merchant, and Lars Johnoen, fisherman at Smolen. The latter said he had frequently seen it, and that one afternoon in the dog-days, as he was sitting in his boat, he saw it twice in the course of two hours, and quite close to him. It came, indeed, to within six feet of him, and, becoming alarmed, he commended his soul to God, and lay down in the boat, only holding his head high enough to enable him to observe the monster. It passed him, disappeared, and returned; but, a breeze springing up, it sank, and he saw it no more. He described it as being about six fathoms long, the body (which was as round as a serpent's) two feet across, the head as long as a ten-gallon cask, the eyes large, round, red, sparkling, and about five inches in diameter: close behind the head a mane like a fin commenced along the neck, and spread itself out on both sides, right and left, when swimming. The mane, as well as the head, was of the colour of mahogany. The body was quite smooth, its movements occasionally fast and slow. It was serpent-like, and moved up and down. The few undulations which those parts of the body and tail that were out of the water made, were scarcely a fathom in length. These undulations were not so high he could see between them and the water.

"In confirming this account Mr. Soren Knudtson, Dr. Hoffmann, surgeon in Molde, Rector Hammer, Mr. Kraft, curate, and several other persons, testified that they had seen in the neighborhood of Christiansand a sea serpent of considerable size.

A Norwegian artist painted this version of a sea monster during the early 1800s to illustrate a book called Norwegian Folk Tales. Photo, Nasjonalgalleriet, Oslo, Norway.

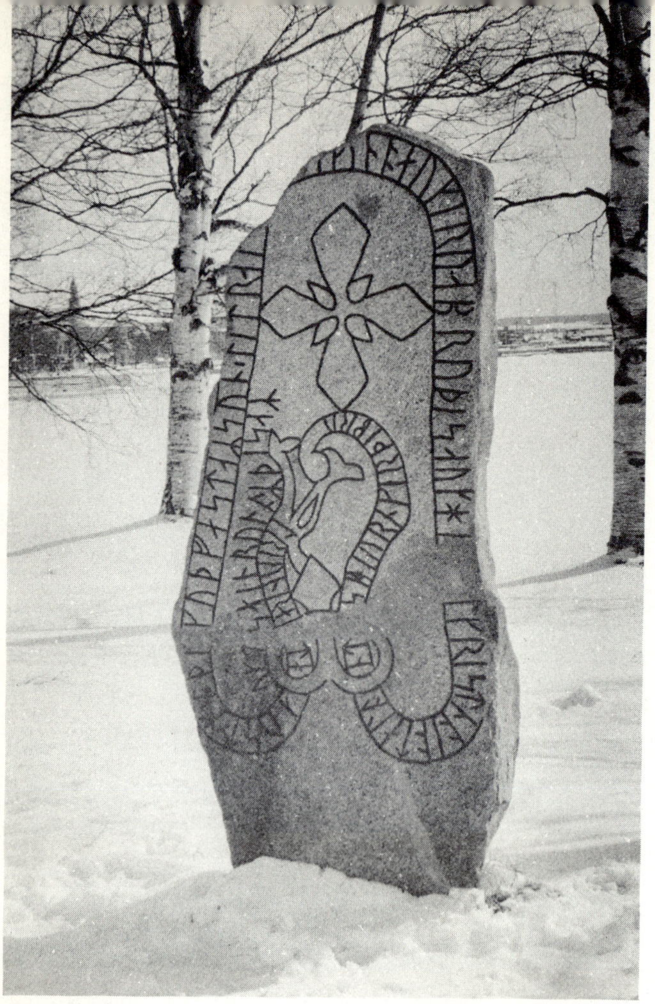

Mr. Goran Rosander, museum director, Foreningen Heimbygda, Østersund, Sweden, wrote that the northernmost runic monument is situated along the shore of Storsjön Lake, Froson Island. This ancient slab tells the story of a sea monster that was so long it reached around the entire island and was tied with a ribbon at head and tail. If the ribbon was broken, so went the story, the island would sink. This runic stone dates from about the year 1050.

"Mr. William Knudtson, and Mr. Bochlum, a candidate for holy orders, also gave their account of it, much to the same report; but some of these remarks are worthy of note for future comment. They say, 'its motions were in undulations, and so strong that white foam appeared before it, and at the side, which stretched out several fathoms. It did not appear very high out of the water; the head was long and small in proportion to the throat: as the latter appeared much greater than the former, probably it was furnished with a mane.'

"Sheriffe Gottsche testified to a similar effect. He could not judge of the animal's entire length; he could not observe its extremity. At the back of the head there was a mane, which was the same colour as the rest of the body."

This same account is given in Lee's book of 1883, *The Great Sea Serpent.* The account was confirmed for this book (using Lee's version as a basis for verification because it is eighty-eight years closer to the facts) by the Ministerie Van Landbou en Visserij, Sectie Bibliotheek en Alg, and by the librarian, Documentation Department, Museum Voor De Nederlandse Zeevisserij, both of the Netherlands. These confirmations stated that several versions of the report were in existence but that the basic facts, as given, are correct.

Although in some cases, it will not be stated because of space limitations, the same care will be taken throughout ensuing pages to verify each historic sea monster case cited.

Sea serpent, from a 1660 book on the history of Holland. Photo, Rijksmuseum, Leiden, Holland.

Many people were convinced that the ancient seas were populated with sea monsters like this one, shown in a book published in the 1800s. Photo, Library of Congress, Washington, D.C.

A marine monster depicted in an early book (about 1660) on Holland. Photo, Rijksmuseum, Leiden, Holland.

This early sixteenth-century woodcut from Cosmographia Universalis, *by Sebastian Munster, shows a wide variety of sea monsters.* Photo, Maritime Museum, Haifa, Israel.

Coelacanth. Photo, Bureau of Commercial Fisheries (Bob Williams).

This drawing of Jonah and the sea monster, which is in the Catacombs in Rome, appeared in a book published in 1883. Photo, Navy Library, Washington, D.C.

This piece of art depicts a giant squid that supposedly attacked a man who happened into a partially submerged cave. Photo, The Mariners Museum, Newport News, Virginia.

A giant squid (sometimes called a kraken), *or cuttlefish, is shown attacking an oceangoing ship.* Photo, Forbes Collection, Massachusetts Institute of Technology.

This early weather map shows man's awareness of the existence of sea monsters. Photo, TRW, Inc., California.

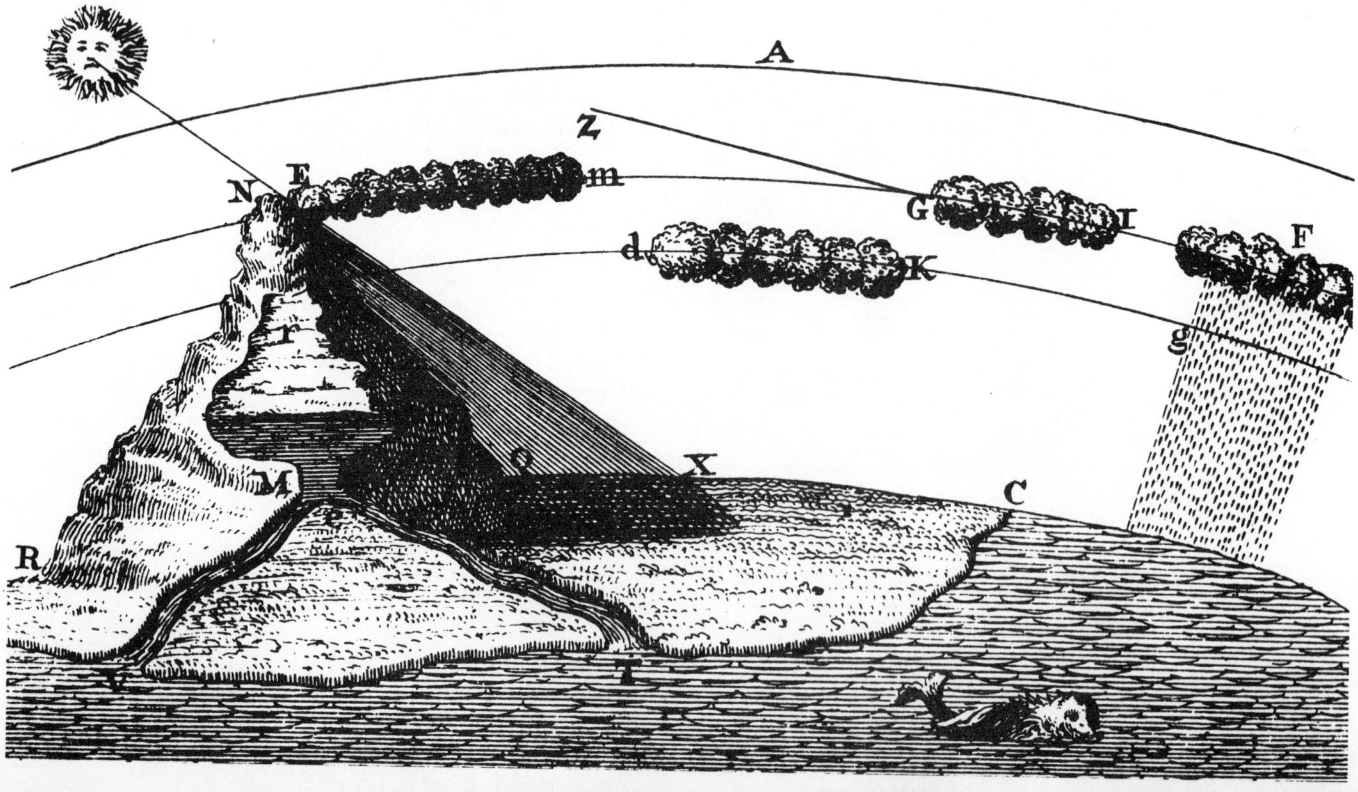

This runic stone from Sweden, dating back to about 1000 A.D.*, shows the story of man's fight against sea serpents. It presently stands in the Altuna Church, Uppland, Sweden.* Photo, ATA, Stockholm, Sweden.

These Zeuglodons, or primitive whales, might still have existed during man's early years on earth. Photo, Field Museum of Natural History, Chicago, Illinois.

2

From Mythology to A.D. 1500

Skeptics frequently attempt to explain away one of the great mysteries of the world by simply saying that sea monsters are *myths*. Ironically, in a historical sense, they might be more correct than is usually realized, for mythology is a collection of ancient stories describing gods, demigods, and great beasts possessing supernatural capabilities. Mythology is full of stories about sea monsters and sea serpents.

One such story concerns Perseus, son of the divine figure Zeus and Danaë. When Danaë was pregnant with Perseus, her father tried to kill her by placing her in a chest, which was then thrown into the sea. She somehow survived this ordeal and gave birth to Perseus, whom the gods loved. One day, in his wanderings, he came upon a fair maiden named Andromeda, who was about to be fed to a sea monster. Perseus "slew the sea monster by cutting off its head and leaving the headless body fall back into the sea."

This story was told by Apollodorus, an Athenian painter and sometime poet who lived from 430 to 400 B.C. And even in those times, it was considered an ancient narrative. Later, the Latin poet Ovid (43 B.C.–A.D. 17) put together a more detailed story of Perseus's sea monster episode, which is

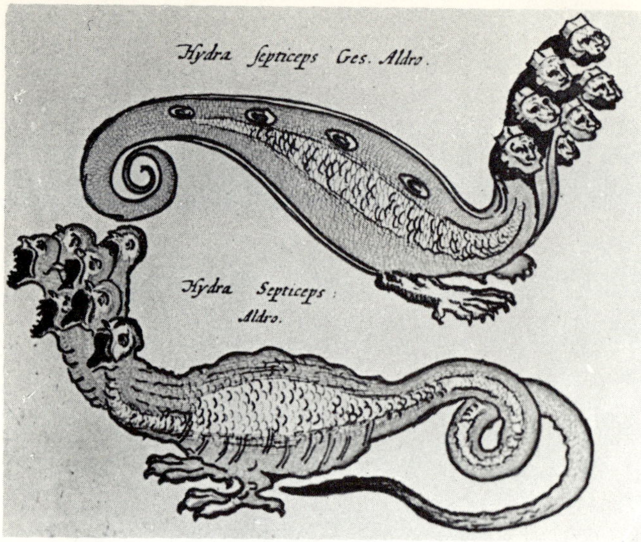

In Holland, in 1660, a document was published showing these seven-headed sea monsters that were long believed to have existed. Photo, Rijksmuseum, Leiden, Holland.

said to have been taken from several different sources.

The Homeric hymns are thought to have been composed sometime around 800 B.C. Homer took his materials from existent mythology. He relates a highly complicated love story about a beautiful girl named Scylla, who was turned into a horrible sea monster by a jealous goddess named Circe. Homer tells this story in Book 12 of the *Odyssey*.

"Her [Scylla's] voice indeed is no stronger than the voice of a newborn whelp, but a horrible sea monster is she, nor would anyone look on her gladly, not if it were even a god that met her. She hath twelve feet all dangling down, and six necks exceedingly long, and upon each a hideous head, and therein three rows of teeth set thick and close, full of black death. Up to her middle is she sunk far down in her hollow cave, but forth she holds her heads from the dreadful gulf, and there she fishes, swooping around the rock, for dolphins or sea-dogs, or what so greater beast she may find anywhere. . . . Thereby no sailors boast that they have fled scathless ever with their ship, for with each head she carries off a man, whom she has snapped-up from out the dark-prowed ship."

Homer follows up his sea monster story by telling of another horrible marine beast: "And beneath it (a specific section of shore line), mighty Charybdis sucks down the black water, for thrice a day she spouts it forth, and thrice a day she sucks it down in terrible wise. Never shouldst thou be there when she sucks the water down, for there are none who might save thee from thy bane. . . . But thou shouldst take heed and swiftly drawing nigh to Scylla's rock drive the ship past, since it is far better to mourn but six of thy company in the ship, than all in the selfsame hour."

It might well be that Homer was alluding to what we now commonly refer to as a giant octopus, which has eight arms, a bulbous head, and large, glaring eyes. Our ancient ancestor could well have been relating a warning against the presence of just such a 60- or 80-foot sea monstrosity.

Herodotus, a Greek historian who lived in the fifth century B.C., has often been cited for his

Plesiosaurs of the Mesozoic era. Photo, Field Museum of Natural History, Chicago, Illinois.

Herodotus spoke of these and other partly human monsters that lived in the sea. Photo, Mitchell Library, Australia.

objectivity in reporting. His works also reflect a good deal of skepticism concerning older stories that were related to him as fact. In other words, he possessed all the requirements for a good reporter. Although he displayed these sterling qualities, he told of numerous sea monsters or monsters related in some fashion to the sea as he wandered about in Persian and Egyptian lands. He began one passage by saying, "I don't believe it," and then went on to write about a one-eyed sea monster that was part fish and part lion, a symbol that long endured on ancient sea charts. He then displayed further misgivings in the matter by quoting tales of native Ethiopians (in whose country he appears to have been at that moment) about seafaring people who were dogheaded, about others who were no-headed with a single eye in the middle of their chests, and still again, about men who were part sea serpent and part eagle.

The Greek poet Hesiod, who lived several hundred years before Herodotus, also wrote of such beasts. Consequently, it can be assumed that as a student, Herodotus had closely studied the writings of Hesiod and accepted them as truth. Furthermore, a contemporary of Herodotus, a Greek philosopher named Empedocles, who is considered to be the founder of what we now know as physics, related the amazing beliefs that parts of gigantic animals came out of the sea and wandered about. Heads, tails, and massive body sections floundered around until finding themselves compatible parts. All this took place in a land where human parts, such as legs, hands, heads, and torsos, also flopped about. As a result, some very strange combinations were formed: sea monsters with human heads, humans with sea monster heads, or perhaps sea serpents with human hands and feet.

In Idyll XXI, thought to have been written by Theocritus (about 270 B.C.), since labeled "The Mariner's Dream," we discover a story (as translated in 1889) which clearly portrays man's awareness that extraordinary beasts existed in the sea. The episode commences in this fashion. "Once there were two fishermen who lived in a hovel, and for a bed they had nothing but dried sea moss.

This sea monster is from an old chart (c. 1560). It was said to have resembled the one described by Theocritus in the story "The Mariner's Dream." Photo, The Mariners Museum, Newport News, Virginia.

In the sod-covered hut where they lived were only the barest necessities for plying their trade: a few badly fabricated ropes, a pair of crude oars, mended nets, and a small selection of lobster pots woven from reeds.

"Here was all their toil and all their wealth," so the verse ran, and one of the mariners, named Asphalion, asked his friend to interpret the horror of his most recent dream.

"As I lay sleeping, amidst my labors of the sea, I found myself busy upon a rock, and there I squatted and observed the fish and kept casting the bait with my rod.

"And suddenly a fat fish nibbled at my bait; for in sleep, dogs dream of chasing massive bears, but I dream only of fish. So he was firmly hooked, and his blood was being let run, and my rod was bent double from the weight of him.

"So with my two arms and hands I struggled, and was hard put to hold the rod. The question was in my mind as to how I was going to land him with such slender gear? Then my line slackened and I commenced to take in the catch.

"My toil came to an end with the gleeful sight of my prize. Lo! I drew in a golden sea monster! a sea animal all plated over thickly with gold. At once fear gripped me, lest he be some monstrous fish belonging to Poseidon, or perhaps some jewel of the Amphitrite. With great care I unhooked my monstrous catch, for fear that the hooks might remove some of the gold from its mouth. Then I pulled him on to the shore with ropes and pledged that never again would I put to sea, but would forever live on land and rule over my vast amount of gold."

The friend called upon to interpret Asphalion's dream, warned the poor fisherman to continue seeking "ordinary fish of the flesh," but to keep an eye out for his golden sea monster. Obviously, here was the beginning of a sea monster story that very likely became more factual, more authentic, and more physical with each telling.

It therefore comes as no surprise to learn that Berossus, a Babylonian priest-historian, who was prominent a short time after Theocritus, wrote that "at a time when men were sunk in barbarism, there came out of the Erythrean Sea [now the Gulf of Persia], and landed on the shore of Babylon, a golden sea monster named *Oannes*, which had the body and head of a monstrous fish. But above the monster was the head of a man, and below the tail of the fish were human feet. It also had human arms, a human voice, and human language. This strange monster sojourned among the common people during the day taking no food, but retiring to the sea at night; and it continued for some time thus to visit them, teaching them the arts of civilized life, and instructing them in science and religion."

In a book written in 1883, by Henry Lee, called *Sea Fables Explained*, it is noted that this sea monster Oannes was also known by numerous other names, one of the more common being Dagon, who was the great fish-god of the Phoenicians. In ancient Hebrew, *dag* means a "male fish"; *aun* and *oan* were two of the words meaning "Noah." Regardless of biblical connotations, Dag-aun, or Da-oan, was portrayed in two ways.

"The more ancient image," writes Henry Lee, "was that of a man issuing from a fish, as described of *Oannes* by Beros[s]us; but in later times it was varied to that of a man whose upper half was human, and the lower parts those of a fish. The image of *Dagon* which fell upon its face before 'the ark of the God of Israel,' was probably of this latter form, for we read (1 Samuel v. 4) that in its fall, 'the head of *Dagon* and both the palms of his hands were cut off upon the threshold: only the *stump* of *Dagon* was left to him. This

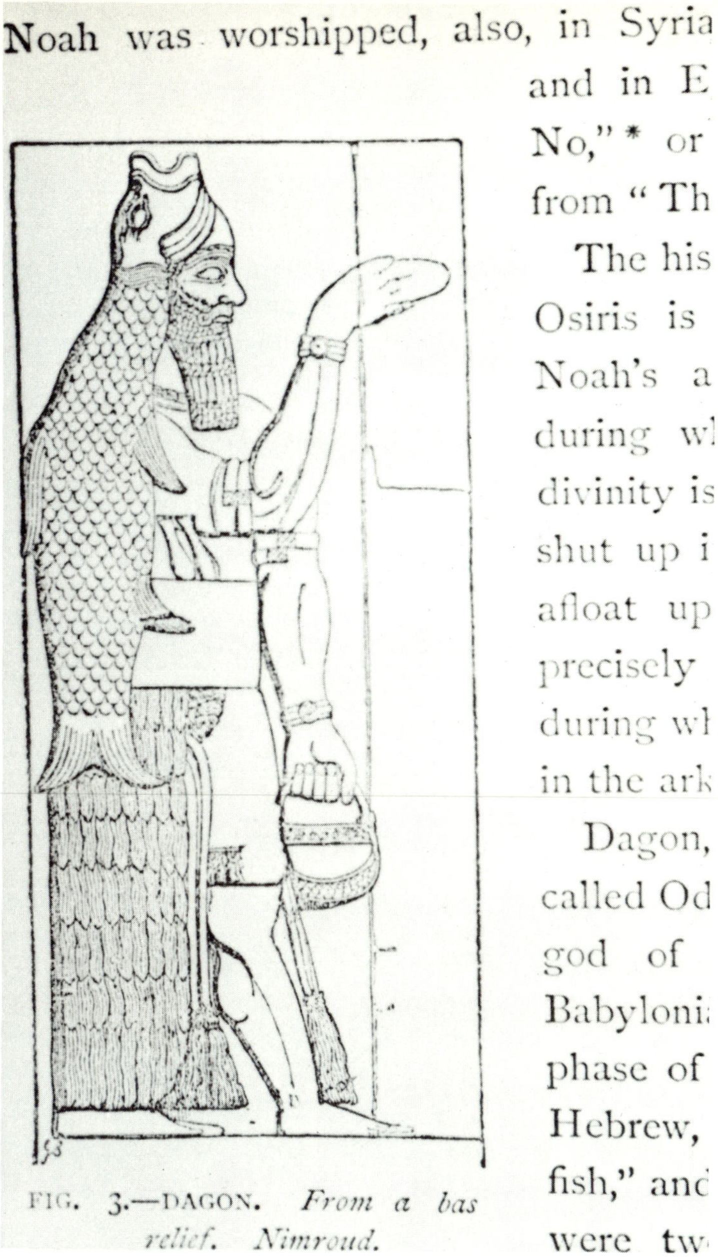
Noah was worshipped, also, in Syria
and in E
No,"* or
from "Th
The his
Osiris is
Noah's a
during wl
divinity is
shut up i
afloat up
precisely
during wl
in the ark
Dagon,
called Od
god of
Babyloni
phase of
Hebrew,
fish," and
were tw

FIG. 3.—DAGON. *From a bas relief. Nimroud.*

The great fish-god Dagon, as shown in a book published in 1883, now the property of the United States Navy Library, Washington, D.C.

was evidently Milton's conception (Paradise Lost, Book 1. 1.462) of him:

> " 'Dagon his name; *sea monster,*
> upward man
> And downward fish.' "

In India, a supreme Hindu god, Vishnu, is represented as coming out of the mouth of a monstrous fish. He is revered for having saved the Vedas (the sacred religious records of that ancient faith) when the entire world was drowned. An amazing fact in this story is that although the theme is quite similar to the story of Noah, the names *Vishnu* and *Noah* are identical. *Vishnu* is the Sanskrit form of *Ish-nuh*, which means "the man Noah." In fact, much of the mythology of Greece and Rome, as well as Asia, is full of the deeds of Noah.

Thereafter, in ancient Greece, a sacred sea monster image was sometimes created by combining a half-human, half-fish figure into a symbol of heavenly power. The goddess Artemis was created in this manner, and it is from her grotesque figure that the first generation of mermaids is said to have evolved. However, people called Phigalians, who lived in one of Greece's earlier

fish. The image of Dagon which fell upon its face to
ground before "the ark of the God of Israel," was
obably of this latter form, for we read * that in its fall,
he head of Dagon and
th the palms of his hands
re cut off upon the thres-
ld: only the *stump* (in the
argin, "*the fishy part*") of
gon was left to him. This
s evidently Milton's con-
ption of him:

Dagon his name; sea-monster,
upward man
And downward fish." †

In some of the Nineveh
lptures of the fish-god,
the head of
the fish forms
a kind of
mitre on the
head of the
man, whilst
the body of
the fish ap-
pears as a
cloak or cape
over his
shoulders and
k. The fish varies in length; in some cases the tail

. 5.—DAGON. *rom an Agate ignet. Nineveh.*

FIG. 4.—DAGON. *After Calmet.*

Another artist's conception of Dagon, as shown in a book published in 1883, now the property of the United States Navy Library.

cultural periods, are known to have worshipped a monsterlike wooden statute of a sea goddess named Eurynome. This effigy was human in the upper portion and fish in the lower section and was of such value to those who adored it that it is said to have been held in captivity with chains of gold.

Ancient Egypt also had her mythical sea monsters. Many are depicted on the massive columns that were once part of the temple at Luxor. This great edifice was built sometime around 1570 B.C., and is replete with serpents shown coming out of the sea to help or harm the people of that country. Again, it is evident that the people of those ages either knew about, or thought they knew about, monsters that lived in the depths of the sea.

In China, the sea monster is clearly evident in some of that country's most ancient art. However, artists generally symbolized sea monsters in the form of sea dragons. Ironically, the forces of good that controlled any dangers from evil sea dragons were depicted as being attractive young women. Sometimes, when a great emperor stood in danger, a massive sea turtle or porpoise or whale would be called upon to act as a sea horse, thereby carrying the dignitary to safety.

A work of Chinese literature, probably written shortly after the time of Confucius (479 B.C.), called *The Botanical and Medical Works of Li She Chan*, tells its readers the following (from a translation made about 1890): "Dr. So Tsung says that in the autumn a certain fish changes itself into a Dragon, and leaves its original bones, which are of five different colours, and are used by men as medicine. In Shanshi is the dragongate, through which, when the fish leaps, it becomes a dragon. Dr. Kai Tsung Shik says that it is well known that the Dragon is invisible to man. If this were the case, how could we see his bones?"

It is interesting to note that Chinese sea dragon bones, according to the same authoritative writing, "cures heart-ache, drives away ghosts, cures colds, and dysentery, cures fainting in children, irregularities of the digestive organs, heart or stomach, paralysis, nocturnal alarm, etc., and increases the general health."

The ancient Hindu god Vishnu, as shown in a book published in 1883, now the property of the United States Navy Library, Washington, D.C.

In Africa, monstrous sea serpents were closely associated with numerous myths. Since this gigantic landmass is entirely surrounded by water, it is understandable that many of its superstitions and early beliefs were tied to the sea. One very early carving shows a gigantic sea serpent doing battle with two undersea warriors. In many of the early tapestries showing various aspects of African life, sea monsters or sea serpents play a dominant part in man's struggles upon earth.

All early civilizations appear to have depicted monsters living in the sea. Even Polynesians, who were then, and still are, people of the sea, paid

This detail from an ancient Chinese tapestry shows an emperor riding a massive sea turtle. Photo, Library of Congress, Washington, D.C.

homage to gods who were either sea monsters themselves or possessed the power to ward off the evils of sea monsters.

Australia's aboriginal art depicts many great beasts who lived in the sea. According to ancient rock drawings, many of these sea monsters were known to have attacked men and to be related to cruel gods. Some few, however, were characterized as being both kindly and useful. The sing-sing man, or village storyteller, amongst the aborigines, still relates yarns of mythical characters that live in the sea. These legends probably originated thousands of years ago and have been handed down by word of mouth without change.

There are many examples in history where sea monsters are specifically mentioned. A case of this sort can be found in writings concerning activities in the Circus Maximus during the days of Nero (Claudius Caesar), the Roman emperor. There an arena of about 1,500 feet long would be flooded and transformed into an "enormously large aquarium full of *sea monsters*." Greek sponge divers, imported for the occasion, would dive, with knives held between their teeth, into the great tank and do battle with these sea monsters. What did they mean by "sea monster"? No one

A Mediterranean seagoing galley, shown in a stern-on view. It is interesting to note that even in this piece of art (1684), there are also sea monsters. Photo, Deutsches Museum, Munich, Germany.

can really say because no detailed descriptions have been left for study. We must therefore assume that writers of the period were referring to large sharks and not to the type of sea monster we are seeking. This is regrettable, for in fact they might really have captured some monstrous eel or massive sea serpent worthy of the name "sea monster."

Very early Canaan inscriptions, found mainly on household instruments, religious articles, and weapons, show that many had been dedicated to a sea serpent goddess named Ba'alat, portrayed as a snake charmer whose reptiles were really monsters of the sea. Ideographics, or picture symbol writing, was not adequate to express size, ferocity, or species of these monsters, but we know that something big and animalistic was frequently taken from surrounding oceans. Details, however, are lacking.

Psalm 104 states that the sea teems "with things innumerable, living things both small and great. There go the ships, and leviathan which thou didst form to sport in it."

The word *leviathan* is Hebrew and means an aquatic monster thought to have been a gigantic seagoing crocodile, an especially large whale, or even a dragon of sorts. In the Bible, the word was used as a symbol of massive animal life that inhabited the sea and had to be defeated by the power of good, (Job 41, Psalms 74:14, 104:26, Isaiah 27:1). We can only surmise that all these descriptions were representative of some one thing and were presented to enlighten mankind of dangers that dwell in the deep. Several clear examples are to be found. In Isaiah: "In that day the Lord with his hard and great and strong sword will punish Leviathan the fleeing serpent, Leviathan the twisting serpent, and he will slay the dragon that is in the sea." And in the Book of Job: "He [Leviathan] makes the deep boil like a pot; he makes the sea like a pot of ointment. Behind him he leaves a shining wake; one would think the deep to be hoary. Upon earth there is not his like, a creature without fear. He beholds everything that is high; he is king over all the sons of pride."

It should be said here that in this book, we use

the terms *sea monster* and *sea serpent* interchangeably. By the word *monster* we mean any enormous animal or thing. By *serpent* we mean any large-sized marine animal having serpentine movements, such as winding, twisting, coiling, hissing, or rearing of the head on a long neck. Thus, in this context, a sea monster could readily be a sea serpent, or vice versa.

In many mythological writings, beautiful women and sea monsters or sea serpents have been closely associated. It appears as if the exquisite loveliness of a female was easily transformed into the murderous figure of a sea monster. We have seen how Greek mythology turned gorgeous Scylla into a horrible sea monster with six heads. Nor has this close interconnection been confined to ancient Greece or Rome, for the same thing is found in the legends of Norway, Germany, Denmark, Sweden, and other European countries. France, for example, had its angelic-appearing Melusina, who was in reality a sea monster. Legend relates that she was a beautiful fairy who changed into a horrible sea monster one day each week. Count Raymound, said to be of a great and noble house, fell in love with what appeared to be a stunningly beautiful young girl. Melusina responded and finally agreed to marry the dashing young count, provided he made a promise not to visit her on one specific day of every week. After making the agreement, Count Raymound nevertheless broke his promise and discovered the horrible fact that his wife was in truth a sea monster. Melusina fled and forever thereafter was known to exist in the sea, where her frightening appearance and shrieking voice are said to foretell a ship's disaster.

Orpheus, the celebrated Thracian musician in Greek mythology, was the first to declare that all sea monsters are not ugly beasts. Dangerous, yes, but not necessarily repulsive. One particular species he warned the marine world against was the *siren,* a beautiful woman that sometimes took the form of a mermaid who possessed a magical voice. She it was who sat on a rock and lulled sailors into coming so close that they were smashed upon the shore and killed. Apollonius of Rhodes, a third-century Greek poet, relates this story.

"Orpheus sailed with Jason on the Argo. When all hands were weary of rowing, they suddenly heard, from far over the sea, singing that was so enchantingly sweet it lured them to come ever

The terms sea monster *and* sea serpent *became interchangeable. When erect, the creature is often called a sea monster; but when swimming, it becomes a sea serpent.* Photo, The Mariners Museum, Newport News, Virginia.

closer and closer. The ultimate end was inevitable, for the attractive young girl sat upon a rock shore. The sailors turned their ship toward the breaking seas near which the siren sat. Orpheus grabbed for his lyre and strummed out a tune so clear and so bright all other sound was lost. The sailors put their ship back on course and sailed away from the dangerous area."

These sirens were the daughters of Phorcys, son of the sea, and Achelous, a river deity. It is said that Odysseus, a leader of Greek forces during the Trojan War, escaped from the enticing clutches of these lovely sirens by having himself lashed securely to a mast and by stopping the ears of his men with rags.

Bestiare d'Amour, which appeared in 1250, written by a man named Richard de Furnival, discussed sirens, who, just as in Greek mythology, lured sailors to their deaths by dint of irresistible songs. This aspect of sea life was so well substantiated that scholar de Furnival set about classifying these dangerous sea animals.

"There are three kinds of sirens. Two are half-woman and half-sea monster, while the third is half-woman and half-bird. All three are known to have possessed a manner of singing that was magnetic. Two out of the three played instruments, one a harp, the other a flute. Sailors found the songs overwhelming and once in the enfolding arms of the siren, fell into peaceful slumber. The siren would then transform herself into a horrible sea monster and slay the defenseless sailor."

Germany's Lorelei has long been an inspiration as a tourist attraction that is still pointed out to foreign visitors. This particular siren is said to sit on a cliff 433 feet above the twisting, dangerously narrow Rhine River. By singing in a voice of great clarity, the siren would momentarily distract the helmsman, and the ship would instantly be smashed against the rocks. Heinrich Heine's poem *Dei Lorelei,* written at a much later date than the *Bestiare,* immortalized sirens.

In 1481, a Frenchman by the name of Gerald Lyon, related his version of the siren in a work called *Créatures Moralisées.* He described his fair lady as performing in the following monstrous fashion: After attracting the attention of a deep-water sailor, as his ship went sailing past where she bathed nude in the sea, she would beckon him into the water. "Young sailor," her voice would come ever so sweetly through the air, "if you love me as you seem to, come to my arms and I will grant all your desires." The story continues, saying that the sailor, mad with physical desire,

In mythology, beautiful women and little cherubs were closely associated with sea monsters. Here, an ancient Scandinavian sea monster, in the form of a siren who lured sailors to their deaths, is waited upon by an entourage. Photo, Nederlandsch Historian Scheepvaart Museum, Amsterdam, Netherlands.

would pitch himself into the sea, only to have his fair lady turn into a voracious monster that devoured him in a single bite.

In our hunt for sea monster facts, we next turn to the contributions of a great Roman naturalist known as Pliny the Elder (A.D. 23–79), who wrote an influential book entitled *Historia Naturalis*. In it he quotes from some 2,000 works then in existence, among which were 326 Greek authors of note and 146 Roman writers of established fame. Unfortunately, as this great source of material available to Pliny has since been lost, we have only his word about what they wrote. However, it has been agreed that Pliny's own observations about monsters of the sea showed considerable accuracy. On the other hand, he assumed that his sources of previously recorded information were as accurate as his own, which was a serious error. Many of those quotations cited in his writings tell of sea monsters, such as the Ethiopian dragons that were 30 or more feet long. There could very well have been some sort of reptile that fit this description. However, Pliny then quotes what other writers had to say about these dragons and is obviously misled. "On the coasts of Ethiopia, it is said that numbers of these dragons arrange themselves into the form of barriers, and with heads raised to form a sail, allow themselves to be carried by the winds to Arabia."

Pliny also gives some very worthwhile material on monstrous sawfish that lived in the Indian Ocean. Once more he errs when he cites other sources and says that they told of encountering sawfish reaching 300 feet in length. This, as we now know, is improbable, for the Indian Ocean species of sawfish has never been found to exceed 28 feet in length. Pliny also tells us of a sea monster that could kill a man with a glance and of marine animals he calls "blowers" that stood higher than a ship's sail while blowing water into the air. It seems obvious that a "blower" is a whale. Thus, so much of Pliny's remarks on sea monsters are obviously founded upon other men's observations that his entire massive work must be viewed with suspicion.

Pomponius Mela, a Roman geographer born in Spain, devoted much of his life to writing a lengthy description of the world as he observed it. His *De Situ Orbis*, written during the time of Pliny, told of part-time sea monsters that changed as the seasons changed. In winter they were massive wolves, but in summer they became great sea serpents, swimming about to gobble fishermen.

Valerius Maximus, a Roman writer, relates problems created by a marine serpent during the Punic Wars, when troops under Atilius Regulus were waylaid by the beast that lived in an underwater cave along the banks of the Bagrados, near Ithaca. He reported that many of the Roman soldiers were swallowed, having first been crushed in the serpent's folds. The sea serpent succeeded in keeping the troops from crossing the river until it was destroyed by large stones hurled from Roman catapults. Valerius Maximus goes on to say that the havoc caused by the sea monster did not end with its death. The water was so polluted with gore and the air so full of sickening stink from putrefying flesh that the Romans were obliged to remove their camp. However, the soldiers did secure the serpent's skin and skull, which were preserved in a temple at Rome. All this action was described by other ancient writers such as Florus, Seneca, Silius Italicus, Aulus Gellius, Orosius, Zonaras, and Pliny. In the main, all their accounts agree.

Diodorus Siculus, a prolific Greek historian and contemporary of Valerius Maximus, tells of a serpent 60 feet long. He said that the sea monster came out of the water at frequent intervals to devour cattle. A group of farmers attacked the sea serpent, but twenty of them were killed and the monster escaped. It was later captured in a strong net and taken alive to Alexander the Great, who then presented it to King Ptolemy II, founder of the Alexandrian library and museum.

Saint Augustine, who lived shortly after Pliny and Mela, noted the existence of sea monsters and sea serpents. In *De Civitate Dei* and other numerous writings, Saint Augustine divided the universe into two camps: one of God and one of the Devil. Thus, as he put it, sea monsters were simply the personification of the Devil, just as angels represented all things good. The uneducated and superstitious fishermen and sailors of those times readily accepted as fact the existence of sea monsters, for the Church, through Saint Augustine, preached just such a fact. Also, they believed that these monsters possessed some very unpleasant characteristics. They were stinking beasts, were carnivorous, ate their own offspring, belonged to no country, believed in no god, and possessed no morals. "They had unto themselves," wrote one commentator of that period, "only a lascivious nature, that was filthy in its promiscuousness, while yet being of unwashed body."

During the twelfth century, there appeared the Latin text of a letter addressed to a Byzantine emperor called Manuel I (Comnenus) and signed by a Christian priest named Prester John. This prelate of the church supposedly ruled a vast empire, either in Africa or Asia, where the largest sea monsters ever heard of existed. This fraudulent letter, appearing so soon after the published concepts of Saint Augustine, gave further impetus to the belief that horrible sea monsters truly existed. It became quite acceptable to believe in sea serpents of great size, and glib-tongued seafarers soon sought to elaborate on the writings of Prester John.

"There is a man," a Latin scholar of those times said in support of the Prester John letter, "that stands 40 ells [approximately 90 feet], who doth ride a swift sea horse that is thrice his own size and who upon occasion plungeth to the bottom of the deepest seas."

Since that would give a highly mobile creature a total length of about 270 feet, and in view of the fact the Prester John letter was later proved to be a complete fabrication, such figures are meaningless. However, they formed a basis for sightings that were to be recorded centuries later.

Ancient fishermen likened their gods to their own image, while giving them sea serpent or sea monster characteristics that no mortal man could ever possess. Glaucus, one of the most powerful gods in Greek mythology, is most often depicted with a serpentine lower body that ends in a fishtail. Unquestionably, there is a certain sea monster aspect about this man-god-fish.

Poseidon was another powerful sea god of Greek mythology. Later, Roman fishermen adopted him for their god and named him Neptune. Our own generations have modified this to Father Neptune, and even today we depict this man-god as being ruler of the seven seas. It is he who controlled sea monsters and used them for his own vengeful purposes. He employed a great sea serpent to ravage the Troad, which was the most important waterfront area of ancient Troy, because he had not been paid for building the walls of Troy. Later he was called upon to punish the Greek fleet for not having paid homage to their gods, and again he used a sea monster to stir up a massive storm that wrecked all the ships and drowned most of the military leaders.

During the so-called Dark Ages, sea monster sightings were negligible because ships, although attaining rather large size ("500 tons of cargo within hatches, in addition to cargo laid upon the decks," according to one authority), still were required to stay within sight of land. There were

Glaucus, Greek mythology's powerful god of the sea, is quite frequently portrayed as being part monster or part sea serpent. Photo, Library of Congress, Washington, D.C.

This picture, taken from Dr. A. C. Oudemans's book The Great Sea Serpent, *clearly indicates that the Scandinavian countries had their myths, legends, and ancient beliefs about sea monsters.* Photo, Library of Congress, Washington, D.C.

few if any navigational instruments, and those in use were crude and unreliable. Mariners hugged the shores and stayed within what was called, in the Mediterranean, "the circle of vision." By keeping land in sight, a ship could circle from Africa to Europe and was therefore quite unlikely to come upon a sea monster or sea serpent.

The far northern nations, such as Norway, Denmark, and Sweden, had their own myths, legends, and traditions. Many of these stories told of enormous sea animals that resembled islands in size. Probably the earliest manuscript describing such a north country sea monster appeared shortly after A.D. 1000 and was attributed to King Sverre of Norway. However, this report has to be in error, since the king was not born until some years after the story is said to have first appeared. Nevertheless, this initial account is important to our future sea monster sightings because it contains the first use of the word *kraken.* In the Scandinavian countries, even at this early date, a kraken was a sea monster of gigantic proportion, with which fishermen were all too familiar. Sea monster reports emerging from these nations made frequent reference to a specific marine animal recognized as a kraken. Eventually this particular species of kraken became identifiable with our own giant squid, which, of course, could be considered a sea monster. However, the word *kraken* unquestionably had its first use in the mythology and ancient folklore of the Scandinavian countries and not, as some historians would have us believe, in the early fourteenth century.

Even when Christianity supplanted mythology, mariners still believed strongly in the presence of

The upper portions of sea monsters were often depicted as having human form. Photo, Rijksmuseum, Leiden, Holland.

This sea monster, with a human head, was said to be able to talk to men of all nationalities in their native tongues. Photo, Rijksmuseum, Leiden, Holland.

Shortly after A.D. *1000, King Sverre of Norway first used the word* kraken *to describe a sea monster. Later, the word came to mean a giant squid or octopus.* Photo, The Mariners Museum, Newport News, Virginia.

sea monsters. On commencing a voyage, it was customary for a sea captain to gather his crew and make a little speech about the hazards that lay ahead. According to historian James Westfall Thompson, one captain is reported to have performed in the following manner: "When but half a day out, the master gathered unto himself his crew of numerous score and said, 'Seeing ourselves now at the mercy of God and the elements, each shall henceforth be held equal to one another, as upon this voyage we are subject to sudden tempests, pirates, and monsters of the deep. . . .' "

However, maritime affairs were changing. Larger and more seaworthy ships were being built. For example, in the year 1268, Louis IX was furnished with a merchant vessel by Venice; it was 108 feet long and carried a crew of 110 men. Even at this early stage, special cargo carriers began to appear. One such ship, constructed to transport horses, had doors in the stern that opened to form a walkway over which the animals could be loaded or unloaded. With such progress in shipbuilding, it was inevitable that some daring master would venture to sail a direct route from the horse-raising center of Arabia to the buying markets of Spain. This meant taking the risk of sailing out of sight of land. In turn, this required better navigational instruments, which, once they were available, enabled the mariner to take even greater risks. With the extension of such travel, reports of sea monsters rapidly increased. Large crews were given to sitting around and telling sea stories; as is always the way with a lively nautical yarn, they begot gigantic sea monsters.

A man named Saewulf is quoted by James Westfall Thompson, in his economic and social history of Europe in the later Middle Ages, about his voyage to Joppa during the Dark or Middle Ages. He describes the ship, its crew, its deck procedures, and its intended markets in accurate detail. In one instance he tells of his ship happening upon a sea monster "many times the length of that vessel wherein we sailed." Another traveler, Ludolph von Suchen, in 1350 wrote that "howbeit, no vessel can ever, with but few exceptions, return without toil, the fear of sea monsters, and tempest."

Saint Brendan, an Irish missionary and man of many travels during the years before 1500, tells of a great fish named Sasconye, upon which his crew alighted and celebrated mass, thinking it to be an island. It was only when they commenced to cook a meal that the heat disturbed the animal and it began to move.

By the end of the fourteenth century, the age of discovery was beginning. Vasco da Gama, a Portuguese navigator, in 1499 concluded a voyage that had taken him around the Cape of Good Hope, up the eastern coast of Africa to Malindi, and across the Indian Ocean to ancient Calicut. This voyage opened up a way for European merchants to trade with the Indies. Commerce was going farther and farther afield, which meant bigger and faster merchant ships, with accompanying war vessels to guard them against pirates and enemy ships. Whalers pursued their quarry into strange waters and brought back the bones of hitherto-unknown sea beasts. Sea monster stories started to circulate wildly and became so numerous that people began to make them a matter of record. One ship's captain is said to have warned the count of Flanders, Guy de Dampierre, a member of the house of Bourbon, that if his ships put to sea, "they must be ready to face monsters of the deep and fend them off, or run from them."

So it is that as the year 1500 arrived, we find many accurately recorded sea monster sightings.

An enormous fish of sea monster proportions was believed to inhabit Scandinavian waters. It had numerous human eyes all over its body. Photo, Rijksmuseum Voor de Geschiedenis, Leiden, Holland.

"And the Lord spake unto the fish, and it vomited out Jonah upon dry land" (Jonah 2:10). This print appeared in an eighteenth-century Bible. Photo, Maritime Museum, Haifa, Israel.

This etching shows a dolphin, which could be considered a sea monster in biblical times. Photo, Maritime Museum, Haifa, Israel.

As late as the 1500s, galleys were being built for use as both warships and merchant vessels. Ships of this design were especially susceptible to attack by giant squid and giant octopus, thereby giving rise to sea monster stories.

Mr. C. A. Repenning, of the United States Geological Survey Laboratory, works on reconstructing the bones of a sea mammal called a paleoparadoxia. This now-extinct beast probably weighed over one ton and measured between 9 and 15 feet in length. Photo, United States Geological Survey, Washington, D.C.

This old French print clearly shows that sea monsters attacked ships and devoured their crews. Photo, Le Clerc Studios, Paris.

3

1500 to 1700

In the 1500s, many exaggerated stories of sea monster sightings were reported; for example, Konrad von Gesner's Heracleoticon was a monstrous animal that would occasionally rise out of the seas bordering the shores of the Scandinavian lands of Finmark and Lapland. Gesner was a knowledgeable Swiss scientist, and it is amazing that he should write about a sea monster of such a great size "that a regiment of soldiers could conveniently maneuver on its back."

It is in Gesner's book *Historia Animalium* that we run across the first chronicled mention of what later was to become known as "Jennie Hanivers," or "Gullion-Poke." These were very realistic sea monster frauds made from the flesh of sea animals. Most frequently the underside of a skate or large ray would be imprinted with the features of a man. When properly treated and dried, they were sold as the deceased offspring of mermaids and were said to be endowed with magical powers. Other such curios were made from the body of an eel and the face of a frog, which gave them an especially ferocious sea monster visage. These were sold as the nurslings of a sea serpent.

Since the Gullion-Poke shams were widely distributed and proved popular items of the day,

Thijs Mol, Arnhem, Holland, says that this engraving was made to portray the true size of the largest marine animal ever washed ashore in that country. It could have been just such a whale that Gesner came to call a heracleoticon.

people became conditioned to accepting readily almost any sea monster story. The net result was that fantastic sea yarns were related as fact.

Gesner was not alone as an exaggerator of sea monster stories. Bartholin, a learned Dane, related how, on a specific occasion, the bishop of Midaros found a sea monster sleeping quietly along the shore. Mistaking the animal for a great boulder, he built an altar upon it and celebrated mass. Apparently the beast awoke partway through the ceremony but, out of respect for the bishop, "waited until the mass was concluded, and everyone safe ashore, before sinking beneath the waves."

Sea monster sightings in the northern seas now were frequently reported. Early in the 1500s, Olaus Magnus, brother of the Swedish Archbishop Johannes Mansson until a revolution caused both to flee to Rome, designed charts of Arctic waters and illuminated them with drawings of hungry sea serpents feasting upon crew members. To substantiate the truth of his pictures, he began to write about sea monsters. In 1555 his *Historia de Gentibus Septentrionalibus* gave the following report, which loosely translated read as follows:

"There is also another serpent of an astonishing size in an island called Moos, in the diocese of Hammer: which portends a change in the kingdom of Norway, as a comet doeth in the whole world, as it was seen, anno 1522, raising itself high above the surface of the water and circling like a spire. Seen from afar this serpent was estimated by conjecture to be fifty cubits long [about 75 feet]; this even was followed by the banishment of King Christiernus and by a great persecution of the Bishops; and it also showed the destruction of the country, as Isidorus tells us of the birds of Diomedes."

Again, writing in 1555, Olaus Magnus states the following:

"They who, either to trade or to fish, sail along the shores of Norway, relate with concurring evidence a truly admirable story, namely that a very large serpent of a length of upwards of 200 feet, and 20 feet in diameter lives in rocks and holes near the shores of Bergen; it comes out of its caverns only on summer nights and in fine weather to devour calves, lambs and hogs, or goes into the sea to eat cuttles, lobsters and all kinds of sea-crabs. It has a row of hairs two feet in length hanging from the neck, sharp scales of a dark colour, and brilliant flaming eyes. It attacks boats, and snatches away the men, by raising itself high out of the water, and devours them: and commonly this does not happen without a terrible

Two well-known objects of Gullion-Poke were the sea monk (left) *and a sea bishop* (right). *These figures appeared in a book written by Schipper and published in 1660.* Photo, Rijksmuseum Voor de Geschiedenis, Leiden, Holland.

Illustrations showing a mass being celebrated on the back of a sea monster were commonplace during the early 1500s. Photo, Museo Naval, Madrid, Spain.

In 1554, a man named Gillaume Rondelet published this drawing of a massive marine centipede. Photo, Mairie de Saint-Malo.

A sea monster as depicted on a sea chart, about 1560. This type of beast was known as "greedy rosmarines."

event in the Kingdom, without a change being at hand, either that the prince will die or be banished, or that a war will soon break out."

The scholar A. C. Oudemans, of the Hague, who was well acquainted with the teachings, temperament, and behavior of Olaus Magnus, appraises his 1555 writing thus: "This narrative tells us that the sea serpent frequents the shores of Norway, that it appears mostly in summer, that it has large dimensions, and a considerable thickness. It has a row of hairs hanging down from its neck, its colour is dark, its eyes are brilliant and flaming. It only appears in fine weather."

Dr. Oudemans continues his evaluation by pointing out that both he and his associates harbored grave doubts about certain aspects of the Magnus's writings. "We consider its [the sea monster] devouring hogs, lambs and calves, and its appearance on summer nights on land to take its prey to be a fable."

On the other hand, Oudemans and his associates assure us that the Magnus's writings telling of a sea monster that eats squids, cuttles, crabs, and

This sea monster is said to have been drawn after one seen near the shores of Bergen. Photo, Schiffahrts Museum, Brake, West Germany.

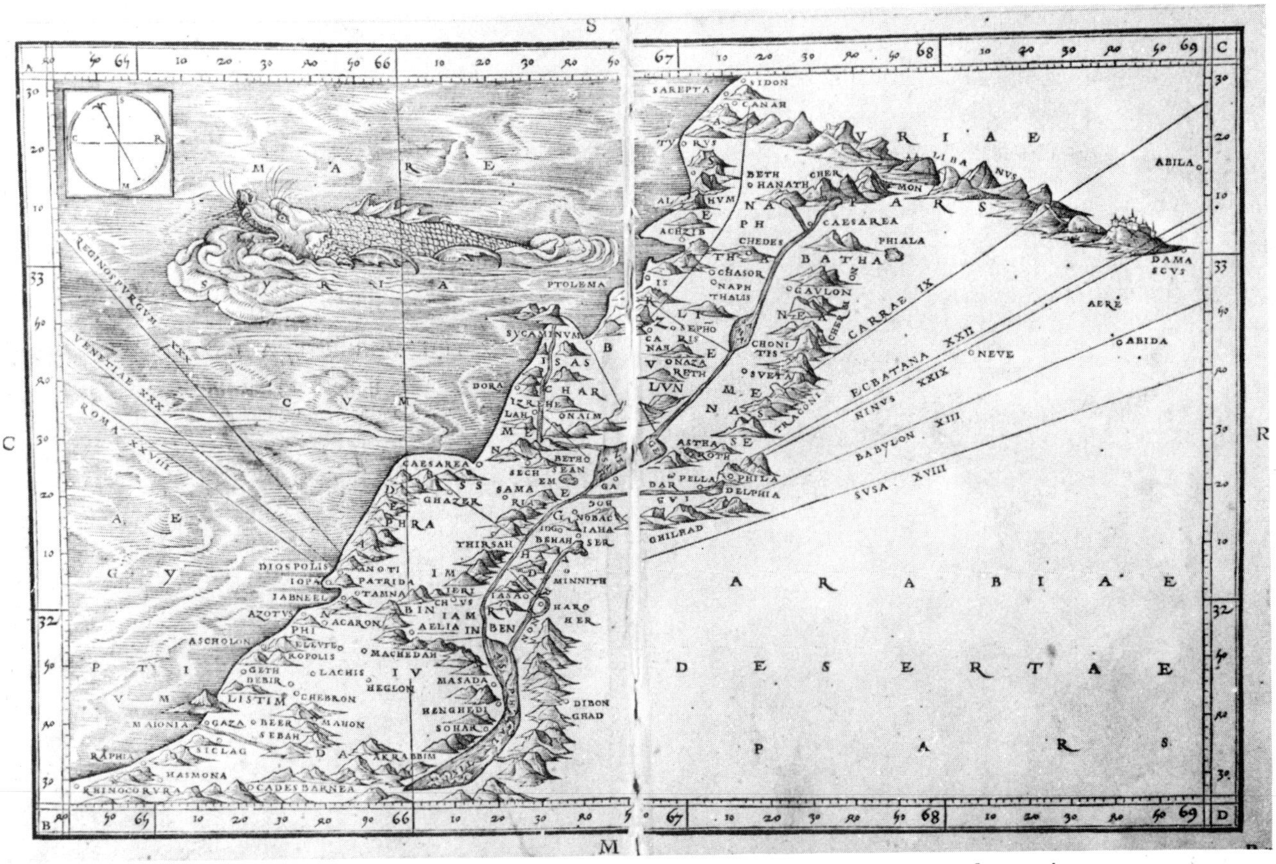

This map of the Holy Land by Jacob Ziegler was printed in Strasbourg in 1532. In the upper left is a sea monster. Photo, Maritime Museum, Haifa, Israel.

Another massive serpent, like the one shown in this ancient sea chart, was said to have been seen near an island called Moos. Photo, Schiffahrts Museum, Brake, West Germany.

42)

This sea chart of 1595 warns mariners to be wary of a wide variety of sea monsters. Photo, National Library of Australia.

lobster may be a fiction or may have been truly witnessed, with "the animal chewing them with its head above water, as seals and sea lions do." However, they reject any notion of a sea monster snatching a man from a ship. They say that Magnus either confused his story with another tale or was making reference to gigantic calamaries (giant squid) that occasionally attack boats and snatch away members of the crew. Oudemans then concludes by stating that "its being covered with scales must be fictitious too, for they who saw a sea serpent at a short distance, are unanimous in stating that it had no scales but a smooth skin."

There are two specific points that should be noted in regard to Oudemans's report on Olaus Magnus. First, at no time did Oudemans or his associates ever deny the existence of a sea monster. Their only quarrel was with certain of the sea monster's attributed physical aspects and behavior. Second, Oudemans uses the word *calamaries* several times without any explanation. It should therefore be noted that the word is an archaic name for a squid and is very likely used to imply just that.

A relatively short period after the Olaus Magnus papers appeared, a story was reported telling of a sea monster experience by a Captain Jean Magnus Dens. Extensive research on the part of the Librairie Larousse in Paris confirms the existence of a Captain Jean Magnus Dens who very probably lived some time shortly after the Magnus report. His story, while well recorded by several French writers, could not be confirmed. Consequently, we present the following information from Henry Lee's *Seamonsters Unmasked*, published in 1883, knowing that the entire affair has never been substantiated.

Captain Jean Magnus Dens was a mariner recognized for his capabilities and his honesty. After making several voyages to China as a master trader, he retired from a seafaring life and settled down in Dunkirk. It is said that whenever he had the opportunity, he told the story that on one voyage, while crossing from Saint Helena to Cape Negro, he was becalmed. He took advantage of the enforced idleness of the crew to have the sides of the vessel scraped and painted. During a time when three of his men were standing on planks slung

Olaus Magnus on several occasions referred to gigantic calamaries (giant squid) that were known to have attacked ships. Photo, Thijs Mol, Arnhem, Netherlands.

A whalelike sea monster depicted on a nautical chart by Olaus Magnus. Photo, Schiffahrts Museum, Brake, West Germany.

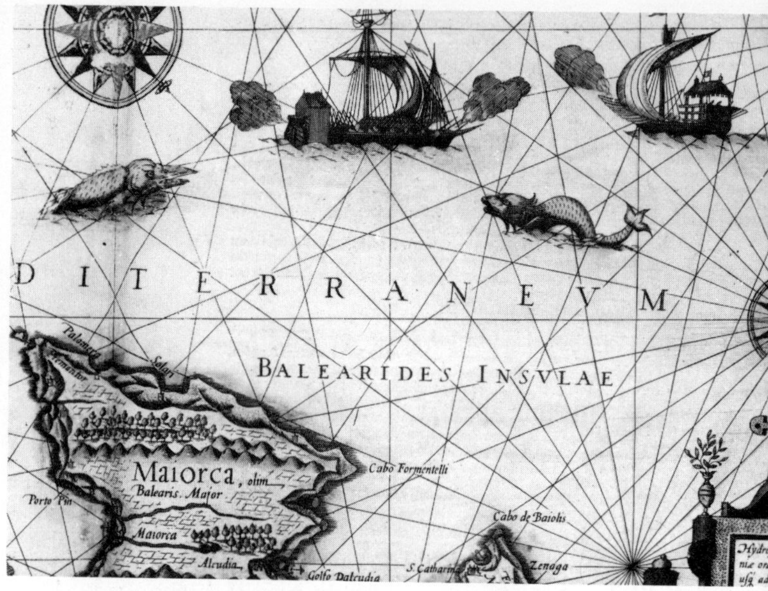

Sea serpents are always in evidence on the charts of ancient mariners. Photo, Schiffahrts Museum, Brake, West Germany.

over the side, an enormous cuttle rose from the water and threw one of its arms around two of the sailors, whom it tore away, along with the scaffolding on which they stood. With another arm it seized the third man, who held on tightly to the rigging and shouted for help. His shipmates ran to his assistance and succeeded in rescuing him by cutting away the creature's arm with axes and knives, but he became delirious and died the following night. Mr. Lee adds this note to his own version of the affair: "I have been told by a friend that he saw in a shop in China a picture of a cuttle embracing a junk, apparently of some 300 tons burthen, and helping itself to the sailors, as one picks gooseberries off a bush."

Other sources of contemporary information, again according to the Librairie Larousse, say that Captain Den's sea monster had one arm cut off in the affair and that this was 25 feet long and "as thick as a mizen-yard," with suckers the size of saucepan lids. At the same time, this monster was said to have been not a giant cuttle, but actually a large land snake, such as a boa constrictor or an anaconda.

It is well known that the African variety of python possesses the ability to swallow a cow whole. What would happen if such a formidable serpent should put to sea? It would certainly be considered a sea serpent in the true sense of the term. And that is just what happens upon occasion, for the African python has been observed swimming in the middle of one of the world's largest bodies of water, the Indian Ocean. These snakes have been known to travel from island to island in search of food. On one occasion, an earthquake caused the trees and foliage of several islands to burn, thereby driving off all the wildlife. Would-be rescuers found nothing but charred desolation. Yet within a relatively short period after the conflagration, grown pythons were reported to be on the islands. Obviously, they reached their destination by swimming.

This leads us to surmise that many of the sea serpents depicted on the charts of ancient mariners and shown enveloping ships while eating members of the crew (as Olaus Magnus liked to picture) might have been inspired by the sight of a tired, hungry python, swimming in the ocean. Should he happen upon a small ship, the obvious thing for the serpent to do would be to climb aboard and devour a few of the crew.

This is an entirely valid assumption because large water snakes, when swimming far from land, will frequently attempt to board a boat or ship. The "attack" is generally misconstrued as an aggressive act, when in reality it is only an attempt to use the ship as a platform that might afford a rest. Nevertheless, any mariner who sees a 40-foot length of snake crawl across his bulwark is bound to be terrified. It is certain the event would result in an oft-repeated yarn, with the aggressor snake growing larger at each telling.

Today we are skeptical concerning Olaus Magnus's words and drawings; but in his own time, he found support. Magnus's contemporary, Gesner, in 1560 substantiated the presence of sea serpents by writing: "In the Baltic or Swedish Ocean are found certain yellow sea serpents of thirty or forty feet in length, which, when not provoked, do not harm anyone."

However, this same writer continues: "There is another sea serpent, a hundred or two hundred feet long, which sometimes appears near Norway in fine weather, and is dangerous to Sea-men, as it snatches away men from their ships. Mariners tell that it incloses ships, as large as our trading vessels, made on our rivers and lakes, by laying themselves around them in a circle, and that the ship is then turned upside down. It sometimes makes such large coils above the water, that a ship can go through one of them."

A treatise called *Les monstres marins*, by an author named Landrin, carried the following information. During the reign (1556–1598) of Philip II of Spain (1527–1598), a massive sea monster was observed in the ocean. It differed from any other animal of the deep hitherto seen. Standing upright, and half out of the sea, it displayed two great fins that enabled it to sail along like a ship. A war vessel caught sight of the beast and fired a cannon, breaking one of the fins. With a gigantic roar and constant howling, the monster turned into the Strait of Gibraltar and died on the shores of Valencia. It had a mouth of such size that seven man could stand upright within its jaws and a rider on horseback could easily enter without having to dismount. The corpses of two men were found in the stomach.

In 1615, one of these same sea monsters was seen swimming in the Elbe River; and in 1638, a fisherman who tried to throw an iron harpoon into a beast of this type was stricken dead by an electric shock. This vague report is of significant interest because it is one of the first, if not *the* first, concerning fish having the ability to generate substantial amounts of electricity. Today we know that certain large eels (specifically, *Electrophorus electricus*) can give off a charge of 600 volts, which is a force equal to the combined power derived from fifty automobile batteries, a shock powerful enough to kill a man.

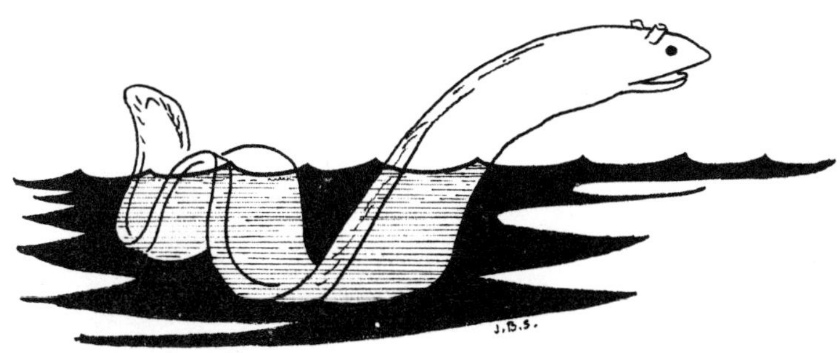

Electric eels, which have sometimes been mistaken for young sea serpents, are capable of giving off a charge equal to the combined power of fifty automobile batteries. Drawing by the author.

In Jewish literature, one chronicler, Rabbi bar bar Hannah, related stories pertaining to his experiences while at sea. He tells a sea monster episode that resembles those told by many of his predecessors, such as Bartholin the Dane.

"At one time while aboard a ship at sea, we came to a gigantic fish at rest on the waters, which we mistook to be an island. We made a fire and cooked our meat. But the fish, on feeling the heat, rolled over, and had the vessel not been nearby, we undoubtedly would have drowned."

The Jewish Talmud contains the following folktale about sea monsters: "One day we were going before the wind in a ship and saw a fish which put its head out of the water and upon which was written: 'I am the meanest creature to inhabit the sea. I am three hundred miles in length, and upon this day I enter into the jaws of the Leviathan.' "

In 1640, a scholar named Adam Olearius published a book called *Gottorfische Kunstkammer*. An English translation was published in 1892. Olearius wrote: "And that this is true has not long been confirmed by a Swedish nobleman at Gottorf, who declared to have heard from the Burgomaster of Malmoi, a trustworthy man, that, whilest standing on a hill on the Norwegian coast, he saw in the calm water a large serpent, which was seen from afar, had the thickness of a wine barrel, and 25 windings. These serpents are said to appear on the surface of the water only in calm weather and at certain times."

The outstanding significance of this statement is simply that it recurs. There are numerous other less-specific accounts, most of which appear to be garbled. But all have one thing in common: They cite a serpent of similar appearance in adjacent waters. There was no doubt in any of their minds that this was the Sound between Sweden and Denmark, and they saw a large animal, which looked like a serpent, and which would appear on the surface only if the weather was calm. Oudemans, in referring to the above quotation from Olearius, said: "I beg the reader to fix his attention on those apparently insignificant statements, as it will be seen that they are given several times independent of one another."

In 1658, Etienne de Flacourt, an admiral of the French navy and governor of Madagascar, published a history of that island entitled *Histoire de la Grande Isle de Madagascar.* In this treatise, he spoke of two animals unique to Madagascar. First, he told of a giant bird, living in the southern areas, which laid enormous eggs. Second, he wrote of a massive sea serpent the natives called a *fanghane*, which apparently meant "dragon" in the local dialect. Science has since proved that the overly large eggs did in fact exist; they were the work of a giant bird, now extinct, scientifically named *Aepyornis maximus*. Unfortunately, no remains of the fanghane have ever been found. However, it is logical to assume that if the admiral was correct in speaking of the presence of one animal, he was equally correct in speaking of the other.

In European waters, sea monsters were still as common as ever. During 1666, the burgomaster of Potom was reported to have seen a sea serpent as thick around as a wine barrel and of "10 loopings." In the year 1687, a substantial sea monster was seen by many people from the village of Dramsfiorden, in Norway. Erik Pontoppidan, a Danish bishop, reported that in that year "a great sea serpent was seen several times by several persons in the Dramsfjord, and once by eleven persons together. The weather was calm, but as soon as the sun set and the wind began to blow, it left the fjord, and like one who runs out a coil of rope can know the length thereof, so one could see how long it was, before it had wound off all its coils, and stretched itself at full length."

In this account by Pontoppidan, it must again be noted that the animal is seen in calm weather and that it shows coils or windings. For the first time, the fact is mentioned that it can stretch itself, evidently in a straight line.

Even the bards of the day commenced to write about the Magnus type of sea serpent. John Milton (1608–1674), in *Paradise Lost*, published in 1667, had this to say:

"Thus Satan talking to his nearest Mate,
With Head up-lift above the wave, and Eyes
That sparkling blaz'd, his other Parts besides
Prone on the Flood, extended long and large
Lay floating many a rood, in bulk as huge
As whom the Fables name of monstrous size,
Titanian, or *Earth-born*, that warr'd on *Jove*,
Briareos or *Typhon*, whom the Den
By ancient *Tarsus* held, or that Sea-beast
Leviathan, whom God of all his works
Created hugest that swim th' Ocean stream:
Him haply slumb'ring on the *Norway* foam
The Pilot of some small night-founder'd Skiff,
Deeming some Island, aft, as Seamen tell,
With fixed Anchor in his scaly rind
Moors by his side under the Lee, while night
Invests the Sea, and wished Morn delays. . . ."
[I. 192–208]

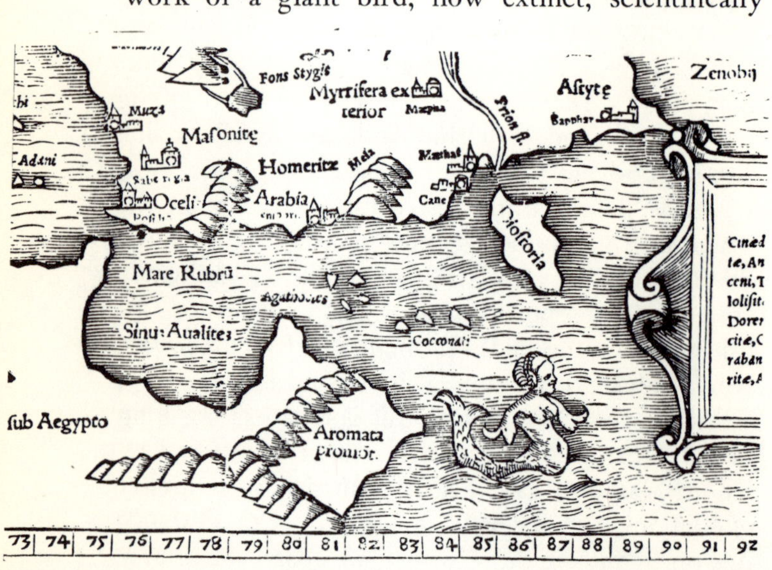

This fragment of a sea chart of the Middle East dates from 1511. It indicates that the artist believed that sea monsters very probably existed. Photo, Maritime Museum, Haifa, Israel.

This, supposedly, was the first recorded sea monster sighting in America. It occurred in 1639. Photo, Forbes Collection, Hart Nautical Museum, Massachusetts Institute of Technology.

As Oudemans states in citing this verse in *Paradise Lost*, "We observe that he mixes here also another story of a large sea monster on which sea men, believing it some island, will anchor, a story told about the Kraken and about the sperm whales."

At a later date, the Reverend Mr. Friis, consistorial assessor, minister of Bodoen in Nordland, and vicar of the College for Promoting Christian Knowledge, was to write the following:

"In the year of 1680, a Krake (perhaps a young and foolish one) came into the water that runs between the rocks and cliffs in the parish of Alstabourg, though the general custom of that creature is to keep always several leagues from land, and therefore they of course must die there. It happened that its extended long arms or antennae, which this creature seems to use like the snail in turning about, caught hold of some trees standing near the water, which might easily have been torn up by the roots; but beside this, as it was found afterwards, he entangled himself in some openings or clefts in the rock, and therein stuck so fast, and hung so unfortunately, that he could not work himself out, but perished and putrefied on the spot. The carcass, which was a long while decaying, and filled a great part of that narrow channel, made it almost impassable by its intolerable stench.

"The Kraken has never been known to do any great harm, except they have taken away the lives of those who consequently could not bring the tidings. I have heard but one instance mentioned, which happened a few years ago, near Fridrichstad, in the diocess of Aggerhuus. They say that two fishermen accidentally, and to their great surprise, fell into such a spot in the water as has been before described, full of a thick slime almost like a morass. They immediately strove to get out of this place, but they had not time to turn quick enough to save themselves from one of the Kraken's horns, which crushed the head of the boat, so that it was with great difficulty they saved their lives on the wreck, though the weather was as calm as possible; for these monsters, like the sea snake, never appear at other times."

Before the end of the century, other serpents were reported, usually a creature with serpentine coils and undulations, making an appearance in calm waters. In all these there are several points of importance that should be stressed. First, all the sightings occurred in places where people's lives were closely connected with the sea. Most were associated to some degree with commercial fishing. Hence, they would readily know the difference between a large shark, a school of playful porpoises, and a long, serpentine beast. Furthermore, there is a strong implication here that the brute stayed around for some time, until "the wind began to blow," affording observers a good long look. Combine this lengthy period with weather that was "calm," and you have an extremely favorable set of circumstances for people to look, study, and agree or disagree on what they are seeing. In many of these cases, numerous witnesses seem to have agreed.

In Japan, the 1500s and 1600s were important years for sea monsters. Although that nation, like other inhabited lands, had its myths, legends, and early fables about dragons of the deep, these stories did not become known to the European world until the mid-1600s. Dr. Tokujero Kanamori, director of the Japanese Diet Library, explained that the first Europeans visited Japan in about 1543. These people were Portuguese who landed on the island of Tanegashima, off the southern coast of Kyushu, and did not find their way home for many years. As a result of this time lapse, all Japan's authenticated beliefs in sea monsters, much of which relates to mythology, were not translated into one or another of the European languages until sometime after the year 1650.

Japanese mythology has many tales of sea serpents, sea monsters, and sea dragons. Baku, one of their more ancient sea monsters, gradually swallows each full moon, thereby saving mankind from complete destruction. However, this sea monster is also the "dream eater" for fishermen who suffer from nightmares. The cures for these horrible ordeals may be brought about by painting a picture of baku on a slip of paper and placing it under the pillow. As related by Professor Saki Kichitaro Atsuharu, in his book entitled *Japan*, published many years ago, the fisherman, "then no more will have dreadful dreams that produce sweat and tremble in sleep, because 'baku the sea dragon' eats up dreams and leaves man unmolested."

The earliest reference to sea monsters in Japanese literature, which appears in Portuguese records of about 1650, is in the *Koziki*. The *Koziki* is considered the most treasured and accurate source of information concerning the origin and foundation

The arms of a giant cuttlefish are here exhibited in a Japanese fishmonger's stall. The print is from a missionary's report dated 1883. Photo, United States Navy Main Library, Washington, D.C.

This scene is from a missionary's report dated 1883. It shows a Japanese fisherman being attacked by a cuttlefish. Photo, United States Navy Main Library, Washington, D.C.

Professor Torao Mozai, of Tokyo University, states that this picture shows umibohzu, *meaning "a priest of the sea." Many umibohzu were horrible, giantlike monsters.* Photo, Tokyo University, Tokyo, Japan.

of the Japanese Empire. It reports that *Hoori-no-Mikoto*, a heavenly prince skilled in the use of bow and arrow, borrowed a large fishhook from Hoderi-no-Mikoto, who was having all the "sea luck," and tried his fortune with rod and line. But he never caught anything and only managed to lose the valuable hook that was blessed with luck, which greatly angered its owner. The hook was eventually found sticking in the throat of a large sea serpent and so all ended well, for Hoderi-no-Mikoto sold the sea beast for a great sum of money and became very wealthy.

Another Japanese classic, the *Nihon-syoki* (The Chronicles of Japan), contains this passage: "Zingu Kogo, or Empress Zingu, bent a needle and made it into a hook. She took grains of rice and used them as bait. Pulling out the threads of her garment, she made them into a line. Then she stood on a stone in the middle of the river and cast the hook. Pulling up her rod she found that she had indeed caught a great sea dragon."

The South Pacific also had its sea monster stories. The Reverend Wyatt Gill, a missionary who spent a lifetime amongst the small islands of that area, once related a legend told him by natives of Aitutaki Island. At that time, he said, he was well aware that the story had existed for over 200 years and therefore probably originated in about 1650. It seems that there once was a powerful fisherman, who was also a great explorer, by the name of Rata. He built himself an especially strong canoe, decked it over, and rigged it to catch a great amount of wind. He then set out on what was to be a lengthy voyage of exploration. "At the prow," the Reverend Gill wrote, "was stationed the dauntless Nganaoa, armed with a long spear and ready to slay all monsters.

"One day when speeding pleasantly over the ocean, the voice of the ever vigilant Nganaoa was heard: 'Oh Rata, yonder is a terrible enemy starting up from the ocean depths.' "

The legend goes on to relate that the "terrible enemy" was actually a sea monster of massive dimensions. As the story unfolds, the beast appears to develop into a hybrid species of underwater animal that was part octopus, a good deal whale, and mostly wild-eyed imagination. "Its huge tentacles encircled the vessel in their embrace, threatening its instant destruction. At this critical moment Nganaoa seized his spear and fearlessly drove it through the head of the creature. The tentacles relaxed and the dead monster slowly floated off on the surface of the ocean."

Ireland, an island reputedly without snakes, was indeed surrounded by seas alive with serpents. Before the Battle of the Boyne in 1690, imaginative Irish Celts related their sea monster stories in the Gaelic tongue. Native writers such as O'Rahilly, McMurphy, Collins, and O'Curry put these stories into ballads that told of those great beasts ruling

A sacred Micronesian sea monster is shown with three sacred sea gods. Photo, Maritime Museum, Kingston-on-the-Hull, England.

the waves. Here is a translation of the beginning of one old Gaelic ballad:

We started out for Dublin town
The Captain steered us straight
But when we reached the ocean
The poor oul' ship was bate

The ballad goes on to relate how the crew and passengers called upon the saints to save them from a watery grave after their tiny ship was blown upon the rocks. Their prayers were answered in the form of a mile-long sea monster that formed a bridge from the broken hull back to the solid shores of the Irish Sea. Across this walked every living soul.

During the period from 1500 to 1700, it was usual to classify any nonrecognizable animal remains from the sea as parts of sea monsters. Obviously, marine life was badly in need of a more understandable classification system. Early in the 1700s, it became quite apparent that science was aware of its own shortcomings. Slowly, facts began to be compiled. Sketches were made with considerable care in measurement. These drawings were accompanied by descriptions. Gradually, out of a morass of misunderstanding, there evolved a clear picture of what we might expect a true sea monster to look like.

But there were still greatly exaggerated and tall tales of sea monsters. An excellent example of such a story is to be found in Hans Egede's *Natural History of Groeland* (Greeland), written just before the turn of the eighteenth century.

"Fishermen also relate another tale about a sea monster. They say that a great ghost-like sea monster now and then puts in an appearance in the main sea, which they call a *Kraken,* and is without any doubt that same one the islanders call *Hafgufa,* or *Drew.* They claim that its body reaches several miles in length, and it is most often seen in calm weather. When it comes out of the water, it appears to cover the whole surface of the sea. All kinds of fishes flock around it, as upon a bank of the sea, and that many fishing boats come to this point to catch fish, not aware of the fact they rest upon a dreadful sea monster. This they finally understand by the entangling of their hooks and angles in its body. The sea monster, feeling the latter, rises softly from the bottom and seizes them all. They can, however, most easily avoid such destruction, as they only have to call it by name, which it no sooner hears than it sinks down again as softly as it did rise."

In concluding the examination of this important period in the history of sea monsters, there are several reportorial characteristics that should be highlighted. First, numerous occasions are cited wherein a holy mass was celebrated on the back of a sea monster. Typical of such reports is noted by Eric Falkendorff, bishop of Nidros, in a letter which he sent to Pope Leo X. He stated that one Sunday, when traveling at sea, he was considerably disturbed that he was unable to celebrate mass on land with all due solemnity. "Suddenly there emerged out of the seas an island of considerable size and upon this the party was able to celebrate mass. Hardly was the sacred ceremony completed and the party safely returned to its ship, than the sea monster vanished into the depths."

Second, there are many gross exaggerations reported by responsible persons, such as a story in "Gummer's Ore." This is another one of the sea monsters that were so often mistaken for islands.

"Amongst the rocks about Stockholm there is sometimes seen a certain tract of land which at other times disappears, and is seen again in another place. Buraeus had placed this as an island on his map. Peasants claim that it is not always seen, and that it resides out in the open ocean. Now Baron Grippenhielm related that on a certain Sunday, when he was walking amidst the rocks, he spotted something that appeared as three spots of land in the surrounding waters, which surprised him somewhat. He sought a peasant to inquire if this were Gummer's Ore, but when they returned to the spot nothing could be seen of it. The peasants said that this forecast the coming of a great storm, or great quantity of fish. Gummer's Ore, however, is nothing else than the Kraken, keeping himself always about that area, and frequently rising to be seen amidst the rocks and cliffs."

Called the Dingle I-Cosh squid, *this sea monster was seen in Ireland during 1673.* Photo, British Museum of Natural History, London, England.

Hans Egede helped several artists reconstruct this version of a sea monster that he called a hafgufa. Photo, Forbes Collection, Hart Nautical Museum, Massachusetts Institute of Technology, Boston, Massachusetts.

4

1700 to 1800

Early in the 1700s, several reported sea monster sightings were mentioned briefly in print. A physician named Christian Paulinus of Eisenach and a Danish priest reported an incident with a kraken. By now, this Old Norse word was applied to any unknown animal of the sea that was large or was thought to be large. Later the word *kraken* came to represent any one of the more than 350 known species of giant squid, some of which have been reported to range up to 100 feet in length. This assertion about length has never been proved, but certainly these monsters measured up to 60 and 80 feet. In this reporting by Paulinus, there is obviously an exaggeration, for the doctor claims that his companion, like so many priests before him, celebrated mass upon the animal's back. Today, the word *kraken* has reverted to its original application, for *Webster's Third New International Dictionary* shows it as meaning "a fabulous Scandinavian sea monster."

Also, at the turn of the century, an old reprobate explorer and buccaneer named William Dampier commanded two English expeditions to Australia. While there, aboard his ship *Roebuck,* he reported having seen several sea serpents of great length that were "as big around as a man's leg."

Reports of celebrating mass and building fires for cooking on the backs of sea monsters were common during the twelfth century. Photo, Museo del Mare, Italy.

Hans Egede was back in print with more fantastic tales of sea monsters. Again referring to the hafgufa, he said that it became completely passive when called by name and was "so awfully terrible and frightening that this writer does not know how to describe it, and this without wonder, because he never had true association with it; its shape, length and bulk seem to be beyond all size and means of measure. Those who would pretend that they have seen it, assert that it appeared more like an island than any fish, or sea animal."

Egede went on to say that there had never been more than two of these seen above the surface of the sea, which would indicate that there could not be a breed of such animals. For, as he reasoned, if such enormous sea monsters should ever succeed in multiplying, all the rest of the fish in the ocean would be destroyed because their large bodies demanded such great quantities of nourishment.

"When the monster is hungry," Hans Egede explained, "it is said to void some essence of a sweet scent, which perfumes the entire sea around; and by such means does it allure and attract all sorts of fishes and animals, even to the whales, that come close to it. They flock to this point and swim into the wide open mouth of the hideous monster, as if caught in a whirlpool, till its belly is well freighted with a generous load of all sorts of fishes and then it closes the mouth, and has for an entire year sufficient to live on; for it is claimed that it eats but one meal each year."

About 1741, Egede's son Povel published this somewhat altered drawing of a sea monster that his father had seen in Greenland. Photo, Naval Historical Foundation, Washington, D.C.

The Reverend Mr. Egede, in his book *The Natural History of Greenland*, also told of another sea monster which he said fishermen called a Drew. This peculiar beast is not known to have kept any particular shape or figure. It frequently was heard and seen just before a shipwreck or other nautical misfortune.

During these early years of the eighteenth century, the sciences—chemistry, mathematics, physics, medicine, and biology—showed considerable advancement. Many men of learning became proficient in several sciences. One such man was Carolus Linnaeus (1707–1778), who was born in Rashult, Sweden. The son of a minister, he studied botany at the University of Uppsala. Later he switched to medicine and received his M.D. degree from the University of Harderwijk, Holland. His education represents the best that the era had to offer. During his lifetime, he had 180 books published, including several pertaining to Lapland and Sweden.

Carolus Linnaeus, by his own admission, was a firm and vocal believer in sea monsters and sea serpents. In his first edition of *Systema Naturae*, Linnaeus uses the old term *kraken* as a label for

An illustration from an old book telling how to outwit a sea monster. Photo, Musée Maritime, Marseilles, France.

all marine oddities of great bulk and writes that "our fishermen unanimously affirm, and without the least variation in their accounts, that when they row out several miles to sea, particularly in the hot summer days, and by their situation (which they know by taking view of different points of land) expect to find eighty or a hundred fathoms of water, it happens that they do not find above twenty or thirty, and sometimes less. At these places they generally find the greatest plenty in fish, especially cod and ling. Their lines, they say, are no sooner out than they may draw them up with the hooks all full of fish. By this they know the kraken is at the bottom. They say this creature causes those unnatural shallows mentioned above, and prevents their sounding."

According to Linnaeus, these fishermen are always glad to find such a sea monster, for it guarantees them a plentiful catch. He says they operate as a fleet, with twenty or more boats in a line. All trail their lines astern, and if the depth stays the same for all boats, the men busy themselves fishing. However, if the water begins to turn shallow, "they know that the kraken is raising himself nearer the surface, and then it is not time for them

Two versions of a sea monster as depicted by I. I. Schipper of the Netherlands. Photo, Rijksmuseum Voor de Geschiedenis der Natuurwetenschappen, Netherlands.

Erik Pontoppidan described his first sea monster as looking something like this. Photo, Forbes Collection, Hart Nautical Museum, Massachusetts Institute of Technology.

to stay any longer; they immediately leave off fishing, take to their oars, and get away as fast as they can."

When out of danger, Linnaeus goes on to say, the fleet stands by to watch. In a few minutes, a great monster rises to the surface; however, although a large portion becomes visible, an even larger share remains submerged. "Its back, or upper part," he writes, "which seems to be in appearance about an English mile and a half in circumference (some say more, but I chuse the least for greatest certainty), looks at first like a number of small islands surrounded with something that floats and fluctuates like sea-weeds. Here and there a larger rising is observed like sandbanks, on which various kinds of small fishes are seen continually leaping about until they roll off into the water from the sides of it; at least several bright points or horns appear, which grow thicker and thicker the higher they rise above the surfaces of the water, and sometimes they stand up as high and as large as the masts of middle-sized vessels. It seems these are the creature's arms, and it is said if they were to lay hold of the largest man of war they would pull it down to the bottom."

Not content with describing the kraken's great size in somewhat general measurements, Linnaeus says that when the sea monster decides to submerge again, so massive a bulk going down causes a whirlpool that can easily suck the entire fishing fleet to the bottom of the sea.

Scientist Linnaeus, after classifying this kraken with other "polype, or of the starfish kind," describes the sea monster by saying that "the great Creator has also given this creature a strong and peculiar scent, which it can emit at certain times, and by means of which it beguiles and draws other fish to come in heaps about it. This animal has another strange property, known by the experience of many old fishermen. They observe that for some months the Kraken or Krabben is continually eating, and in other months he always voids his excrements."

Bishop Pontoppidan attributed divine powers to sea serpents, as can be seen in this early drawing. The devil, especially, was in reality a sea monster. Photo, Glasgow Museum, Glasgow, Scotland.

This piece of art clearly tells the story of what may happen to any ship wrecked in the Indian Ocean. Photo, Rex Nan Kivell Collection, National Library of Australia, Canberra, Australia.

During these evacuations, Linnaeus writes, the water is colored with excrement, which appears to be quite thick and turbid. This muddiness, he says, is so agreeable to other fish they come in great schools to gather in a spot directly over the kraken. "He then opens his arms or horns, seizes and swallows his welcome guests, and converts them after due time, by digestion, into a bait for other fish of the same kind.

During this early period of the new century, Pontoppidan also reverted to the earliest meaning of *kraken* and told of a sea monster in the following words:

"Many authors tell of certain islands which suddenly appear and suddenly vanish. This was something that no one could comprehend; so that one should not wonder at the common people, and even if those who were a degree above them, for looking upon those moving islands to be inhabited by evil spirits, which appeared sometimes in such places where the fishermen, by daily experiences, knew that there was no such thing as a rock, far

A sea chart printed in France about 1555, clearly warns the mariner against sea monsters. Photo, National Library of Australia, Canberra, Australia.

Pontoppidan's sea serpent, as depicted in a magazine article. Photo, Library, City of Liverpool, England.

less an island. They often, however, discovered something at sea which bore the resemblance of land, and consequently were confounded, made false reckonings, and were taken off course, and brought into greater dangers."

Bishop Pontoppidan, like Linnaeus and Egede, shows no uncertainties about his sea monster. The information he has at hand is the absolute truth, at least to his own satisfaction. He says that numerous mariners give accounts of such appearances of land and of their sudden vanishing, and particularly in the North Sea. These lands, he states, in an "ocean so boisterous," cannot be misconstrued as being of those *real floating lands* that are observed, on stagnant water, in and about Norway. These real floating islands, so he tells us, could not stand as firmly against the massive waves of the deep ocean.

Bishop Pontoppidan, like so many devout clerics before him, then rationalized the entire matter by attributing it to Beelzebub. "Our seamen have, therefore concluded that this delusion could stem from no other than the great deceiver, the Devil. But I think that we should not charge so apostate a spirit without cause. I believe that this devil who so suddenly makes and unmakes such pieces of land is nothing but the *kraken,* which many seafaring people call *Sol-draulen,* of sea-mischief."

At about this time Pontoppidan brought out another book, entitled *Det forst Forsog paa Norges naturlige Historie* (the Natural History of Norway). In it he said, "Thorlack Thorlackson has told me that in 1720 a sea serpent had been shut up a whole week in a little inlet, into which it came by high tide through a narrow entrance of seven or eight fathoms deep, and that eight days afterwards, when it had left the inlet, a skin of a snake or serpent was found. One end of the skin had entirely sunk into the water of the inlet, and no one could guess how long it was, the inlet in which the skin partially lay, being several fathoms deep. The other end of this skin was washed ashore by the current, where everybody could see it; apparently it could not be used, for it consisted of a soft slimy mass. Thorlackson was a native of the harbor of Kobbervueg."

In 1740, Hans Egede's book entitled *Det Gamle*

116 THE VARIOUS ACCOUNTS, [N°. 5.]

News, Oct. 28, 1848, and in Mr. LEE's *Sea Monsters Unmasked*, London, 1883. A reduced copy of it also appeared in the *Illustrirte Zeitung* of February 3, 1877.

Fig. 21. — The drawing of Bing, as reprinted and altered in the Illustrated London News of 1848.

In the Danish work of HANS EGEDE *Det gamle Groenlands nye Perlustration* we read: "that it was seen at 64° lat. before the colony". "Its body was as thick as the ship and three or four times longer". Moreover the description of the animal is the same as in PAUL EGEDE's *Continuation* of the Journal.

In the German edition of this work, entitled *Des alten Groenlands neue Perlustration*, Copenhagen, 1742, we read: "that it was seen before the Danish Colony, the Good Hope, that it had

Egede's sea monster, as drawn by Bing was reprinted in the Illustrated London News *in 1848.* Photo, Gallop International Photo Service.

Grönlands nye (perlusträtion eller Naturel-Historie) was published. In it, he said that "on the 6th of July, 1734, there appeared a very large and frightful sea monster which raised itself so high out of the water that its head reached above our main-top. It had a long, sharp snout, and spouted water like a whale; and very broad flappers. The body seemed to be covered with scales, and the skin was uneven and wrinkled, and the lower part was formed like a snake. After some time the creature plunged backwards into the water, and then turned its tail up above the surface, a whole ship-length from the head. The following evening we had very bad weather."

In 1745, he published another book, *A Description of Greenland,* in which he told of either a new monster or another version of the one just described. As quoted by Henry Lee in 1883, Egede now saw it in the following light: "This monster was of so huge a bulk, that when it came out of the water, its head reached as high as the mast-head; its body was as bulky as our ship, and of about three or four times our own length. It possessed a long pointed snout, and spouted like a whalefish; great broad Paws, and the body appeared to be covered with shell-work, its skin very rugged and uneven. The underside of the body was shaped like an enormous Serpent, and when it dived to go beneath the water, it plunged backwards into the Sea, and raised its tail aloft in so doing, which appeared to be at least one ship's length distance from the more bulky part of the body."

Henry Lee, curator of the Brighton Aquarium, editor of a publication called *Land and Water*, and author of *Seamonsters Unmasked*, had this evaluation to make, some years later, of both Egede and his sighting: "Egede was a truthful, pious, and single-minded man, possessing considerable powers of observation, and a genuine love of natural history; his statements are modest, accurate, and free from exaggeration. His illustrations bear the unmistakable signs of fidelity." Lee then went on to say that from what had been said of the animal seen by Egede, it must be gathered that it appeared on July 6, 1734, in fine weather before the Danish colony of Good Hope, Davis Strait, Greenland. Henry Lee continues his comment in these words, "Egede says . . . the body *seemed* to be covered with hard skin. For truth sake Egede wrote *seemed,* which is well done; for a hard skin or crust would not have been *wrinkled* when the animal bends its body." In his critical analysis of Egede's various reports, Lee was forced to conclude that they were sincere and truthful.

At this point it would be well to note that scientists of the United States Naval Oceanographic Office, one of the largest oceanographic organizations in the world, have made several astounding observations that relate to the written works of Egede. For example, as recently as November, 1963, scientifically schooled observers witnessed an island rear up out of the sea just south of Iceland. The emergence of this volcanic island, Surtsey, could well be a repetition of what might have happened many centuries before, thereby resulting in the belief in supersized sea monsters. Again, during 1969 oceanographers of this same office recorded the jumpings and cavortings of whales that, in many instances, because of the presence of several, could easily have been mistaken for a single sea monster that performed exactly as Hans Egede described.

While Hans Egede and Bishop Pontoppidan unquestionably held the center stage during these years, lesser reports of sea monster sightings can

Copy of a sketch released by Henry Lee, naturalist in charge of the Brighton (England) Aquarium, pertaining to the sea monster seen by Egede. Drawing by the author.

Dagfind Korsbeck caught this sea monster prior to 1744. It was placed on view for the public. Photo, Thijs Mol, Arnhem, Netherlands.

be found. A few of these instances will be noted here. In the pages of a book dated 1742, written by Antonino Mongitore, and entitled *Items of Interest Taken from a Sicilian Trip*, we find a brief report from the Mediterranean. "There along the sandy shores of the Sahara Desert, we were informed of an imposing sea monster—spotted in numerous colors and, ... having in many ways the shape of a serpent—that constantly tore up fishing nets, thereby letting a horde of fish escape." Such occurrences were confirmed by many witnesses, for net fishing in this area had always been a function of an entire town. The housewives of each village joined together to weave especially strong nets, while the town's fathers worked them as a single unit. Everyone—men, women, and children—was present on special days to help string out the one great net and pull it in, thereby receiving shares in the profit. As a result of this communal effort, a net being ripped apart by a gigantic monster of the sea was a well-documented event.

Another obscure report comes from a fisherman named Dagfind Korsbeck, who, prior to 1744, caught a sea monster in the parish of Sundelvems on Sundmoer. The beast was seen by many villagers, and the descriptions given by these illiterate people were recorded in several Scandinavian languages. There are various translations available, but the most comprehensible is the one especially prepared for this book by *Dansk Sportsfiskerforening*. It states in concise form that the sea monster in question was a "monstrous fish, as tall and round as a good sized mast, which many people were able to view at Dagfind Korsbeck's home. The head of the beast was like that of a cat and it purred in the manner of a feline. About its body was a hard shell and when it was touched by a stick, it would bight [bite] at the offending object. The oddity of all of this frightened the superstitious people and they returned the monster to the sea."

Those who observed sea monsters and other dangerous creatures of the deep inevitably used their own vessels as yardsticks of comparison for both size and speed. Hans Egede measured a sea monster's height by stating that "the animal reached above our main-top." A little later on he said that the tail was "an entire ship length from the rest of the body." Immediately the question arises: How tall and how long was the ship? Unfortunately, he has not provided us with a description of the vessel from which he made these observations, but we can get an approximate idea of what he meant by looking at the type of ship that sailed the seas during this particular century.

A typical merchant ship of the 1700s. Photo, Glasgow Museum, Glasgow, Scotland.

Commercial ships were round-bottomed and measured up to 70 feet, with masts that projected upward for almost 85 feet. Generally speaking, the hulls were approximately three times the length of their beam, with heavy suited sails towering higher above the deck than the entire distance from bow to stern. Such dimensions created a slow and unstable ship.

Usually, only one man, the ship's master, or at best a few privileged persons, saw the encountered sea monster well enough to make a report. There is a logical explanation for this. A square-rigged ship was literally a mass of rigging, stays, braces, lines, lifts, booms, spars, tackle, yardarms, and gammoning, all hidden behind immense masts and billowing sheets of canvas. Such surroundings could hardly be deemed ideal settings from which to make accurate observations. On the other hand, the poop deck, that raised section over the stern, was "officers' country"; and if a seaman was required to go into this area, the section where the helm was located, he was careful to use the lee ladder, the weather side of the poop being reserved for the ship's master or whatever officer might be in charge. Since the weather side of the poop offered the best view of the area surrounding the ship and the officer of the watch was the only man not having his hands occupied, it was generally he who caught a glimpse or two of such oddities as a unique sea beast.

There are many reasons why more specifics about sea monsters—such as exact location, state of the seas, depth of the bottom, and proximity to shoreline—were not logged. The answer rests in the crude tools with which a ship's master had to work. He possessed a sextant, compass, chronometer, and barometer, plus the ability to take deep-sea soundings with a weighted line, provided the ship was not canted over while making a fast clip. In other words, he knew his approximate position, likely direction, and hoped-for depth. If he were

Commercial ships were round-bottomed and poorly constructed, as can be gathered from this etching showing a boatyard of this period.

Reporting the location of a sea monster sighting was at best haphazard because nautical instruments in those days were not accurate. Here a mariner is depicted shooting the sun with a quadrant. Photo, Spanish Archives, Madrid, Spain.

to be suddenly confronted by a sea monster, it is hardly likely that he would scramble his already-busy watch just to lay main sails aback in order to stay his forward progress so that all hands could make observations.

In about 1744, we find a brief but intriguing notation by Pontoppidan that "it is also told that a sea serpent stranded near the Isle of Karmen and that the stranding of sea serpents took place in more localities."

The casualness of this remark was later explained by Dr. Oudemans, who said that indeed there was nothing strange in the stranding of sea serpents. Unluckily, the Norwegian people's fear of sea serpents, he explained, was great enough to keep them far away from them, even from their carrions, and so these accounts do not include the results of closer investigation.

During 1745, Pontoppidan recorded: "A fisherman relates to me that, on Sunderland, two miles from Bergen, he once saw a long, large and strange animal so close to his boat, that the water, brought in motion by the animal, dashed against it, but immediately it disappeared again under water. The head resembled that of a seal, its skin was also as woolly, but the body was as thick and as long as a yacht of fifty tons, and the tail, which seemed to be about thirty-five feet long, tapered towards the end which was as pointed as a boat-hook."

In 1754, a German named Johann G. O. Richter proved that catfish could grow to massive size and could easily kill one or more men. A few years later, explorer R. P. Gumilla reported finding giant stingrays in the Orinoco Basin of Venezuela. His works were published in Paris in 1758 under the title *Histoire naturelle, civile et géographique de l'Orénoque, et des principales rivières qui s'y jettent.*

It is important that such discoveries or conclusions be noted because it might now be fairly assumed that in many cases such oddities were mistakenly identified as sea monsters.

In the same year as Richter's announcement, a French scientist named François-Doissier de la Crois de Sauvage introduced to the world an interesting study on giant sharks under the title of *Venematis Galliae Animalibus.* At once, news about mammoth sharks began to appear in headlines from various sources. A Frenchman wrote into the town log of his fishing village that "the headless body of a man encased in armor, has been found in the stomach of a monstrous shark."

In themselves, these things mean little in our search for sea monster sightings, except for one important point: People were slowly being educated to the fact that every sea lion, shark, or whale was not a sea monster. As a result, reported sightings became more accurate.

Pontoppidan released another writing, in which he gives the following fully authenticated and thoroughly witnessed sighting of a sea monster:

"Last winter I happened to meet the Royal Commander and Pilot-General at Bergen, Mr. Lorenz Von Ferry, and we spoke about this subject. He told me that for a long time he had doubted the existence of the sea monster, but that at last his experiences in 1746 had convinced him. And though I could not say anything of importance against it, he offered to my satisfaction and that of others, two seamen, who were with him in his boat, and had seen the animal and its blood which coloured the water red after a shot of Von Ferry at it, to appear before the public court of justice at Bergen. What those men confirmed on oath may be found in the following instrument which I received in original, and which I therefor think valuable enough to communicate in extenso. . . ."

There then follows a name-packed document, replete with titles, seals, and legal jargon such as "begging that these men might be admitted to a declaration on oath. . . ." After all signatures had been affixed, the following statement was duly recorded:

"In the latter end of August, in the year of 1746, as I was on a voyage, on my return from Trundheim, on a very calm and hot day, having a mind to put in at Molde, it happened that when we had arrived with my yacht within a mile of the aforesaid Molde, being at a place called Jule-Naess, as I was reading in a book, I heard a kind of murmuring voice from amongst the men at the oars, who were eight in number, and observed that the man at the helm kept off from the land. Upon this I enquired what was the matter, and was

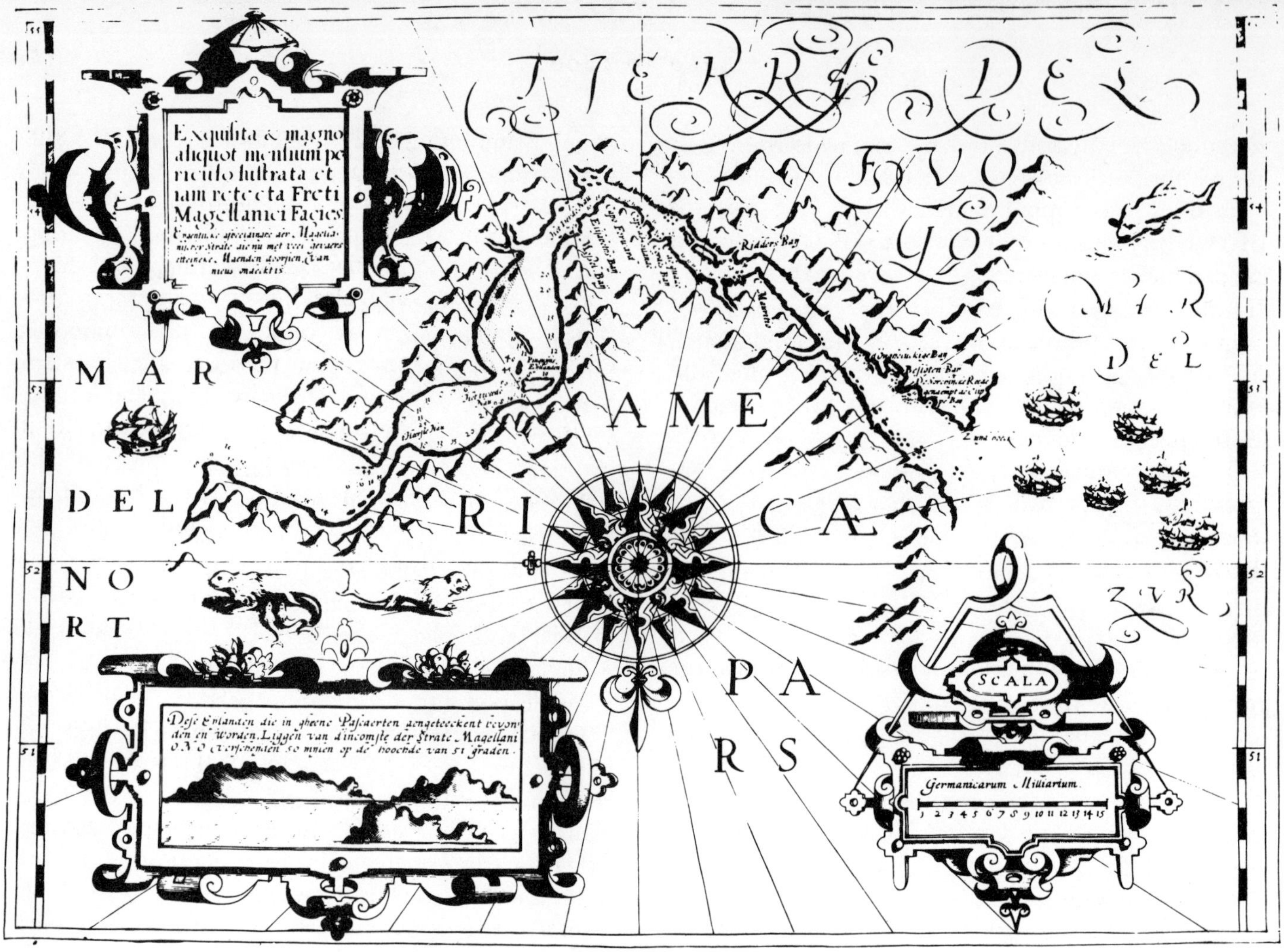

This is part of a sea chart drawn by Gerhardus Mercator in 1613 and illustrating several sea monsters. Photo, National Library of Australia, Canberra, Australia.

informed that there was a sea serpent before us. I then ordered the man at the helm to keep the land again, and to come up with the creature of which I had heard so many stories. Though the fellows were under some apprehension, they were obliged to obey my orders. In the meantime the sea snake passed by us, and we were obliged to tack the vessel about in order to get nearer to it. As the snake swam faster than we could row, I took my gun which was loaded with small shot, and fired at it; on this he immediately plunged under water. We rowed to the place where it sank down (which in the calm might easily be observed) and lay upon our oars, thinking it would come up again to the surface; however it did not. Where the snake plunged down, the water appeared thick and red; perhaps the small shot might have wounded it, the distance being very little.

The head of this sea serpent, which it held more than two feet above the surface of the water, resembled that of a horse. It was of a grayish colour, and the mouth was quite black, and very large. It had black eyes, and a long white mane, which hung down from the neck to the surface of the water. Besides the head and neck, we saw seven or eight folds, or coils, of this snake, which were very thick, and as far as we could guess there was a fathom's distance between each fold. I related this affair in a certain company, where there was a person of distinction present, who desired that I would communicate to him an authentic detail of all that happened; and for this reason two of my sailors who were present at the time and place where I saw this monster, namely, Niels Petersen Kopper, and Niels Nielsen Angleweigen, will appear in court, to declare on oath every particular herein set forth; and I desire the favour of an attested copy of the said descriptions."

In 1749, Pontoppidan said that a Mr. Tuchsen of Herroe told him that he too had seen the animal several times when he went to church in his boat. To this Pontoppidan then adds that he could mention "many other persons of the same high rank and trustworthyness. The same Mr. Tuchsen is the only one who told me that he distinctly saw the difference in thickness between the trunk and the tail of the animal, viz., the trunk is not gradually growing smaller where the tail begins, but becomes smaller at once and very distinctly. The body is as thick as a barrel of two hogsheads. The tail is tapering towards its end, which is very pointed."

Although Pontoppidan was the most prolific writer on sea monsters at this time, there were other sea serpent reports turned in by quite reliable sources. Aboard the H.M.S. *Intrepid*, in 1751, a Joseph Kent reported sighting a sea serpent; in 1779, a Stephen Tuckey claimed to have seen a sea monster; and in 1780, off Meduncook, another monster was sighted. None of these were well substantiated, and only brief mention can be found of each in any of the archives.

Again, in 1750, Pontoppidan reported that "one of these north-sailors tells that he was once so close to the serpent that he might have touched its smooth skin." In 1751, he stated that "an uncertain rumor tells me, that some peasants of Sundmoer have lately captured in their nets a serpent eighteen feet with four paws under its belly; which they brought ashore. Thus, it resembled a crocodile. The peasants in their terror fled from their nets, and by doing so they gave an opportunity to the serpent to do the same."

After hearing hundreds of eyewitness accounts concerning sea monsters, Pontoppidan finally came to these conclusions: "The sea serpent, *serpens-marinus*, by some people also called Aele Tust, is the second wonderful and frightful sea monster which ought to be studied by him who looks with delight on the great deeds of the Lord, and which is considered as the greatest wonder next to the Kraken, which will be described hereafter. Before describing its habit and shape, I feel again obliged to prove the real existence of the serpent, as I did before with the mermen."

It must be noted here that Pontoppidan did not mean to imply that he believed in mermen. He long ago had concluded that both mermen and mermaids were "nothing else but sea cows or manatees." As is said by one of his numerous biographers, some of the mermen Pontoppidan wrote about were nothing else but bladdernose seals. However, Pontoppidan goes on to say, "If it were not the wise and careful arrangement of the Creator, that this sea animal perpetually lives in the depths of the sea, except in July and August, its pairing-time, during which it appears, when the sea is quite calm, but dives as soon as the wind ruffles the surface of the water; if this arrangement, I say, were not thus made for man's safety, the existence of the sea serpent would want fewer proofs, than even in Norway, thanks to God! is the case, the shores of which are the only ones of Europe, frequented by this monster."

Pontoppidan obviously is in error here, for sea serpents and sea monsters were previously sighted in the Davis Strait and the Baltic Sea, as well as in certain far reaches of the Orient, all of which must have been known to him. In any event, it becomes obvious that Pontoppidan never considered that the beast could be migratory. Furthermore, once his mind was made up, he was inflexible. He betrays this characteristic when he writes: "Like all who are enemies to credulousness I too doubted of the existence of the sea serpent, when at last my doubt was dispelled by incontestable proofs. Amongst our ablest navigators and fishermen of this country there are many hundreds who prove the existence of the sea serpent as eyewitnesses, and they agree pretty well in their descriptions, though there are many others who declare that they know the sea serpent only from the tales of their neighbors. I, however, in my inquiry hardly met with a person who when born in the Northern provinces, did not answer immediately with the greatest certainty and assurance. Nay, some so-called north-sailors, who are here (in Bergen) every year for commercial interests, even consider it a shame to be earnestly questioned on that subject. They consider this question as superfluous as that one, whether there exists a cod-fish or an eel."

In 1751, in a letter from Captain George Little to the Reverend Alden Bradford, which appeared in the *American Journal of Science and Arts*, there is this comment: "A sea monster of the above description was seen in the same place (Round Pond in Broad Bay near Muscongus Island), by Joseph Kent, of Marshfield. Kent said he was longer and larger than the main boom of his sloop, which was 85 tons. He had a fair opportunity of viewing him, as he saw it within ten or twelve yards of his sloop."

A Mr. Knud Leems, in 1765, wrote in his book entitled *Beskrívelse over Finmarkens Lappen:* "The Finmark Sea also produces the hydra or sea serpent, a huge monster, forty 'passus' long [200 feet], with a head resembling in size that of large sea fishes, in shape that of a snake. This beast has a neck with a mane like a horse's, a grey back and whitish belly. In the dog days, when the sea is free from wind, the sea serpent will come to the surface, bend itself in several coils, of which some are partially visible above the water, whilst others remain hidden under it. The seamen greatly fear this monster, and they do not trust themselves on the sea, when the animal is on the surface."

As the century came to an end, reports of sea monster sightings become more frequent. One significant but scoffed-at report was contained in a letter written by the Reverend William Jenks, which in part reads: "He (Mr. Staples of Prospect) told me also that about 1780, as a schooner was lying at the mouth of the river, or in the bay, one of these enormous creatures leaped over it between the masts—that the men ran into the hold for fright, and that the weight of the serpent sunk the vessel 'one streak' or plank. The schooner was of about eighteen tons."

In a subsequent letter, addressed to the Honorable Judge Davis, of Boston, the Reverend Mr. Jenks said of a sea serpent: "The British supposed the length of that which they saw to be 300 feet, but this Mr. Cummings imagines to be an exaggeration." It would appear as though *both* of Mr. Jenks's stories were exaggerations.

In the records of the *American Journal of Science and Arts*, we find an interesting letter written

The monster shown in this detail of an old chart is said to have been the type that, according to the story told by a Mr. Staples of Prospect, leaped over a ship. Photo, Schiffahrts Museum, Brake, West Germany.

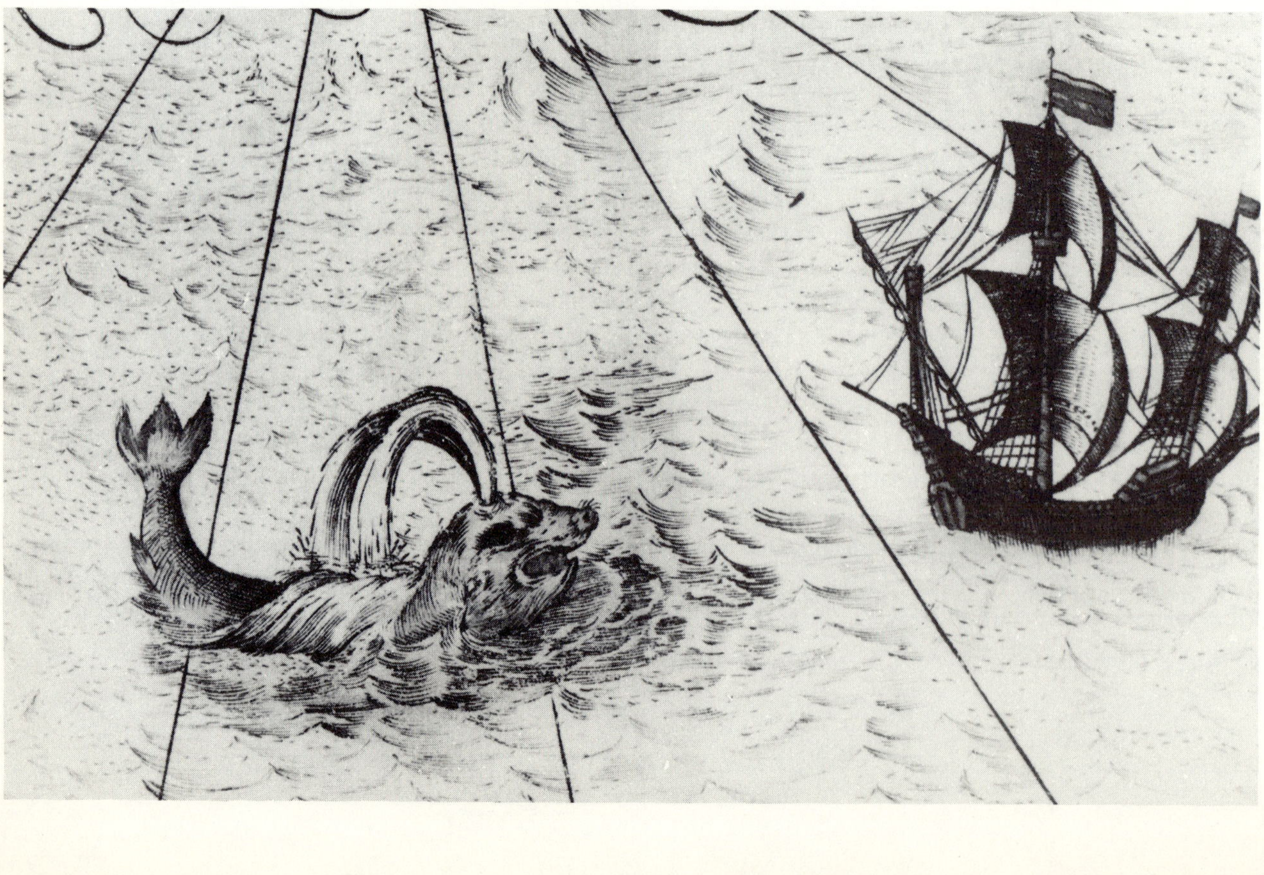

by the previously mentioned Captain George Little about the sightings he reported during 1780. His letter had been prompted by a request for an affidavit.

"In answer to yours of 30th. of January last, I observe, that in May, 1780, I was lying in Round Pond, in Broad Bay, in a public armed ship. At sunrise, I discovered a large Serpent, or monster, coming down the Bay, on the surface of the water. The cutter was manned and armed. I went myself in the boat, and proceeded after the serpent. When within a hundred feet, the mariners were ordered to fire on him, but before they could make ready, the Serpent dove. He was not less than from 45 to 50 feet in length; the largest diameter of his body, I should judge, 15 inches; his head nearly of the size of that of a man, which he carried four or five feet above the water. He wore every appearance of a common black snake. When he dove he came up near Muscongus Island. . . . we pursued him, but never came up within a quarter of a mile of him again."

In a contemporary magazine called *Zoologist*, there appeared this very interesting letter: "Having seen much notice taken in the *Zoologist* of the question of the great sea serpent, allow me to subjoin an extract from the log-book of a very near relative, dated August 1st, 1786, on board the ship '*General Coole*,' in lat. 42 degrees, 44 minutes N., and long. 23 degrees, 10 minutes W.—

"A very large snake passed the ship; it appeared to be 16 or 18 feet in length, and 3 or 4 feet in circumference, the back of a light ash-colour and the belly thereof yellow.

"According to the log the ship was becalmed at the time. You may rely on the correctness of this, and any one desiring to satisfy himself may see the original log. [Signed], S. H. Saxby; Bouchurch, Isle Of Wright."

Regarding the incident reported by Saxby, Dr. Oudemans makes this personal evaluation: "Only the length is given of the visible part, else it would be impossible that an animal of 16 feet in length and 3 or 4 feet in circumference made the impression of being a serpent or snake; the whole trunk and tail must have been hidden under water. As the colour of the animal's back is noted as a light ash-colour, I suppose that the animal having swum a long time in the sun without diving under water, the skin had become dry and showed the ash-colour; the colour of the belly (read throat) is stated to be yellow. This statement already mentioned we shall see repeated more than once."

During these years between 1700 and 1800, others evaluated the known facts. One recognized Norwegian authority writes as follows:

"We see hereby that in Norway the belief in the existence of the sea serpent was as firm as possible amongst the sea-faring people.

"Though no one has ever been able to measure this animal, many witnesses agree in telling that the serpent must be as long as a cable, viz., 100 fathoms [600 feet], or 300 ells, whilest it lay on the surface of the water, so that only here and there behind the head, which is held upwards, some parts of the back were visible, which were also held upwards, whilest the serpent bent; and from afar one would have believed that he saw some tuns [a large cask] or hogsheads, which floated in a line, so that there was space between each of them.

"Though the length of a cable or six hundred feet given to the serpent is exaggerated, it may be more than 100 feet, why not? For there are other sea animals, such as the whale, which measures more than 88 feet, and the fin-fish (*Balaenoptera boops*) which sometimes attains a length of 105 feet.

It has been stated to Pontoppidan by most of the eye-witnesses that the animal shows by its vertical undulations several coils above the water, and that these coils resemble from afar tuns or hogsheads floating in a line. It is very remarkable that these facts are repeatedly stated by witnesses who are independent of one another, even by persons who never heard of a sea serpent.

"The head of all of these animals has rather a high and broad forehead; some, however, have a sharp snout, others a quadrangular beak as cows and horses have, with large nostrils, and on the side there are a few stiff hairs, or bristles, as other animals have with a good nose. And that the sea serpent has a good nose, is proven by its flying away at the smell of castoreum, which the people who go out in the summer to fish on the great bank, will never forget to take with them."

A widely published picture depicting the Gloucester sea serpent. Photo, Naval Historical Foundation, Washington, D.C.

5

1800 to 1850

By the beginning of the nineteenth century, there was a substantial increase in the number of sea monster sightings and more evidence that they were authentic.

Let us look at a few well-substantiated cases of this period. There was a letter about a sighting that occurred in July, 1802, written by the Reverend William Jenks, of Bath, to the Honorable Judge Davis, of Boston. "The Rev. Mr. Abraham Cummings who has been much employed in Missions in the District of Maine, and navigated his own boat among the islands, etc., in the discharge of his duty, informs me, in conversation, which was immediately written from his lips, that in Penobscot bay has been occasionally seen within these thirty years, a sea serpent, supposed to be about 60 feet in length, and of the size of a sloop's mast. Rev. Mr. Cummings saw him, in company with his wife and daughter, and a young lady of Belfast, Martha Spring; and judged that he was about three times the length of his boat, which is twenty-three feet. When he was seen this time he appeared not to notice the boat, though he was distant, as nearly as could be ascertained, but about fifteen rods."

Correspondence continued between Mr. Cummings, the American Academy of Arts and Sciences, and Judge Davis. One of these letters, written by Cummings himself about his sighting of the sea monster, is well worth quoting from here.

"It was some time in July 1802 that we saw the extraordinary sea monster, on our passage to Belfast, between Cape Rosi and Long Island. His first appearance was near Long Island. I then supposed it to be a large shoal of fish with a seal at one end of it, but wondered that the seal should rise out of the water so much higher than usual; but, as he drew nearer to our boat, we soon discovered that this whole appearance was but one animal in the form of a serpent."

Cummings states that the sea monster's mode of swimming was exactly the same as had been described to him by some people of the Fix Island group who had sighted the animal before. This fact, according to Cummings, confirmed what he had seen. Cummings then went on to say that the creature had not "the horizontal but an ascending and descending serpentine motion. This renders it highly probable that he never moves on land to any considerable distance and that the water is his proper element. His head was rather larger than that of a horse, but formed like that of a serpent."

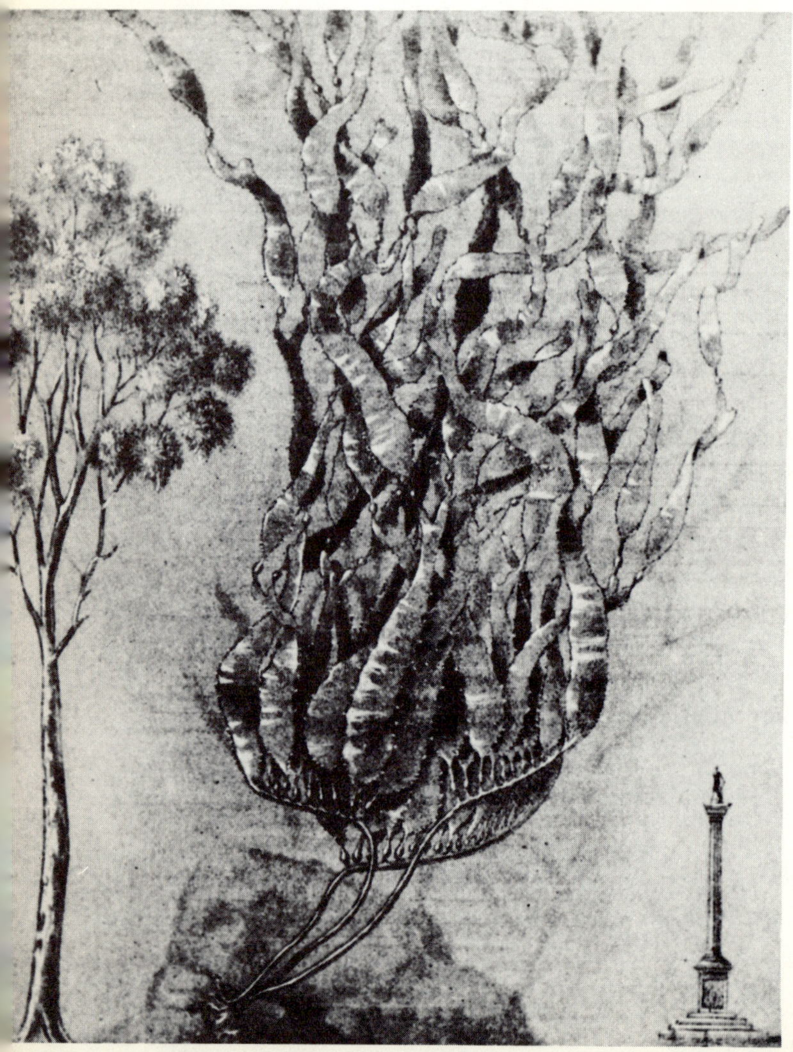

Continuing, Cummings stated that the body was 60 feet in length and that the head, plus the exposed part of the anatomy, was blue, the exception to this solid color being a black circle around each eye. The motion of the sea monster was at first moderate, but when it decided to move along the sea, it did so at a very rapid pace. Cummings then notes that "this monster is the sixth of the kind, if our information is correct, which has been seen in this bay within the term of eighteen years."

The Reverend Mr. Cummings said that his daughter and a Mrs. Martha Spring were with him in the boat at the time of the sighting and that they could be called upon to attest to his facts as stated.

It might seem strange that no indication was given in regard to these sightings that what they saw was actually seaweed, for it was well known that in the area of certain sightings, there existed seaweed plants that grew to 1,000 feet in length and weighed hundreds of tons. If a cannonball were to be fired at any one of these enormous, cylindrical plants, and the sea savagely disrupted by the action, it is likely that they would bob, weave, and act much like a serpent in the act of making an escape.

About 1805, a Mr. Rafinesque-Schmaltz, in his writings entitled *Additions*, says that "Mr. W. Lee has brought to notice another Sea-Snake, seen by him many years ago near Cape Brenton and Newfoundland, which was over 200 feet long, with the back of a dark green: it stood in the water in flexuous hillocks, and went through it with impetuous noise. This appears to be the largest on record and might well be called *Pelamis monstrosus;* but if there are other species of equal size, it must be called then *Pelamis chloronotis*, or Green-back Pelamis."

One of the most publicized sea monster sightings in history involved a creature that has come to be known as the "animal of Stronsa." It is first reported in an 1808 letter printed in a work entitled *Life and Letters of Campbell.* These statements were also picked up and reprinted by the magazine called *Zoologist.* "A snake (my friend

It is claimed by some mariners that seaweed can grow to 1,000 feet in length and take on a resemblance to a sea monster. Photo, Seamen's Church Institute of New York, New York City.

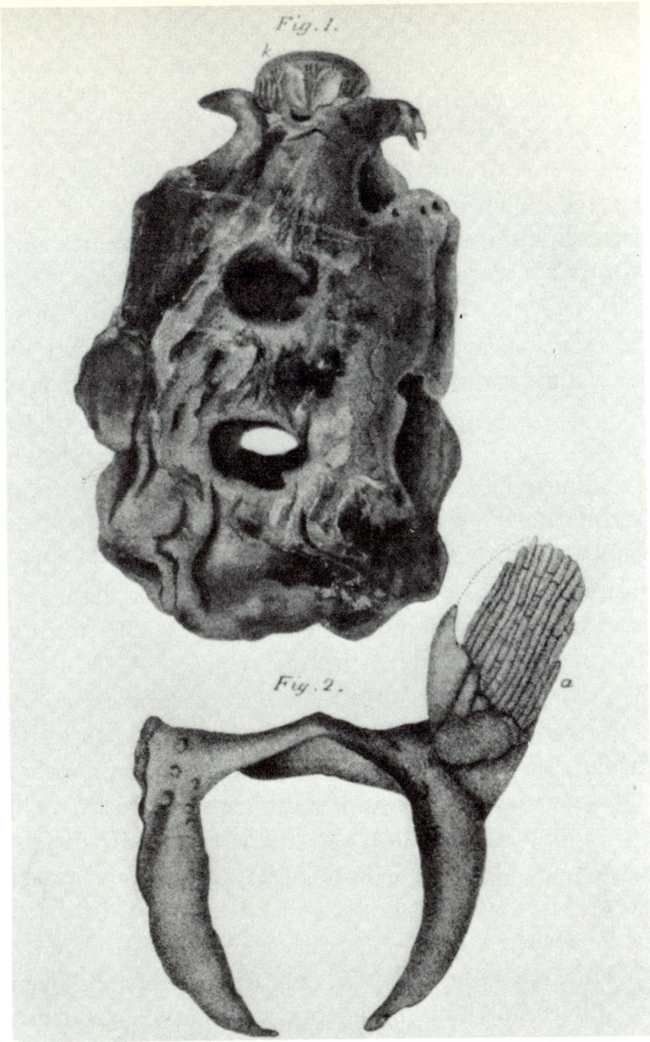

Skull and sternum of the "animal of Stronsa," from a book entitled Memoirs of the Wernerian Society, *published in 1808.* Photo, Naval Historical Foundation, Washington, D.C.

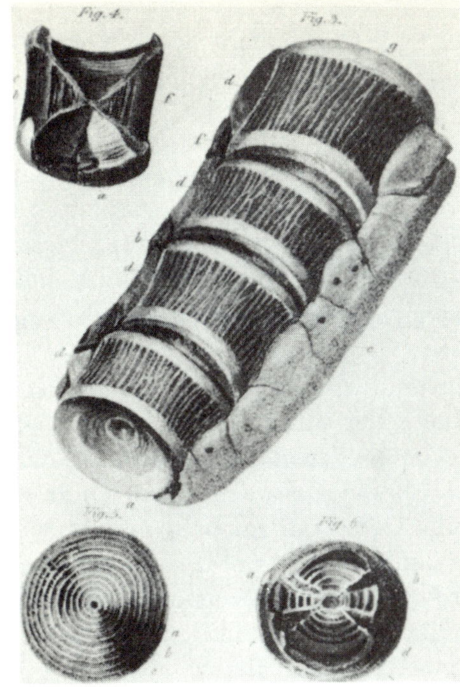

Vertebrae of the animal of Stronsa, 1808. Photo, Naval Historical Foundation, Washington, D.C.

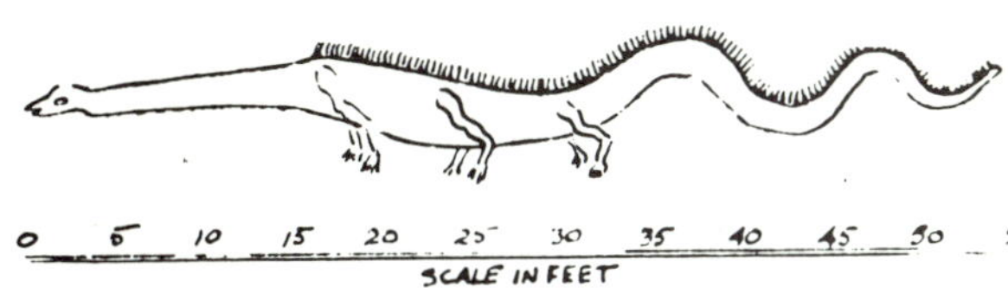

The animal of Stronsa from the report, "Philosophical Transactions Of 1809." Photo, Naval Historical Foundation, Washington, D.C.

Telford received a drawing of it) has been found thrown on the Orkney-Isles, a sea-snake with a mane like a horse, 4 feet thick and 55 feet long, this is seriously true. Malcolm Laing, the historian saw it, and sent a drawing of it to my friend."

The next recording of the animal, as well as its demise, is found in *Proceedings of the Meeting of the Wernerian Natural History Society*, dated November 19, 1808, and printed in the *Philosophical Magazine*.

"At this meeting Mr. P. Neill read an account of a great sea snake, lately cast ashore in Orkney. This curious animal, it appears, was stranded in Rothiesholm Bay, in the Island of Stronsa. Malcolm Laing, Esq., M.P., being in Orkney at the time, communicated the circumstances to his brother, Gilbert Laing, Esq., advocate at Edinburgh, on whose property the animal has been cast. Through this authentic channel Mr. Neill received his information. The body measured 55 feet in length, and the circumstance of the thickest part might be equal to the girth of an Orkney pony. The head was not larger than that of a seal and was furnished with two blow holes. From the back a number of filaments (resembling in texture the fishing tackle known by the name of silk-worm gut) hung down like a mane. On each side of the body were three large fins, shaped like paws, and joined. The body unluckily was knocked to pieces by a tempest; but the fragments have been collected by Mr. Laing, and are to be transmitted to the Museum at Edinburgh."

Another report read: "Kirkwell, Nov. 19, 1808 —Compeared John Peace, tenant in Dounatoun in Rothiesholm; and being interrogated, solemnly declares: That on the 26th day of September last, he went a fishing off the east part of Rothiesholm-Head, when he perceived as he imagined, a dead whale, on some sunk rocks, about a quarter of a mile from the Head: That his attention was first directed to it by the sea-fowl screaming and flocking about it, in his boat, he found the middle part of it above the water: That he then observed it to be different from a whale, particularly in having fins or arms, one of which he raised with his boat-hook above the surface of the water: That this was one of the arms next the head, which was larger and broader than the others nearer the tail; and at that time the fin or arm was wedged all around from the body to the extremity of the toes, with a row of bristles about ten inches long, some of which he pulled off, and examined in the boat: That about ten days afterwards, a gale of south east wind came on, and the surge drove the fish ashore on Rothiesholm-Head: That he measured it by fathoms, and found it about fifty-four or fifty-five feet in length. . . ."

The testimony continued to give a great deal of unscientific trivia about the animal and concluded on this note: "And all this he solemnly swears to be truth. And declares he cannot write." In other words, this man Peace was a workaday fisherman and knew a strange monster when he saw one.

Many people are inclined to view sea monster sightings as isolated episodes. This is not so. Frequently, the same type of monster was seen in places often many hundreds of miles apart, within a reasonably short period of time. In June, 1808, in the Hebrides off the northern coast of Scotland, there was another sighting. It was recorded in a letter addressed to the Wernerian Natural History Society by a man who signed his name as Donald Maclain. In part, the letter states: "I saw it in June of 1808 not on the coast of Eigg, but on that of Coll. Rowing along the coast, I observed, at about the distance of half a mile, an object to windward, which gradually excited astonishment. At first view it appeared like a small rock. Knowing there was no rock in that situation, I fixed my eyes on it close. Then I saw it elevated considerably above the level of the sea, and after a slow movement, distinctly perceived one of its eyes."

Alarmed at the unusual appearance and magnitude of the animal, Donald Maclain guided the craft in order to be close by the shore. Just as he succeeded in getting between the monster and the shoreline, the animal appeared to notice the boat for the first time and turned in its direction. Maclain put the tiller hard over and headed for shore. When the boat grounded, he leaped out and ran to a high vantage point, from which he could view the sea monster as it came under the stern of the now-deserted boat. However, finding itself in exceedingly shallow water, the beast "raised its monstrous head" and then began to struggle back to deeper water. Once back in deep water, the great animal appeared satisfied with what it had seen and moved off, with its head above water, until at last it disappeared in the distance.

Donald Maclain then wrote a very accurate description of what he had seen. "Its head was rather broad, of a form somewhat oval. Its neck somewhat smaller. Its shoulders, if I can so term them, considerably broader, and then it tapered toward the tail, which last it kept pretty low in the water, so that a view of it could not be taken so distinctly as I wished."

Maclain went on to tell that the sea serpent swam in an up-and-down fashion, was close to 80 feet in length, and at no time raised its head entirely out of water. Although, as he said, the neck was mostly submerged, there seemed to be no scales or armorlike skin. The sea monster had great speed, Maclain noted, which was only true as long as the head remained underwater. When the head was elevated, it seemed as if the sea monster had paused to take a look.

Could this have been the same beast as the one later found by John Peace? There are many differences in shape, but these are understandable. The animal that Peace saw had begun to decompose; whereas Maclean saw his sea serpent alive, but only once and from a distance.

Brief accounts of numerous other sea monster sightings can be found. Names are given, sometimes even places, but little else. Captain Lillis, in 1809, sighted a sea monster along the eastern coast of the United States; the first mate of the brig *Cox Neck*, in 1810, claimed to have seen a mighty sea serpent; and a fisherman named Klimb, in 1811, is reported to have netted a strange animal. Matthew Gaffney, of Gloucester, took an oath that on August 14, 1817, between four and five in the afternoon, he saw a strange marine animal resembling a sea serpent. This was while he was aboard a small craft in the main harbor, and he was able to get within 30 feet of the animal. Later, Gaffney said that the sea monster's head appeared fully as large as a 4-gallon keg, his body as large as a barrel, and his overall length somewhere beyond 40 feet. The top of the animal's head was of a dark color, and the underpart of his head and several feet of its belly were almost white. Gaffney, like so many before him, hauled out a gun and fired at the sea monster. The animal turned toward the boat but, instead of attacking, passed beneath the keel. The monster surfaced one more time, at about 100 yards, and was gone.

Then a sudden upsurge in sea monster sightings occurred. A ship's master named Solomon Allen III signed a sworn statement that he too had seen a sea monster during August, 1817, and in about the same locale. His description of the sea monster was that it appeared to be about 90 feet in length, was about the size of a half barrel around, and apparently had joints from its head to its tail. Allen stated that the head looked like the head of a rattlesnake and was the size of a horse's. He claimed that its color was dark brown and without any noticeable spots.

So many sea monster sightings were reported in and about the harbor of Gloucester, Massachusetts, that the Linnean Society of Boston formed a special investigative committee. The group was limited to three members of some prominence: Messrs. John Davis, Jacob Bigelow, and Francis Gray.

The three men set to work in a thorough and energetic fashion. In a relatively short while, they had compiled an impressive number of sworn statements concerning what had been seen by whom. It is said that these men did exceedingly well as long as they stuck to reporting what others had seen. When, however, they took it upon themselves to explain what none of them had ever seen, they arrived at numerous false conclusions that did much to negate the impact of the report.

One obvious and irrefutable factor did remain, nevertheless: the actuality of a sea monster. Unfortunately, it is not possible to cover each and every case here because of space limitations. But sampling will be sufficient to make the point.

In the harbor of Cape Ann, which lies just to the north of Gloucester, two women watched in horror as a sea monster swam into the harbor. This was in August, 1817. It later turned out that the event had been witnessed by numerous fishermen as well. Also during August, a seaman named Amos Story viewed the beast from a vantage point onshore. This was in the entrance to the harbor of Gloucester, and two days later the monster was seen again by another man. Also, a few days later, twenty people saw it for a considerable period of time. Mr. London Nash, justice of the peace in Gloucester, was one of this group.

There were others, but the first man to be interrogated by the committee was Amos Story.

"I, Amos Story of Gloucester, in the County of Essex, mariner, depose and say, that on the tenth day of August A.D. 1817, I saw a strange marine animal, that I believe to be a serpent, at the southward and eastward of Ten Pound Island, in the harbour of said Gloucester. It was between the hours of ten and one o'clock when I first saw him, and he continued in sight for an hour and a half. I was sitting on the shore, and was about 20 rods [330 feet] from him when he was the nearest to me.

"His head appeared shaped much like the head of the sea-turtle, and he carried his head from ten to twelve inches above the surface of the water. His head at that distance appeared larger than the head of any dog that I ever saw. From the back part of his head to the next part of him that was visible, I should judge to be three or four feet. He moved very rapidly through the water, I should say a mile in two, or at most, in three minutes. I saw no bunches on his back. On this day I did not see more than ten or twelve feet of his body."

The three-man committee, to do them complete justice, did not limit themselves simply to taking statements. They questioned many of the witnesses with great care. An example of this is to be found in the testimony of Solomon Allen III.

"Q. How many distinct portions of it were out of the water?"

"A. I should say fifty distinct portions."

"Q. Did it appear smooth or rough?"

"A. I should say rough and scaly."

A Monstrous Sea Serpent,

The largest ever seen in America,

Has just made its appearance in Gloucester Harbour, Cape Ann, and has been seen by hundreds of Respectable Citizens.

The Editor of the Salem Gazette, says:—We have in our possession an extract of a letter from John Low, E q. to his son in this town, dated Gloucester, Thursday afternoon, August 14, 1817.

"There was seen on Monday and on Tuesday morning playing about our harbor, between Eastern Point and Ten pound Island, a SNAKE with his head and body about eight feet out of water, his head is in perfect shape as large as the head of a horse, his body is judged to be about FORTY-FIVE OR FIFTY FEET IN LENGTH, it is thought he will girt about 3 feet round the body, and his sting is about 4 feet in length.

While writing the above a person has called in, who says that there are two to be seen, playing from the Stage-head into the harbor inside of Ten pound Island.

The spectators are Mr. Charles Smith, Mr. John Proctor and several others. A number of our sharp shooters are in pursuit of him but cannot make a ball penetrate his head, Another party is just going in pursuit with guns, harpoons &c. Our small craft is fearful of venturing out a fishing.

The above can be attested to by twenty different people of undoubted veracity."

In addition to this account the Salem Register states, that the Serpent is extremely rapid in its motions which are in all directions, that it shews a length of 50 feet; that a man who discharged his musket within 30 feet of the Serpent, says its head was partly white and that he hit it, that a large sum had been offered for it; that "it appears in joints like wooden buoys on a net rope almost as large as a barrel, that musket balls appear to have no effect on it, that it appears like a string of gallon kegs 100 feet long."

The editor of the Register quotes an account of a Sea Serpent seen on the coast in 1746, something like it. It had a head like that of a horse, and as he moved he looked like a row of large casks following in a right line.

The Boston Daily Advertiser in speaking of this *Monstrous Serpent*, says—We have seen several letters from Gloucester, which describe a prodigious Snake that has made its appearance in Cape-Ann Harbour. It was first seen by some fishermen, 10 or 12 days ago, but it was then generally believed to be the creature of the imagination. But he has since come within the harbor of Gloucester, and has been seen by hundreds of people. He is declared by some persons who approached within 10 or 15 yards of him, to be 60 or 70 feet in length, round, and of the diameter of a barrel. Others state his length variously, from 50 to 100 feet. His motions are serpentine, extremely varied, and exceedingly rapid. He turns himself completely round almost instantaneously. He sometimes darts forward, with his head out of water, at the rate of a mile in 3 minutes, leaving a wake behind, of half a mile in length. His head, as large as the head of a horse, is shaped some what like that of a large dog, is raised about 8 feet out of water and is partly white, the other part black. He appears to be full of joints and resembles a string of buoys on a net rope, as is set in the water to catch herring. Others describe him as like a string of water casks. His back is black. Various attempts have been made, without success, to take him. Four boats went out on Thursday, filled with adventurous sailors and experienced gunners, armed with muskets, harpoons, &c. Three muskets were discharged at him from a distance of 30 feet, two balls were thought to hit his head, but without effect. He immediately after plunged into the water, and disappeared for a short time, after which he moved off to the outer harbor, and was seen no more that night. A number of persons are employed in making a net of cod-line, of sufficient size to take him. It is conjectured that he has resorted to this harbor for the purpose of preying upon a very numerous shoal of herrings, which have lately appeared there. If he has been instrumental, as is supposed, in driving these herring into the harbor, he has rendered an essential service to the town.

The Salem Gazette of the 19th inst. says, "We are informed, that on Sunday this creature was seen playing sometimes within 15 or 20 feet of the shore, affording a better opportunity to observe him than had before occured. Gentlemen from Gloucester state, that he appeared to them even of greater magnitude than had before been represented, and should judge from their own observation, that he was as much as 150 feet in length, and as big round as a barrel. They saw him open an enormous mouth, and are of opinion that he is cased in shell.

Aug. 22, 1817.

Printed and Sold by Henry Bowen, Devonshire-Street, Boston.

The Gloucester sea serpent was big news. This full-page story was printed and sold by Henry Bowen, of Boston, during the peak of public interest. Photo, George M. Cushing.

"Q. Had it ears, horns, or any other appendages?"

"A. I perceived none."

"Q. How did its tail terminate?"

"A. He seemed to taper toward his tail, though I had no distinct view of his tail."

Sightings of the Gloucester sea monster continued to be reported at a remarkable rate. A merchant named James Mansfield told of a sighting; Colonel T. H. Perkins and a Mr. Lee saw it at another time; a ship's captain and his entire crew saw it. A few days later, a ship passed within an oar's length of the monster, and then reports were turned in by numerous ships and people.

In the end, it was generally concluded that what had been seen was indeed a transient sea monster that frequented Gloucester's harbor from time to time from August 1 till August 23. Upon leaving the Gloucester area, it was again seen heading in a northerly direction on August 28; however, it was observed on October 3 and 5 in Long Island Sound. Numerous people had an opportunity to observe the Gloucester sea monster under a wide variety of circumstances. Some saw it in the early morning at rest; others caught sight of it in the afternoon. Sometimes it was swimming with vigor; at other periods it moved with a certain quiet dignity.

Summing up what had been seen, attested to, and witnessed by the very young, the elderly, and the hard-bitten old mariners, with their coldly calculating eyes, the Gloucester sea monster could be described as snakelike in appearance, with smooth skin that showed no scales, dark brown to black in color, and very likely about 65 feet in length. The visible part of the torso was about 12 inches in diameter, and the head was carried 6 to 12 inches out of the water. When it surfaced, it appeared as a series of small humps. It was very agile when making sharp turns. When it submerged, it dropped like a stone, disappearing from sight all at one time. The top speed of the sea monster was estimated to be from 40 to 70 knots, no one person being able to say for sure. The head was said to resemble a snake's or a turtle's and was about the size of a horse's skull.

The Gloucester sea monster was seen by several hundred people, each viewing it differently under a wide variety of conditions. It was impossible for the committee to arrive at a valid conclusion because there were so many conflicting reports. As a result, the final report was ridiculed.

Another case of a likely sighting that was derided was one submitted in May, 1818, by Joseph Woodward, master of the schooner *Adamant*.

"I, the undersigned, Joseph Woodward, captain of the *Adamant* schooner of Hingham, being on my route from Penobscot to Hingham, steering W.N.W., and being about ten leagues from the coast, perceived late Sunday, at two PM something on the surface of the water, which seemed to me to be about the size of a large boat. Supposing that it might be part of the wreck of a ship, I approached it; but when I was within a few fathoms of it, it appeared, to my great surprise, and that of my whole crew, that it was a monstrous serpent. When I approached nearer, it coiled itself up, instantly uncoiling itself again, it coiled itself a second time, and placed itself at the distance of 60 feet at most from the bow of the ship."

The captain immediately had his crew break out heavy armament and fire off a cannonball. Having aimed at the clearly visible head, both he and his crew saw the missile rebound and heard its impact. The serpent appeared to suffer no visible effect; however, it did turn toward the ship and, with open jaws, swim in that direction. His report continued: "I had caused the cannon to be reloaded, and pointed at his throat; but he had come so near, that all the crew were seized with terror, and we thought only of getting out of the way. He almost touched the vessel; and had I not tacked as I did, he would certainly have come on board. He dived, but in a moment we saw him appear again, with his head on one side of the vessel, and his tail on the other, as if he was going to lift us up and upset us. However, we did not feel any shock. He remained five hours near us, only going backward and forward."

Captain Woodward says that during this period of inaction, their fears subsided and they were able to make some keen observations. The captain estimated the sea monster's length to be twice that of his own ship, or about 130 feet. His head was 12 to 14 feet long, and the diameter of the body was about 6 feet. The sea serpent was blackish in color, with "ear holes (ouies), about 12 feet from

A view of the harbor of Gloucester showing the sea monster sighted in August, 1817.
Photo, George M. Cushing.

the extremity of the head. In short, the whole had a horrible look."

After additional details concerning the sea monster's characteristics and body movements had been expanded upon, the captain signed and attested to the truth of his report under oath. Furthermore, two of his crewmen, Peter Homes and John Mayo, joined him in making the affidavit.

This report was promptly branded a hoax by many who should have been at least a little tolerant. Characteristic of the mad howl that went up is the lengthy protest written by an armchair authority named W. W. Cooper, of Worcester. After ripping Captain Woodward's report from one end to the other, the letter of protest sarcastically concludes: "So here we have an animal sixty feet from the ship, moving with astonishing rapidity towards the ship, which it appears was also moving towards the animal, and yet allowing him to load a cannon, point it at the throat, and afterwards to tack to get out of its way. Truly a most accommodating serpent!"

Many sea monster sightings, such as the one reported by the war vessel *Elephant* in 1818, smacked of fraud. Its credibility was based on a letter written by a sailor to a friend. In it the writer said that "at dawn we saw more than 20 of these [sea monsters] surrounding our vessel."

Another reported sighting that went unsubstantiated was turned in by the master of the brig *Wilson,* off Cape Henry, Virginia. He saw what he thought to be the remnants of a sunken ship and lowered a boat to have a closer look. It turned out to be a sea serpent that quietly moved away at the approach of the captain's boat. At about the same time, the *London Literary Gazette* reported that a sea monster of massive proportions had been located off the coast of Calabria. The story does not seem valid, for it stated that some fishermen had observed a fire in the sea and, thinking it to be a coasting vessel in need of assistance, approached the monster, whose movements resulted in a phosphoric light, which was what the sailors had taken for fire. They noted thick smoke, heard a bellowing sound, and were obliged to flee posthaste.

The *Philadelphia Magazine*, in 1819, carried an interesting description of a sea serpent. It was written by a man considered an eminent authority in such matters, Mr. Rafinesque-Schmaltz. In a lengthy article he says that "immediately we recognized our well-known sea serpent, an individual of which the length is estimated at 58 feet, which held its head and very long neck upright whilst swimming. . . . It was surrounded by porpoises and fishes. Evidently the animal swam between them with the purpose to snatch one of them. It had a smooth skin, no scales, and eight gills under its neck. Dark brown above (i.e., on the dorsal part of the head and neck), muddy white beneath (i.e., on its throat); head obtuse (read rather obtuse, seen from above or below, or in front; just from aside it is rather pointed). The head was two feet long, the mouth fifteen inches (of course estimated dimensions), and the eyes over the jaws similar to those of a horse (this definition was caused by the heavy eyebrows and by the little bunch above each eye). The whole description is exactly that of our animal in the above-mentioned position and seen from a certain direction. For 'eight gills' we may safely read 'eight gillsplits,' or eight splits caused by and lying between nine folds or wrinkles, which in in their turn are caused by the animal bending its head rectangularly towards the throat. Such folds or wrinkles are also seen in sea-lions, when they make the same motion, and stout, corpulent persons will know what is meant by a double chin!"

Shortly after these questionable sightings, a sea monster report came from Sag Harbor, Long Island. Then a sea serpent, supposedly the same one seen in Sag Harbor, was sighted off Cape Ann by the merchant ship *Delia*, captained by Shubael West. Another sighting, again quite similar to the others, was reported off Massachusetts by the schooner *Mary*, and on and on go the reports.

It is said that there are two sides to every story. Let us take an incident during this period and report the two versions of an attack by a sea monster. For this we go to no less an authoritative source than the *Times* of London of the early 1800s. The following story was presented under the headline "A Successor to the Sea Serpent":

"We had left Colombo in the steamer *Strathowen*, had rounded Galle, and were well in the bay, with our course laid for Madras, steaming over a calm and tranquil sea. About an hour before sunset on the 10th of May we saw on our starboard beam and about two miles off a small schooner laying becalmed. There was nothing in her appearance or position to excite remark, but as we came up with her I lazily examined her with my binoculars, and then noticed between us, a long, low, swelling lying on the sea, which from its color and shape I took to be a bank of seaweed. As I watched, the mass, hitherto at rest on the quiet sea, was set in motion. It struck the schooner, which visibly reeled, then righted. Immediately thereafter the masts swayed sideways, and with my glass I could clearly discern the enormous mass and the hull of the schooner coalescing—I can think of no other term. Judging from their exclamations, the other gazers must have witnessed the same appearance. Almost immediately after the collision and coalescence the schooner's mast swayed toward us, lower and lower; the vessel was on her beam ends, lay there a few seconds, and disappeared, the masts righting as she sank, and the main exhibition being a reversed ensign struggling toward its peak."

The writer, thought to have been the skipper of the *Strathowen*, went on to say that a cry of horror arose from his passengers, and he immediately headed his ship toward the scene, which was now marked by the survivors battling for their lives. As soon as those saved could tell their story, they astounded the passengers of the *Strathowen* by asserting that their vessel, named the *Pearl*, had been submerged by a gigantic cuttlefish or octopus. Each narrator had his version of the story, but in the main, all tallied reasonably well under such circumstances. A short time after the disaster, the captain of the *Strathowen* prevailed upon the captain of the *Pearl* to give the facts as he knew them.

The other side of the story was then presented by the *Times* of London. They printed the views of the master of the schooner, which generally tallied with that given by the master of the *Strathowen*. In any event, the entire statement was sworn and attested to by "James Floyd, late Master, schooner *Pearl*."

Immediately protests arose based on a single

factor: How was it possible, and who would take the time, to reverse the ensign and hoist it to the peak while the schooner was under such attack? No one, not even the master of the schooner *Pearl,* could answer these questions. As a result, it was considered a hoax and no further investigation appears to have been undertaken. However, it is worthy of some note that the newspaper *Mystic Press,* of West Mystic, Connecticut, during the same period, carried this same story, only in an abbreviated form.

New Englanders were acutely aware of the existence of sea monsters or sea serpents. In 1818, a report was circulating that one such beast was haunting the waters in and about Boston. According to the noted historian Dr. A. C. Oudemans, "many rewards were offered to whalers to catch the animal." One reward totaled $5,000 for that whaler who would bring the animal, dead or alive, to port.

Meantime, a few years prior to this, a committee to investigate such nautical matters had been formed in the maritime city of Göttingen, Germany. They systematically collected every scrap of information available on the subject of sea monsters. Within a relatively short period of time, they could boast of a very thorough file on this topic. Even today, the records on sea monsters preserved in the library of the University of Göttingen, are some of the most complete to be found anywhere in the world. Consequently, it is not surprising that we are able to unearth an original report written by a Captain Rich, who launched an expedition to hunt out the monster and thereby earn himself a windfall.

"Squam River, Aug. 20th, 12 o'clock. After several unsuccessful attempts, we have at last fastened to this strange thing called the Sea Serpent. We struck him fairly but the harpoon soon drew out. He has not been seen since, and I fear the wound he received will make him more cautious how he approaches these shores. Since my last, yesterday, we have been constantly in pursuit of him; by day he always keeps a proper distance from us, to prevent out striking oars. But a few hours since, I thought we were sure of him, for I hove the harpoon into him as fairly as ever a whale was struck; he took from us about 20 fathoms of warp before we could wind the boat, with as much swiftness as a whale. We had but a short ride when we were loose from him to our sore disappointment." This report carried the signature "Rich'd. Rich."

The addendum to this report is informative. It is dated Gloucester, Aug. 20, and says that Captain Rich was then away on a trip, but the writer had a brother who, during the incident, was in command of one of Rich's whaleboats. It was through this source the writer received a letter saying that the party had consisted of one large whaleboat and two smaller ones. All boats were well manned when they met the sea serpent off Squam and promptly closed with him. The great beast passed directly under the bows of Captain Rich's boat and was struck by a harpoon. The iron went about two feet into the monster. Immediately the powerful beast made a run for it. They traveled a considerable distance at a fast speed, only to have the boat break apart and the harpoon come loose. There followed the only brief description given of the sea monster, which is worth quoting here.

"He has no scales on him, and no bunches on his back, but his skin is smooth, and looks similar to an eel. In the attack Capt. Rich had one of his hands wounded. These particulars I have in a letter from my brother, signed, Sam'l Dexter."

Close upon the heels of this report, another was made public concerning a sighting in Lake Erie during the year 1819. The story was first presented in the *Philadelphia Magazine* that year, under the title "Dissertation on Sea-Snakes," by Rafinesque-Schmaltz. "It appears as though our large lakes have huge serpents or fishes, as well as the sea. On the 3rd of July 1817, one was seen in Lake Erie, three miles from land, by the crew of a schooner, which was 35 or 40 feet long, and one foot in diameter; its colour was a dark mahogany, nearly black. This account is very imperfect, and does not notice if it had scales; therefore it must remain doubtful whether it was a snake or a fish. I am inclined to believe it was a fish, until otherwise convinced: It might be a gigantic species of eel, or a species of the above genus *Octipos.* Until seen again, and better described, it may be recorded under the name of *Anguilla gigas* or Gigantic Eel."

Sea serpent reports are not exclusive to men. A number of sightings have been reported by

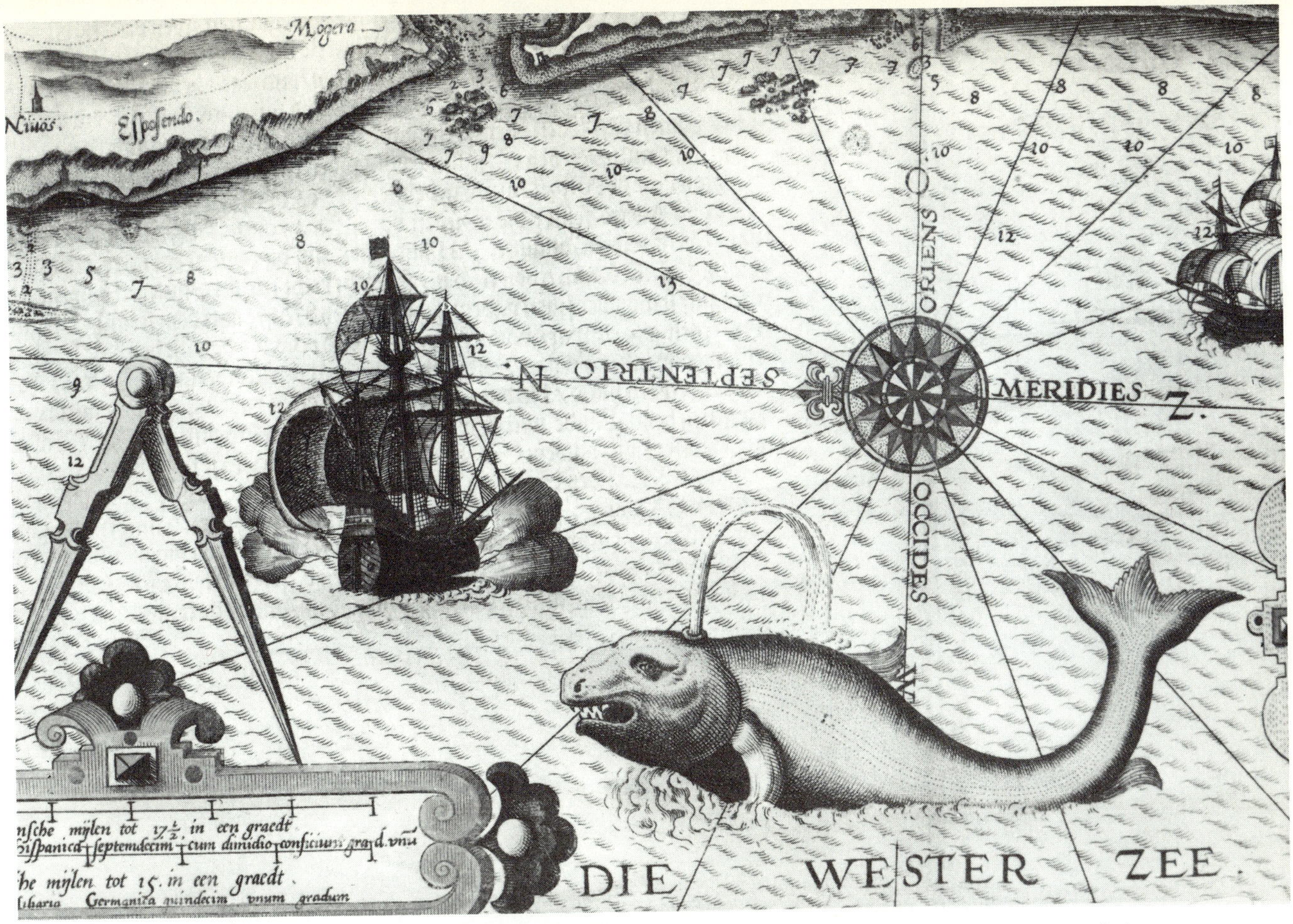

German chart makers of the late 1800s were still showing sea monsters on their maps. Photo, Schiffahrts Museum, Brake, West Germany.

women. Typical of those from reliable sources is the account of Mrs. Ida Mathilde Helmer, wife of Captain H. C. Helmer. It has been translated especially for this book by staff members of Norsk Sjofartsmuseum, Oslo, Norway. The report reads as follows:

"Yesterday both the Pilot, Helmer and myself saw a sea serpent. . . . I saw the tail, it was like a fish-tail, the other two saw the head, it was like the head of an eel. It was of about 90 fathoms [540 feet] long, as thick as the mizzen boom where it is at its thickest.

"It squirmed slowly close to the ship, just at the surface. We saw it amidships and sailed along beside it, so we had an excellent view of it. Long afterwards, I was sad to think that if we had caught it, we might perhaps have been rich and immortalized our name."

In 1830, American newspapers carried this story about a sea serpent. It originated in Charleston, South Carolina. "Captain Deland with the schooner *Eagle* ran into Charleston on the 27th of March, from Turtle River, and with his crew is willing to confirm by oath the truth of the following declaration: On the 23rd of March, at 11 o'clock A.M., at about a mile from Simons Bay, we perceived at the distance of about 300 yards a large body, resembling an alligator, which sometimes moved with the vessel, sometimes lay motionless on the surface. Capt. Deland perceived that he approached the animal, loaded a musket with a ball, and steered so that he approached it within 20 or 25 yards at a moment that it lay quite still and apparently careless. Capt. Deland aimed with great sagacity at the hindpart of the head, the only part that was just visible, and the ball evidently struck."

The report continued to say that the massive animal now attacked the vessel. It came down on the ship and, in passing, dealt several blows with its tail that severely rattled everyone aboard. On seeing that the animal was making for their ship, the captain and his crew scrambled to find a safe haven. They all had an excellent opportunity to

A painting of a lake monster by an artist named Theodor Kittelsen. Photo, Nasjonalgalleriet, Oslo, Norway.

Jonah and the whale depicted in an engraving by J. Vernet (1714–1789). Photo, Maritime Museum, Haifa, Israel.

The skeleton of a sea monster called Hydragos Sillimanii *displayed, 1845, by a Dr. Koch. It was later proved to be a fraud.* Photo, Naval Historical Foundation, Washington, D.C.

observe the sea monster at close range and later agreed that its length was about 70 feet. They said that the body was as thick as a 60-gallon keg, or thicker, of a gray color, eel-shaped, without visible fins, and apparently covered with scales. They also said that the back was full of joints or bunches, the head and back resembled an alligator's, and the skull was about 10 feet long and as big around as a hogshead. A smaller sea serpent was observed at a greater distance. This second animal disappeared at sound of the shot, but later it was seen again as the two disappeared in the distance.

The report immediately resulted in a meeting of experts on the subject of sea monsters. After due and weighty deliberation, the entire affair was proclaimed a hoax because, as one authoritative source put the matter, "it is not the habit of the sea serpent to attack a ship after having been struck by a ball, but to plunge down and to disappear."

In the year 1848, the now-famous *Daedalus* case became front-page news. Its impact still echoes in the halls of serpentine history. The incident has been so analyzed, dissected, and publicly debated that it has lost proper perspective. The simple facts are these:

H.M.S. *Daedalus* put in at the port city of Plymouth, England, after completing a voyage from the East Indies. Immediately, a somewhat brief report appeared in a local newspaper the *Times*.

"Plymouth, October 7, 1848—When the *Daedalus* frigate, Captain M'Quhae, which arrived here on the 4th inst., was on her passage home from the East Indies, between the Cape of Good Hope and St. Helena, her captain, and most of her officers and crew, at 4 o'clock one afternoon saw a sea serpent. The creature was twenty minutes in sight of the frigate, and passed under her quarter. Its head appeared to be about four feet out of the water, and there was about 60 feet of its body in a straight line on the surface. It is calculated that there must have been under water a length of 30 or 40 feet more, by which it propelled itself at the rate of 15 miles an hour. The diameter of the exposed part of the body was about 16 inches, and when it extended its jaws, which were full of large jagged teeth, they seemed sufficiently capacious to admit of a tall man standing upright between them. The ship was sailing north at the rate of eight miles an hour. The *Daedalus* left the Cape of Good Hope on the 30th of July, and reached St. Helena on the 16th of August."

The ship's captain, her navigation officer, and a host of other responsible mariners saw the sea serpent. The following is part of the captain's report: "Our attention being called to the object, it was discovered to be an enormous serpent, with head and shoulders kept about forty feet constantly above the surface of the sea. . . . It passed so rapidly, but so close under our lee quarter, that had it been a man of my acquaintance, I should have easily recognized his features with the naked eye. . . ."

During October, 1848, the *Illustrated London News* carried the following report: "*Mary Ann*, of Glasgow, October 19, I have just reached this port, on a voyage from Malta and Lisbon, and my attention having been called to a report relative to an animal seen by the master and crew of her Majesty's ship *Daedalus*, I take the liberty of communicating the following circumstances:

"When cleared out of the port of Lisbon, on the 30th of September last, we spoke the American brig *Daphne*, of Boston, Mark Trelawaney master. He signaled for us to heave-to, which we did; and standing close round her counter, lay-to while the mate borded us with the jolly boat, and handed a packet of letters to be dispatched per first steamer for Boston on our arrival in England. The mate told me that when in lat. 4 degrees 11′ S., long. 10 degrees 15′ E., wind dead north, upon the 20th of September, a most extraordinary animal had been seen: from his description it had the appearance of a huge serpent or snake, with a dragon's head."

GLEASON'S PICTORIAL DRAWING ROOM COMPANION.

APPEARANCE OF THE SEA SERPENT WHEN FIRST SEEN FROM H. B. M. SHIP DÆDALUS.

THE GREAT SEA-SERPENT.

You have not seen his snakeship, perhaps? [W]ell, let those who have eyes to see, see him [h]erewith—a bona fide representation of a most [re]markable creature, whose natural propensities [s]eem to lead him annually—just about this period [o]f the year—to the shores of Nahant and Cape [A]nn. Last year he visited us, and year before [la]st; and he will, beyond a doubt, be here again [th]is year. Of course, every captain that sails out of Boston, from the master of a fishing smack to a regular liner, has seen the monster, and very many are the wonderful stories they relate. That there is such a creature, however, there can be but little doubt, as his appearance has so often been alluded to. One of the best descriptions, and on the very best authority, is that which we have seen in the report of an English officer to the war department of his own country. When the *Dædalus* frigate, Captain M'Quhæ, which arrived at London, not long since, was on her passage home from the East Indies, between the Cape of Good Hope and St. Helena, her captain, and most of her officers and crew, at four o'clock one afternoon, saw a sea serpent. The creature was twenty minutes in sight of the frigate, and passed under her quarter. Its head appeared to be about four feet out of the water, and there was about sixty feet of body in a straight line on the surface. It is calculated that there must have been under water a length of thirty or forty feet more, by which it propelled itself at the rate of fifteen miles an hour. The diameter of the exposed part of the body was about sixteen inches; and when it extended its jaws, which were full of large jagged teeth, they seemed sufficiently capacious to admit of a tall man standing upright between them. The ship was sailing north at the rate of eight miles an hour. The Dædalus left the Cape

A story on the H.M.S. Daedalus *at the time of the sea monster report.* Photo, George M. Cushing.

Again we find that this ship's company went through the same routine as most of the others. They immediately broke out a gun. One of the deck cannons was brought to bear on the sea serpent. The gun, having been loaded with pieces of iron, was fired at a range of about 40 yards. The serpent, obviously struck by the deadly pellets, reared its head high into the air, then plunged violently down. The *Daphne* was to leeward at the time, but was put about on the starboard tack and stood toward the brute, which was seen foaming and lashing the water. However, when the brig drew near, the serpent, although apparently badly wounded, made off at the rate of 15 or 16 knots. The *Daphne* pursued for some time, but with the night coming on, the master was obliged to put about and continue his voyage. From the description given by the mate, the sea serpent must have been 100 feet in length. The newspaper story concludes by stating that the report carried the signature of the ship's master, Henderson.

Zoologist magazine, in the March 26, 1849, issue, carried the following announcement: "A strange marine animal, of great size and strength, was captured off Cullercoats, near Newcastle. By the enclosed handbill which has been forwarded to me, it appears to be quite unknown to the neighboring savants. The honest fishermen who drew the struggling monster to land are not, however, overscrupulous about the name, provided it be attractive enough to extract from the pockets of 'ladies and gentlemen 6d.; working people 3d. each': they therefore boldly announce him as 'the great sea serpent caught at last.' My correspondent very

THE SEA SERPENT PASSING UNDER THE STERN OF THE DÆDALUS.

Another newspaper print popular at the time of the Daedalus *report*. Photo, George M. Cushing.

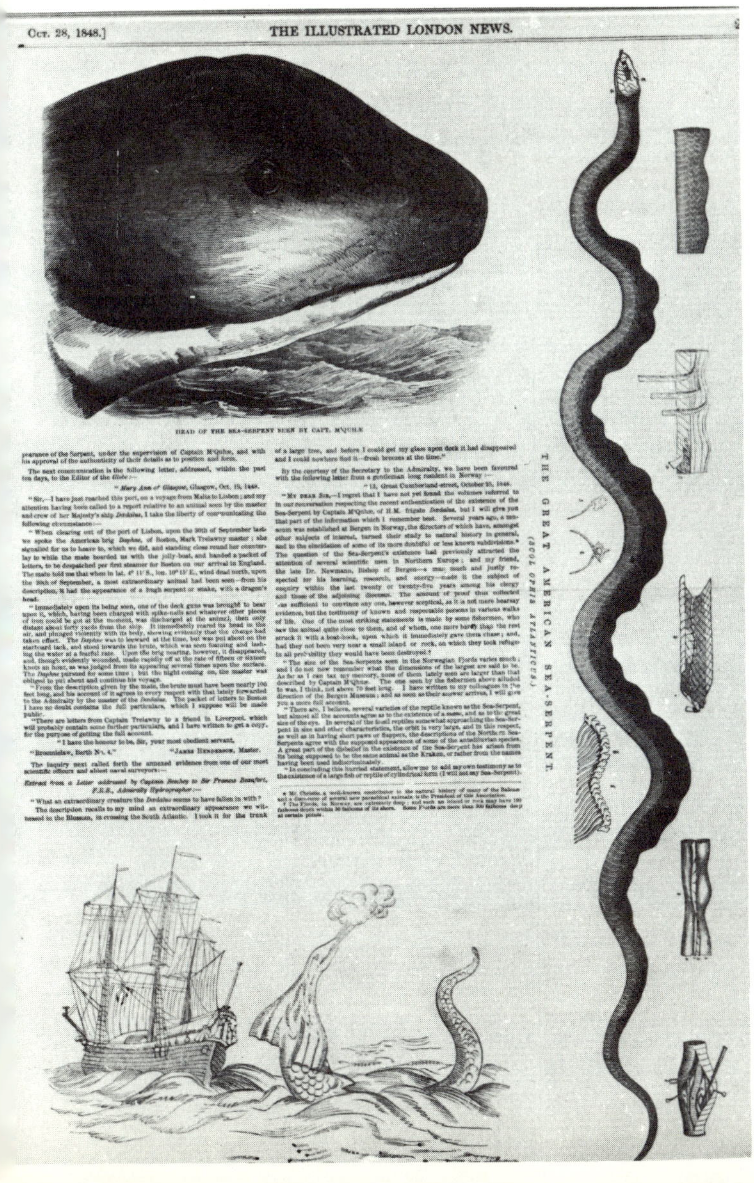

OCT. 28, 1848.] THE ILLUSTRATED LONDON NEWS.

HEAD OF THE SEA-SERPENT SEEN BY CAPT. M'QUHAE

pearance of the Serpent, under the supervision of Captain M'Quhae, and with his approval of the authenticity of their details as to position and form.

The next communication is the following letter, addressed, within the past ten days, to the Editor of the *Globe*:—

"*Mary Ann of Glasgow*, Glasgow, Oct. 19, 1848.

"Sir,—I have just reached this port, on a voyage from Malta to Lisbon; and my attention having been called to a report relative to an animal seen by the master and crew of her Majesty's ship *Dædalus*, I take the liberty of communicating the following circumstance:—

"When clearing out of the port of Lisbon, upon the 30th of September last, we spoke the American brig *Daphne*, of Boston, Mark Trelawny master; she signalled for us to heave to, which we did, and standing close round her counter, lay to while the mate boarded us with the jolly-boat, and handed a packet of letters, to be despatched per first steamer for Boston on our arrival in England. The mate told me that when in lat. 4° 11′ S., lon. 10° 15′ E., wind dead north, upon the 20th of September, a most extraordinary animal had been seen—from his description, it had the appearance of a hugh serpent or snake, with a dragon's head.

"Immediately upon its being seen, one of the deck guns was brought to bear upon it, which, having been charged with spike-nails and whatever other pieces of iron could be got at the moment, was discharged at the animal, then only distant about forty yards from the ship. It immediately reared its head in the air, and plunged violently with its body, showing evidently that the charge had taken effect. The *Daphne* was to leeward at the time, but was put about on the starboard tack, and stood towards the brute, which was seen foaming and lashing the water at a fearful rate. Upon the brig nearing, however, it disappeared, and, though evidently wounded, made rapidly off at the rate of fifteen or sixteen knots an hour, as was judged from its appearing several times upon the surface. The *Daphne* pursued for some time; but the night coming on, the master was obliged to put about and continue his voyage.

"From the description given by the mate, the brute must have been nearly 100 feet long, and his account of it agrees in every respect with that lately forwarded to the Admiralty by the master of the *Dædalus*. The packet of letters to Boston I have no doubt contains the full particulars, which I suppose will be made public.

"There are letters from Captain Trelawny to a friend in Liverpool, which will probably contain some further particulars, and I have written to get a copy, for the purpose of getting the full account.

"I have the honour to be, Sir, your most obedient servant,

"Broomielaw, Berth No. 4." "JAMES HENDERSON, Master.

The inquiry next called forth the annexed evidence from one of our most scientific officers and ablest naval surveyors:—

Extract from a Letter addressed by Captain Beechey to Sir Francis Beaufort, F.R.S., Admiralty Hydrographer:—

"What an extraordinary creature the *Dædalus* seems to have fallen in with? The description recalls to my mind an extraordinary appearance we witnessed in the Blossom, in crossing the South Atlantic. I took it for the trunk of a large tree, and before I could get my glass upon deck it had disappeared and I could nowhere find it—fresh breezes at the time."

By the courtesy of the Secretary to the Admiralty, we have been favoured with the following letter from a gentleman long resident in Norway:—

"13, Great Cumberland-street, October 25, 1848.

"MY DEAR SIR,—I regret that I have not yet found the volumes referred to in our conversation respecting the recent authentication of the existence of the Sea-Serpent by Captain M'Quhae, of H.M. frigate *Dædalus*, but I will give you that part of the information which I remember best. Several years ago, a museum was established at Bergen in Norway, the directors of which have, amongst other subjects of interest, turned their study to natural history in general, and to the elucidation of some of its more doubtful or less known subdivisions.* The question of the Sea-Serpent's existence had previously attracted the attention of several scientific men in Northern Europe; and my friend, the late Dr. Newmann, Bishop of Bergen—a man much and justly respected for his learning, research, and energy—made it the subject of enquiry within the last twenty or twenty-five years among his clergy and those of the adjoining dioceses. The amount of proof thus collected was sufficient to convince any one, however sceptical, as it is not mere hearsay evidence, but the testimony of known and respectable persons in various walks of life. One of the most striking statements is made by some fishermen, who saw the animal quite close to them, and of whom, one more hardy than the rest struck it with a boat-hook, upon which it immediately gave them chase; and, had they not been very near a small island or rock, on which they took refuge, in all probability they would have been destroyed.†

"The size of the Sea-Serpents seen in the Norwegian Fjords varies much; and I do not now remember what the dimensions of the largest are said to be. As far as I can tax my memory, none of them lately seen are larger than that described by Captain M'Quhae. The one seen by the fishermen above alluded to was, I think, not above 70 feet long. I have written to my colleagues in the direction of the Bergen Museum; and as soon as their answer arrives, I will give you a more full account.

"There are, I believe, several varieties of the reptile known as the Sea-Serpent, but almost all the accounts agree as to the existence of a mane, and as to the great size of the eye. In several of the fossil reptiles somewhat approaching the Sea-Serpent in size and other characteristics, the orbit is very large, and in this respect, as well as in having short paws or flappers, the descriptions of the Northern Sea-Serpents agree with the supposed appearance of some of the antediluvian species. A great part of the disbelief in the existence of the Sea-Serpent has arisen from its being supposed to be the same animal as the Kraken, or rather from the names having been used indiscriminately.

"In concluding this hurried statement, allow me to add my own testimony as to the existence of a large fish or reptile of cylindrical form (I will not say Sea-Serpent).

* Mr. Christie, a well-known contributor to the natural history of many of the [illegible] and a discoverer of several new parasitical animals, is the President of this Association.

† The Fjords, in Norway, are extremely deep; and such an island or rock may have 100 fathoms depth within 30 fathoms of its shore. Some Fjords are more than 300 fathoms deep at certain points.

THE GREAT AMERICAN SEA-SERPENT.
(SCOLIOPHIS ATLANTICUS.)

A full-page story on sea serpents as it appeared in an October, 1848, issue of the Illustrated London News. Photo, Naval Historical Foundation, Washington, D.C.

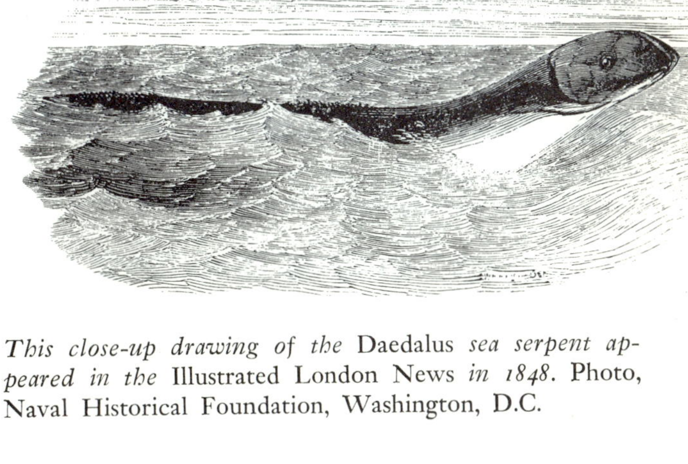

This close-up drawing of the Daedalus *sea serpent appeared in the* Illustrated London News *in 1848*. Photo, Naval Historical Foundation, Washington, D.C.

judiciously observes, that whatever the animal may be, it adds another to the many evidences constantly occurring that there are more things in heaven and earth, than are drempt of by the experienced practical observers. Some thirty-five years since, the distinguished anatomist Dr. Barclay, was fain to reproach his contemporaries with the folly of affecting to suppose that they knew everything. What additions have five and thirty years not given to Science! As the animal in question must be at least a local visitor, may we not hope that some resident naturalist will favor us with a notice of it?"

Then the *Zoologist,* in a follow-up story at a later date, printed this commentary on the great sea serpent caught off Cullercoats. "This most wonderful monster of the deep was discovered by a crew of fishermen about six miles from the land, who, after a severe struggle, succeeded in capturing this, the most wonderful production of the mighty deep. This monster has been visited by numbers of the gentry and scientific men of Newcastle, and all declared that nothing hitherto discovered in Natural History affords any resemblance to this. As an object of scientific inquiry, this 'great unknown' must prove a subject of peculiar interest. Many surmises as to its habits, native shores, etc., have already been made, but nothing is really known. The general opinion expressed by those that are best able to judge, is, that this is the great sea serpent, which hitherto has only been believed to have a fabulous existence, but which recent voyagers have declared they have seen. Now exhibiting, at the shop, 57, Grey Street, opposite the High Bridge."

The *Illustrated London News* appears to be the information source most keenly aware of sea monsters and of public concern with sightings. The April 14, 1849, issue carried a story from H.M.S. *Plumper*. We quote from the ship's official report as given by the newspaper: "Being nearly due west of Oporto, I saw a long black creature with a sharp head moving slowly, I should think about two knots, through the water, in the north westerly direction, there being a fresh breeze at the time, and some sea on. I could not ascertain its exact length, but its back was about twenty feet if not more above water; and its head, as near as I could judge, from six to eight. I had no time to make a closer observation, as the ship was going six knots through the water, her head E. half S., and wind S.S.E."

According to the *Plumper*'s officer who made the report, the sea monster moved across his ship's wake toward a merchant bark. The sailors on deck swore that they had never seen the likes of this creature and said that it in no way resembled a whale or giant shark. This same source concluded by saying that there was something on the sea serpent's back that looked like a mane, which kept washing about as it moved through the water.

The following informative article appeared in the *Illustrated London News* on May 19, 1849: "The Sea Serpent . . . a strange and hitherto unknown fish, nearly 13 feet in length, and possessing many of the characteristics which the captain of the *Daedalus* enumerated in his description of the great Sea Snake, has really been caught off the Northumbrian coast, by the Cullercoats fishermen, and has been exhibited in *Newcastle,* where it has created the greatest sensation. The members of the Natural History Society of that town have duly reported upon it, and expressed their opinion,

A sea monster as portrayed on an old German sea chart. Photo, Schiffahrts Museum, Brake, West Germany.

SUPPOSED APPEARANCE OF THE GREAT SEA-SERPENT, FROM H. M. S. "PLUMPER," SKETCHED BY AN OFFICER ON BOARD.

This sketch was made by an officer who was on board the H.M.S. Plumper *when a great sea serpent was sighted.* Photo, Naval Historical Foundation, Washington, D.C.

that it is a young specimen of the genus *Gymnetrus,* only four of which species, and those very rare, are known to ichthyologists, and described by Cuvier and others as inhabiting the Indian, Mediterranean and White Seas. The present specimen has become the property of a Newcastle merchant, who has presented it to the museum of that town; and we understand that, in accordance with a very general wish of most of our distinguished naturalists, it is now exhibiting in the metropolis."

And then in the *Zoologist,* in 1849, a Mr. Albany Hancock and Dr. Embleton declared the new species to be of the genus riband-fish (*Gymnetrus*). These two gentlemen were apparently eminent scientists, and it is at this point, following the publication at their views, that the great sea serpent was laid to rest in the archives.

After a close study of all the reports available in this case of the Cullercoats sighting, it would appear that the captured specimen was in reality an oarfish (*Regalescus glesne*), which commonly grows to a length of 20 feet or more. When caught, an oarfish is quite likely to be displayed as a young sea serpent.

In making an overall evaluation of sightings reported during the years from 1800 to 1850, several could be rated as extremely credible.

Naval chart makers competed with one another to see who could draw the most imaginative sea monsters. Photo, Seamen's Church Institute of New York, New York City.

This monster devil fish, captured in Trinity Bay, Newfoundland, appeared in the Canadian Illustrated News *of October 27, 1877.* Photo, The Public Archives of Canada.

6

1850 to 1900

Lineus longissimus is the scientific name of a zoological invertebrate that is known to grow to almost 100 feet in length. According to P. A. Meglitsch's *Invertebrate Zoology*, there are between 500 and 600 species of *Lineus*, a wormlike creature. Most are dull in coloring, but some few possess bright tones. During the latter part of the nineteenth century, sea serpents were given this Latin name, as though science applied this catchall term in order to purge itself of any association with illusion. There are several interesting examples of sea serpents that are labeled *Lineus longissimus*.

An Irish newspaper called the *Cork Constitution*, on August 29, 1850, ran the following letter from one of its readers: "Sir. . . . when sailing in my yacht, with a slight breeze off shore, about two miles south of the beacon erected to the barrel rocks, one of the party of four gentlemen aboard drew attention towards the structure, with the interrogatory of: 'Do you see anything queer about the barrels?' In an instant the attention of all on board was riveted on an object which at first struck me as the upheaved end of a large mast, but which, as it made out plainer, proved to be the head of some huge fish or monster."

The writer then continued by saying that as the

boat bore down on the object, all on board could clearly see what they were later to describe as an enormous serpent, *Lineus longissimus*, without mane or fur or any similar appendage. The letter said that the portion of the body above water, which appeared to be rubbing or scratching itself against the beacon, was fully 30 feet long and about 1 fathom (six feet) in diameter. With the aid of a glass those on board took turns noting the eyes were of immense size, about 9 inches across the eyeball, and the upper part of the back appeared to be covered with a furrowed shell-like substance. Since the yacht was then within shooting range of the sea monster, although some on board objected strenuously to the act, a rifle was fired at the underside of the creature's body. The instant the piece was fired, the monster shot high into the air as if driven by pain. The party agreed that it reached the height of about 30 feet and then plunged beneath the surface, leaving in its wake a great wave. After the water had subsided, the serpent was seen in the distance heading for the open sea. On coming up to the beacon, those on board found numerous sections of a scaly mass adhering to the supports and assumed that the creature was in the process of changing its old skin for a new one.

"Those interesting objects," the letter went on to say, "can be seen at the Horse Rock Coast Guard Station, and will repay a visit. These particulars I have narrated in the clearest manner I am able, and if others, in other boats, who had not so good an opportunity of seeing the entire appearance of the animal as those in my boat had, should send you a more readable account of it, I pledge myself none will more strictly adhere to the real facts. I am, Sir, your obedient servant, Roger W. Travers."

This drawing of a sea serpent is a typical example of Lineus longissimus. *It was reported seen in English waters at a place called Table Bay.* Photo, Naval Historical Archives, Washington, D.C.

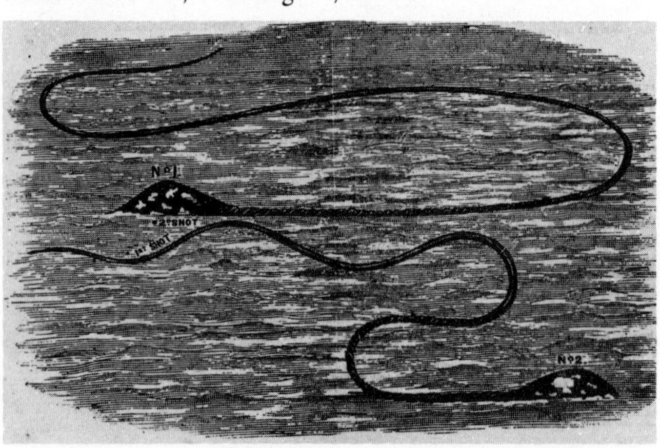

There are several points to be noted in Mr. Travers's story. First, he names a place where some of the remains may be viewed; second, he encourages the public to visit this display; and third, he implies that people, other than his own party, also saw the sea monster and would probably be writing to the newspaper. This then would seem to be an honest story that could be verified.

On September 7, the *Cork Constitution* carried another interesting story that unquestionably seems to be linked to the Roger W. Travers sighting. "On Monday last a party of gentlemen belonging to this city were enjoying a sailing excursion in the *Antelope* yacht, belonging to Mr. Wheeler, along the coast from Glandore to Kinsale. Passing the Old Head of Kinsale, the day unusually fine, they observed an extraordinary commotion in the sea, apparent to everyone on board. The Bay of Kinsale was at the time filled with fish. In a few moments they perceived a large serpent-like fish on the surface, that could not be less than 120 feet in length."

The newspaper report stated that in shape this monster resembled the long funnel of some immense steamer. Unfortunately, they confessed, the yacht was not sufficiently close for those aboard to be able to give a detailed description of the sea monster. After lying on the surface for a few moments, it was reported, the sea serpent suddenly dashed ahead at about 60 miles an hour and, after approximately 2 miles of travel, disappeared. Those on board said that it would seem as if the serpent had been actively engaged in chasing fish.

"It is a singular circumstance," the newspaper went on to say, "that notwithstanding the unusual quantity of fish that was observable, the Kinsale hookers [fishermen] were most unsuccessful, as it was stated that they did not obtain a single take during the evening. The gentlemen who have witnessed the visit of this monster, and whose statement is detailed above, may be relied on as above all suspicion. [Signed] *Cork Constitution.*"

The Irish sea monster as seen from the coast between Glandore and Kinsale. Photo, Irish Tourist Bureau, Dublin, Ireland.

The *Zoologist* carried the following notation, dated September 9, 1850: "A few friends accompanied me on a boating excursion this day whose names are William Silk, John Hunt, George Williams, Henry Seymour, and Edward Barry, and, being off the Souverein Islands, our attention was directed by one of the party to an extraordinary appearance ahead of the boat; immediately all eyes were turned to see what it was, when, to our astonishment and fright, the above monster of the deep was bearing down on us; we were at once thrown into an awful fright, and thought it best to retreat for the shore; on our landing, Mr. W. Silk, who was armed with a double barreled gun, discharged both barrels at the monster, but without effect. I need not describe his appearance, as you are aware of it before, but from inquiries from various boatmen I am told he has been off the harbour the last three days. [Signed] John Good, Kinsale."

All this newspaper copy, plus a great deal more, provided ammunition for those who would cry "hoax." For every written attestation, there were at least fifty armchair authorities who said that no such animal could exist.

Then, once again, appeared *big news!* A headline in the *New York Tribune* carried the word all the world loved to read: "Sea Serpent Caught at Last." The story read: "Ship *Monongahela* at sea, Feb. 6, 1852. A small vessel has just been reported from my mast-head, and as she is apparently bound into some of the northern parts, I intend to speak her, purposely to acquaint, through your widely diffused journal, the people of the United States, of the fact of the existence and capture of the sea serpent—a monster deemed fabulous by many—but the truth of whose existence is now forever settled, and, I trust I shall be excused in saying, by Yankee intrepidity. On the morning of January 13, when in lat. 3 deg. 10 min. south, and long. 131 deg. 50 min. west, the man on the lookout, seated on the foretopmast crosstrees, sang out: 'White water' and in reply to my 'Where away?' said 'Two points on the lee bow.' Supposing it to be made by sperm whales, and being very anxious to obtain oil, I ordered my ship to be kept off, and immediately went aloft with my spyglass."

The writer, who was later to sign himself as "Captain Charles Seabury, Master, Whale-ship *Monongahela*," said he had previously observed that there had been very light winds over the past few days but that on this date the winds had drawn to the south-southwest, become steady, and

Theodor Kittelsen painted this picture to illustrate the book Norwegian Folk Tales. The original now hangs in the *Nasjonalgalleriet in Oslo.* Photo, Nasjonalgalleriet, Oslo, Norway.

threatened to blow a gale. Then he launches into what is probably the all-time record for a detailed description of how it feels to sight, catch up with, and battle a sea monster to the death.

Captain Seabury said that he stayed aloft for nearly half an hour before seeing anything resembling "white water." At that point, he did see something but thought it was probably a school of porpoises. However, wishing to make sure, he ordered the mate to turn up all hands, square in the yards, and send out the port studding sails. He noted that the time was just seven bells (7:30 A.M.), and he returned to the deck, intending to retire for breakfast. A deckhand named Onnetu Vanjau, a Marquesan islander, suddenly gave a call that caught everyone's attention. " 'Oh! Look! Me see—too much—too much!' "

The captain swung himself in the indicated direction, but only just in time to catch a glimpse of "black skin" going beneath the surface of the ocean. The captain luffed his sails and had his crew stand by their whaleboats. For nearly one hour the horizon was scanned before the sea monster was again seen. This time, the captain himself saw the animal and stated that it was the strangest creature he had ever observed.

"It was apparently still," he wrote, "but *shobbing* up and down, as we say of sperm whales. I knew it was not a whale. The head I could not see, but the body had a motion like the waving of a rope when shaken and held in the hand. Every eye in the ship regarded it attentively, and not a word was spoken or sound uttered. In a few minutes the whole length of the body rose and lay in the water; it was of an enormous length. Presently the extremity or tail moved or vibrated, agitating the water, and then the head rose entirely above the water, and moved sideways slowly, as if the monster was in agony or suffering. 'It is a sea serpent,' I exclaimed; 'stand by the boats.' "

The mate, according to Captain Seabury, was against this action, for he said the animal was not a whale, would therefore yield no oil, and as a consequence, could result in nothing more than the loss of valuable time. The captain, seeing the point well taken but still determined to have a try at the sea monster, mustered all hands. When assembled, he made a little speech about the importance of taking a sea monster and said that everyone's courage, in fact their very manhood, was at stake. He called for volunteers to step forward and join him in the hunt.

"Let me say to their credit," the captain wrote, "every American in the ship stepped out at once, followed by all but one native and two Englishmen. I ordered the boat-steerers and officers to examine and see that everything in and about the boats was in perfect order. I had already jumped into my boat when the serpent began to move very rapidly and it was necessary to stand after him."

By now the wind was strong, according to the way the captain described things, and he was forced to give his ship all the sail it could safely carry. The serpent worked to windward of the ship, and shortly after, the fore-topgallant mast carried away. All hands lost sight of the sea serpent. After making necessary repairs, the ship stood in the direction the serpent had last taken and, after about an hour, came within sight of the great animal once again.

The chase began all over again, but the sea monster moved with such rapidity that the ship was forced to tack in an ever-increasing wind. There was almost a gale blowing by now, but abruptly the sea monster came to rest only about a mile from the ship. The captain was determined to close with the animal, rough seas or not. He brought his ship to and then ordered his whaleboats to lower away. Now, here is the account of the actual battle in Captain Seabury's own words.

"I told the boat-steerer, James Wittmore, of Vermont, to 'stand up.' With calm and cool intrepidity he laid hold of his iron [harpoon], and, when I beckoned with a movement of my hand, quick as thought both of his weapons were buried to the socket in the repulsive body before us. I shouted 'stern,' but there was no visible motion of his snakeship. I shifted ends with the boat-steerer, and cleared away a lance as quickly as possible, beckoning them to pull up, that I might get a lance, when a movement of the body was visible, and the head and tail of the monster rushed as it were to 'touch the wound.'

"The frightfulness of the head as it approached the boat, filled the crew with terror, and three of them jumped overboard. I instinctively held out my lance, and its sharp point entered the eye. I was knocked over and felt a deep churning of the water around me. I rose to the surface and caught a glimpse of the writhing body, and was again struck and carried down. I partially lost my consciousness under water but recovered it; when I rose again in the bloody foam, the snake had disappeared, and I shouted, 'pick up the line.' The third mate, Mr. Benson, caught a bight at my line near the end, and bent on his, which in an instant began to be taken out rapidly. The mate picked me up as soon as I rose to the surface, and in a few minutes all were picked up—one was severely bruised and another insensible, but he recovered and both are now well. The snake had taken my line, the third mate's, and was taking the second mate's, when I ordered the mate to bend on and give his line to the ship. The snake was sounding and I cautioned the officers not to hold on too hard, for fear of drawing the irons."

Captain Seabury wrote that at first the line went out rapidly, but he nevertheless was required to get a spare line out of the forehold and bend it on to the others. There were now four boat lines out, each consisting of 225 fathoms, plus other lengths of line that brought the entire lot to 6,000 feet. Apparently the weather was worsening, and very little sail could be carried on the vessel in the face of such wind. The captain brought the line inboard of the ship and made it fast, knowing that the drag of his vessel might draw the irons out of the sea serpent. By 4 P.M., the wind began to shift, which favored the vessel, and by 5 P.M., there was a noticeable drop in the intensity of the gale. Night came on, and by dark, there was a lull in the weather. The line was still taut, and it was evident the sea monster was on the bottom.

"He stayed down a long time," the captain wrote, "but on reflection I considered that was his forte—that he was at home there. At 4 A.M. of the 14th, 16 hours after he went down, the line began to slack, I had it taken to the windlass, when we got nearly two line 'hand over hand,' then there came strain again. This straining continued, I told everybody to bear a hand and get breakfast, and just before we were through, the cook cried out, 'Here he is.' In no time all were on deck, and sure enough he had risen; but all that was visible was a bunch, apparently the bight of the snake, where he had been fastened to."

Continuing his detailed description, the captain said that he lowered three boats and commenced to lance the beast without eliciting any sign of life. Gradually the body floated to the surface, but the lancers kept probing for what life might remain in the great body. When the sea monster suddenly drew himself up, the boats pulled away. Then, according to Captain Seabury, they witnessed the monster's dying struggles. "The evolutions of the body were as rapid as lightning, seeming like the revolving of a thousand black wheels. The tail and head would occasionally appear in the surging bloody foam, and a sound was heard, so dead, unearthly, and expressive of acute agony, that a thrill of horror ran through our veins. The compulsive efforts lasted ten or fifteen minutes, when they suddenly stopped, the head was partially raised—it fell—the body partially turned and lay still."

Captain Seabury said that he took off his hat and led the men in nine terrific cheers. Luckily, the dead sea monster floated, and the ship took him alongside. He then held a consultation with the crew, and since they all agreed it would be impossible to get the body into any port, it was decided they would try to save the skin, head, and bones if at all possible. It was a male sea monster, the men discovered as they worked, and was 103 feet 7 inches in length; 19 feet 1 inch around the neck; 24 feet 6 inches around the shoulders; and the largest portion of the body, which appeared somewhat distended, 49 feet 4 inches. The head was long and flat, with ridges; the bones of the lower jaw were separate; the tongue had an end like the head of a heart. The tail ran nearly to a point, on the end of which was a flat piece of cartilage. The back of the sea monster was black, turning brown on the sides, then yellow, and on the center of the belly, a narrow white streak two-thirds of its length; there were also dark spots scattered over the torso. When the skin was examined, it was discovered that the body was covered with blubber (like a whale) but that it was only four inches thick. The oil was as clear as water and nearly as fast as spirits of turpentine. The blubber was "flensed" (stripped off) but was so elastic it became difficult to handle. A strip of blubber 5 feet long would sometimes stretch to as much as 20 feet when it was being pulled loose by block and tackle.

The head of the sea serpent, said to be a frightful object, was preserved with salt. The animal's bones were cleaned, and they, too, were preserved. One of the serpent's lungs was 3 feet longer than the other, and there were ninety-four teeth in the jaws, very sharp, all pointing backward and each as large as a man's thumb. The sea monster had two spout holes (spiracles), so it was assumed that it breathed like a whale. It also had four swimming paws, or imitations of fins, of loose flesh.

In completing his report, Captain Seabury said that the joints of the back were loose and that it seemed as if, in swimming, the serpent moved two ribs and a joint at a time, almost like feet. The muscular movement of the serpent after it was dead made the body look as if it were encircled by longitudinal ridges. It required three days for the ship's crew to finish cutting up the monster and getting at all the bones, which, the captain pointed out, were very porous and dark in color. The heart, he said, was preserved in liquor, along with one eye; but the head, despite what was done for it with salt, began to give off an offensive odor. Then the longest single report in the annals of sea monster history ends on this note: "2 PM. I have just spoken the vessel; she proves to be the brig *Gipsy*, Captain Sturges, eight days from Ponce, P.R., with oranges and merchandise, bound to Bridgeport. He has kindly offered to put these sheets in the post office when he arrives. As soon as I get in I shall be enabled to furnish you a more detailed account. I am, sir, your obedient servant, Charles Seabury, Master, Whale-Ship *Monongahela*, of Bedford."

As might be expected, this lengthy report drew fire from scientists, who labeled it a "hoax." They could be right, for although there undoubtedly was a *Monongahela*, she never showed up carrying her evidence in any known port nor were any further reports of the episode ever recorded. Of course, the wags of that era concerned themselves with the topic of sea monsters, and a Mr. George

An illustration from Tales of the Ocean, *by John Sherburne Sleeper, published in 1856.*

Guyon, of Ventnor, Isle of Wight, addressed himself to the topic by penning a bit of doggerel:

I've a story to tell—I don't say that it's true—
But just as I heard it I tell it to you.
A ship there was sailing upon the blue sea
With her canvas all set, when the captain, said he
'I feel that the vessel is all of a tremble,
A sort of sea earthquake it seems to resemble;
Send forward the mate to see what is the matter.'
When lo! what he saw would have made your teeth chatter,
An enormous big snake rising out of the sea,
Some three hundred feet it might possibly be,
And in bulk it might equal a 'wide crinoline'
(At least seven yards round that description must mean).
With jaws eight feet long, and with eyes fiercely glaring,
A horn and a mane; he looked horribly daring,
While the bowsprit he shook in his terrible mouth.
Twas in Latitude east and Longitude south,
This is somewhat obscure, but I think on the whole
It occurred th' other side of the Antarctic pole,
The ship making six knots—leaving foam in her wake,
Yet she stopped at the touch of this wonderful snake;
And the jibboom and bowsprit were snapped like a straw;
But his strength was outdone by his marvelous maw;
For he swallowed the stay-sail and also the jib,
Like a boy gulping oysters—they went down to [sic] glib.
With his stay to his stomac he turned about,
And gave with his tail such a vigorous flout,
That some timbers to atoms were crushed by the blow,
And what more might have happened we none of us know,
When an object appeared for the which he set sail,
And both object and story were much like a whale.

In January, 1860, the *Zoologist* printed a sea serpent report that was so well documented and promulgated by such an authoritative source that its accuracy was not questioned. However, as it later turned out, the entire report was in error. The story had its beginning with an article by a Captain Hawtaigne, R.N., and was headlined "A Sea Serpent in the Bermudas." It went on to say that the strange animal captured was no less than the great sea serpent seen by Captain M'Quhae of H.M.S. *Daedalus*. There were two men walking

The sea serpent of Hungry Bay, Bermuda, proved to be a perfect specimen of an oarfish. Photo, Library of Congress, Washington, D.C.

the beach, so the story read, around Hungary Bay, in Hamilton Island, when their attention was attracted by a rushing sound in the water. On reaching the spot from which the commotion arose, they discovered a large sea monster that had thrown itself onto the rocks. It appeared to be dying from a frantic effort to reach the ocean water again. The two men grabbed large forks, which were strewn on the beach as tools with which to gather seaweed, and attacked the monster.

"The reptile was sixteen feet seven inches in length," the article read, "tapering from head to tail like a snake, the body being a flattish oval shape, the greatest depth at about a third of its length from the head, being eleven inches. The colour was bright and silvery; the skin destitute of scales but rough and warty; the head in shape not unlike that of a bull-dog, but it is destitute of teeth; the eyes were large, flat, and extremely brilliant, it had small pectoral fins, and minute ventral fins and large gills."

Captain Hawtaigne continued with his precise description by saying that there was a series of fins running along the back, composed of short, slender rays, united by a transparent membrane, at the interval of something less than an inch from each other. The creature, he stated, had no bones, but only cartilage running through the body. "But its most remarkable feature was a series of eight long, thin spines of a bright red colour springing from the top of its head and following each other at an interval of about an inch; the longest was in the center: it is now in possession of Colonel Munroe, the acting Governor of the Colony. . . ."

The entire report produced quite a stir, with everyone congratulating everyone else on the belated capture of a sea serpent. Unexpectedly, the story, like the fish itself, was quietly left to die. A naturalist named J. Matthew Jones examined the remains and stated with absolute certainty that it was a perfect specimen of an oarfish (*Regalescus glesne*), which is frequently presented to the public as a young sea serpent.

Then the sea monster sighting of the gunboat *Alecton* came into prominence. On November 30, 1861, this French warship happened upon what appeared to be a sea monster swimming on the surface of the ocean. This was about 120 miles northwest of Tenerife, and Lieutenant Bouyer, who was captain of the ship, ordered the animal fired upon as the first step in attempting to capture it. According to his own report, Bouyer identified it as a massive octopus (not knowing the difference between a giant squid and an octopus) that measured somewhere over 24 feet in length. The two long tentacles normal for this type animal seemed to be missing, and the great beast was apparently sick and dying. A noose was finally maneuvered around

Early nineteenth-century painting by Theodor Kittelsen to illustrate the book Norwegian Folk Tales. *Nasjonalgalleriet, Oslo, Norway.*

Sea serpent as seen from City of Baltimore. Photo, Naval Historical Archives, Dublin, Ireland.

the cumbersome body, but it slipped until coming fast on the tail. When it was hoisted inboard, the soft-bodied animal broke apart, and only some remnants were snagged aboard.

The French Academy of Sciences was later informed that when brought aboard, the monster was brick red, had eyes as large as dinner plates, and a mouth 18 inches across. The whole of it possessed a strong stink of musk.

Without any hesitancy whatever, the world of science produced its vocal skeptics; and the captain and the crew of the gunboat *Alecton* were branded liars. Ironically, even while the story of the *Alecton* was being called into question by those who shouted that such an animal could never exist, there were other reports from various areas that indicated an infestation of giant squids. An entire lot of them crawled onto the beaches of Labrador and Newfoundland to die. Ships made numerous sightings, and small fishing boats were attacked. Finally, the giant squid was recognized for what it was, as also was the case with the octopus. Thereafter, both were dissociated from the concept of sea monsters.

In many instances, doubts were well founded. We cite a letter written to an English magazine, *The Natal Colonist*, by Mr. J. Cobbin, of Durban, dated June 13, 1872. "During my last passage from London, I saw no less than three sea serpents, but an account of the last will suffice. On 30th November last, on board the *Silvery Wave*, in lat. about 35′ 10″ S., and long. 33′ 30″ E., at 6:20 PM solar time, an enormous sea serpent passing nearly across our bows compelled the alteration of our

The French corvette Alecton *encounters a giant squid. This drawing was made shortly after the event in November, 1861.* Photo, Seamen's Church Institute of New York, New York City.

A view of the Alecton *squid being captured.* Photo, Library of Congress, Washington, D.C.

course. He was at least one thousand yards long, of which about one third appeared on the surface of the water at every stroke of his enormous fan-shaped tail, with which he propelled himself, rising it high above the waves, and arching his back like a sand snake or a caterpillar."

Mr. Cobbin goes on to say that the shape and proportions of this sea monster led him to liken it to a cobra, especially since it was marked by the same knotty and swollen protuberance at the back of the head and on the neck. He said, however, that the head was more like that of a bull, with large glowing eyes and ears that had circular tips and were level with the eyes. Furthermore, the sea monster's head was surrounded by a horned crest, which he raised and lowered at his own pleasure. He swam with great speed and in doing so lashed the sea into a foam, like breakers dashing over rocks.

"The sun shone brightly upon him," Mr. Cobbin said, "and with a good glass I saw his overlapping scales open and shut with every arch of his sinuous back colored like the rainbow."

In direct contrast with Mr. Cobbin's doubtful tale is another report made public two months later, in August, 1872, by Captain A. Hassel, of the bark *Saint Olaf*, sailing from Newport to Galveston, Texas. "Two days before arrival at Galveston, and about 4:30 PM, on May 13, weather calm, smooth sea, lat. 26′ 52″, long 91′ 20″, I saw a shoal of sharks passing the ship. Five or six came under the vessel's stern, but before we could get out a line they went off with the rest. About two minutes after, one of the men sang out that he saw something on the weather bow, like a cask on its end. Presently another one called out that he saw something rising out of the water like a tall man. On a nearer approach we saw that it was an immense serpent, with its head out of the water, about 200 feet from the vessel. He lay still on the surface of the water, lifting his head up, and moving the body in a serpentine manner. Could not see all of it; but what we could see, from the after part of the head, was about 70 feet long and of the same thickness all the way, excepting about the head and neck, which were smaller, and the former flat, like the head of a serpent. It had four fins on the back, and the body of a yellow greenish colour, with brown spots all over the upper part and underneath white. The whole crew were looking at it for fully ten minutes before it moved away. It was about six feet in diameter." The report was then signed "A. Hassel, master of the Norwegian barque *Saint Olaf;* witnessed by J. Fredk., Walthew."

It is not possible to cover all the reported sightings that came to public attention, either during this period or any other. Some, however, are more important than others because of the trustworthiness of the tellers. One such report is the lengthy account published in the May, 1873, issue of *Zoologist* by two clergymen of established positions, in which they detail their sighting of a sea monster.

The story was headed "Appearance of an Animal, Believed to Be That Which Is Called the Norwegian Sea Serpent, on the Western Coast of Scotland." It seems that the Reverend John Macrae, Minister of Glenelg, Inverness-shire, and the Reverend David Twopeny, vicar of Stockbury, Kent, set out from Glenelg in a small cutter. Along with them were two ladies, another gentleman, and a Highlands lad. The small boat was on a course down the Sound of Sleat, which on that side divided the Isle of Skye from the mainland, the average breadth of the channel being about 2 miles.

A sea serpent, as seen from the steamer Kätie *in May, 1882.* Photo, Berlin Zeitung, Germany.

The day was calm and sunny. There was no wind blowing, and the surface of the water was perfectly smooth. As they rowed the cutter along, the entire group perceived a dark mass about 200 yards astern and just a little to the north. There were three pairs of binoculars aboard, and as many people raised these to their eyes. As they watched, it was noticed that another black lump projected itself out of the water just to the left of the first. Then another and another followed, each showing an interval of water, but all carefully in line. Everyone aboard agreed that it was a single animal. The monster moved across their wake and suddenly disappeared. Presently the first mass, which was evidently the head, reappeared, and was tion: when the lumps sank, other lumps did not rise in the intervals between them. The greatest number we counted was seven, making eight with the head. . . . The parts were separated from each other by intervals of about their own length, the head being rather similar and flatter than the rest, and the nose being very slightly visible above the water; but we did not see the head rise above the surface either this or the next day, nor could we see the eye. We had no means of measuring the length with any accuracy, but taking the distance from the center of one lump to the center of the next to be six feet, and it could scarcely be less, the whole length of the portion visible, including the intervals submerged, would be forty-five feet."

A sea monster, as depicted by French illustrator Gustave Doré for the works of François Rabelais, 1873.

followed by the rising of the other black lumps, as before. Sometimes three lumps would appear; at other times four, five, or six would show; then all would sink out of sight. When the creature resurfaced, the head always came to the surface first, with the lumps slowly rising in regular order. However, when the animal sank, it did so with a single motion, disappearing all at one time.

"It gave the impression," the two clergymen wrote later, "of a creature crooking up its back to sun itself. There was no appearance of undula-

Suddenly, to the horror of all aboard, the sea serpent came directly toward the cutter. It appeared as though it were about to attack. It raised its gigantic head and moved at a furious speed. Most of the body appeared to be immersed as it swam, creating a fine spray as the animal darted through the water. Abruptly, within 100 yards of the cutter, it sank and could be seen moving away in the direction of Skye. The course of the animal could be traced by the waves it caused as it moved beneath the surface. The entire party watched for some time as the animal continued "careening about at a distance," but the animal did not again come close to the boat.

The next day, the party returned to approximately the same spot and found themselves again becalmed. The day, they noted later, was warm and sunny. That afternoon, the creature was again seen, but at some distance. All aboard agreed that it appeared to be longer than it had seemed to be the day before. Its shape also seemed to be somewhat different. On this sighting, it was generally concluded that the sea serpent looked to be at least 60 feet in length. Once again it began careening about and then headed up the Lochburn.

"Later in the afternoon," the two clergymen reported, "when we were still becalmed in the mouth of the Lochburn, and by using the oar had nearly reached the island of Sandaig, it came rushing past us about a hundred-and-fifty yards to the south. It went with great rapidity, its black head only being visible through the clear sea, followed by a long trail of agitated water. As it shot along, the noise of its rush through the water could distinctly be heard on board. There were no organs of motion to be seen, nor was there any shower of spray as on the day before, but merely such a commotion in the sea as its quick passage might be expected to make. Its progress was equable and smooth, like that of a log towed rapidly."

For the rest of the day, until nightfall, as they slowly rowed their way home through the Sound of Sleat, the sea serpent was occasionally within sight of the cutter, rushing about at a distance, showing only its head and a small part of its body on the surface. There was some thought expressed by members of the group that the sea serpent always seemed to stay rather close to the cutter, and since they were always rowing, they were inclined to think the attraction might be the measured beat of the oars.

As it turned out later, other people also saw the same sea serpent, and the *Zoologist* magazine published their observations.

"The ferrymen on each side of Kylerhea saw it pass rapidly through on the evening of the 21st, and heard the rush of water: they were surprised, and thought it might be a shoal of porpoises, but could not comprehend their going so quickly.

"Finlay Macrae, of Bundaloch, in the parish of Kintail, was within the south of Lochourn on the 21st, with other men in the boat, and saw the creature at about the distance of one-hundred-and-fifty yards.

"Two days after . . . Alexander Macmillan, boat builder at Dornie, was fishing in a boat in the entrance of Lochduich, halfway between Druidag and Castledonan, when he saw the animal near enough to hear the noise and see the ripple it made in rushing along in the sea. He says, that what seems its head was followed by four or more lumps, or 'half rounds,' as he calls them, and that they sometimes rose and sometimes sank all together. He estimated its length as not less than sixty to eighty feet. He saw it also on two subsequent days in Lochduich. On all the occasions his brother Farquhar was with him in the boat, and they were both much alarmed and pulled to the shore in great haste.

"A lady at Duisdale, in Skye, a place overlooking the part of the Sound which is opposite the opening of Lochourn, said that she was looking with a glass when she saw a strange object on the sea which appeared like eight seals in a row. This was just about the time we saw it.

"We were also informed that about the same time it was seen from the island of Eigg, between Eigg and the mainland about twenty miles to the south-west of the opening of Lochourn."

The Reverend John Macrae and the Reverend David Twopeny added a long-winded postscript to their history. It did not deal directly with the reporting of hard facts, but rather attempted to refute arguments that they felt certain would arise.

It seems that two years prior to this time, a similar creature of great length had been sighted in the exact same area. Some years before that, a story was related by a well-known gentleman that on crossing from Mull to the mainland, a sea monster came to the surface not more than 50 yards from his boat. His description, according to these men of the cloth, was that it had a sharp, ridgelike back, was about 40 feet in length, and was very dark of color.

"In the meantime," Macrae and Twopeny concluded, "as the public will most probably be dubious about quickly giving credit to our account, the following explanations are open to them, all of which have been proposed to me, *viz:*—porpoises, lumps of seaweed, empty herring-barrels, bladders, logs of wood, waves of the sea, and inflated pigskins; but as all these theories present to our minds greater difficulties than the existence of the animal itself, we feel obliged to decline them."

Between 1850 and 1900, there were literally

Sea serpent sighted from the yacht Osborne *in 1877.* Photo, Navy Library, Washington, D.C.

Another view of the Osborne *sea serpent.* Photo, Navy Library, Washington, D.C.

hundreds of sightings reported. Many were, as is to be expected, based upon hysteria. On the other hand, numerous accounts were given by citizens of solid reputation and honorable position. A ship's logbook, written by Reginald Yonge, aboard H.M.S. *Orontes*, shows that on March 20, 1873, at 3:15 A.M. an engine failure caused his vessel to stop. From the upper bridge, a peculiar object was sighted in the water. It turned out to be the head of an immense monster that projected no less than 5 feet out of the water. The head submerged, only to rise close alongside the ship. It then disappeared, and the creature was seen swimming off in the distance. The entire event was witnessed by a variety of mariners aboard ship, who generally agreed that the creature was about 40 feet in length.

During 1873, another minister, the Reverend James Joass, an archaeologist, saw a sea monster and likened it to a "saurian allied to Plesiosaurus." In the same year, Lady Florence Leveson Gower and the Honorable Mrs. Coke, while driving their carriage beside the sea, also saw the monster.

A story carried in the *Illustrated London News*, dated November 20, 1875, which ran under the headline "The Great Sea Serpent," went on to tell of a massive sea serpent attacking a whale. "Captain Drevar, of the barque *Pauline*, bound with coals from her Majesty's naval stores at Zanzibar, when in lat. 5 deg. 13 min. S., long. 35 deg. W., on July 8 last, observed three very large sperm whales, and one of them was gripped round the body, with two turns, by what appeared to be a huge serpent. Its back was of a darkish brown and its belly white, with an immense head and mouth, the latter always open; the head and tail had a length beyond the coils of about 30 feet; its girth was about 8 feet or 9 feet. Using its extremities as levers, the serpent whirled its victim round and round for about fifteen minutes, and then suddenly dragged the whale down to the bottom, head first. The other two whales, after

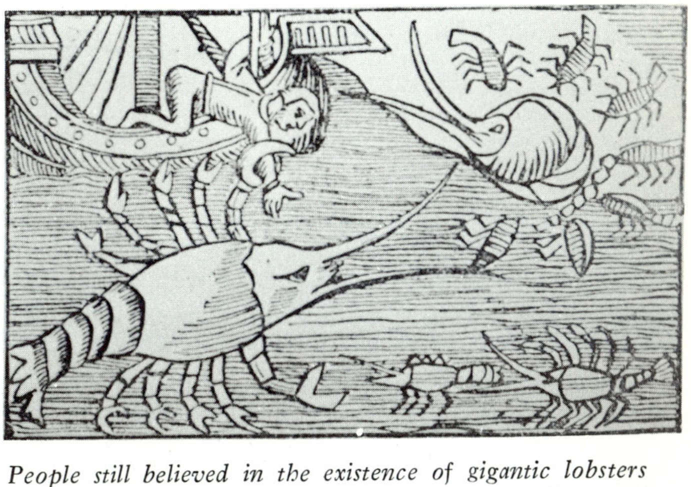

People still believed in the existence of gigantic lobsters capable of dragging a man from a ship. Photo, Library of Congress, Washington, D.C.

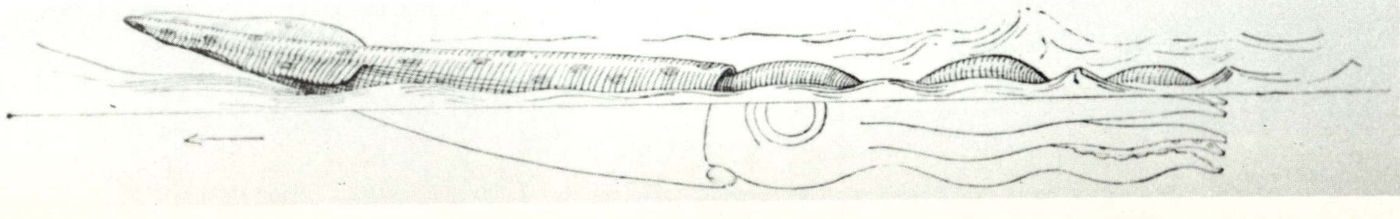

A giant calamary as depicted by Henry Lee, in 1883, for his book Sea Monsters Unmasked. Photo, Naval Historical Archives, Washington, D.C.

A sea serpent seizing a man from on board a ship. Photo, Naval Historical Archives, Washington, D.C.

attempting to release their companion, swam away upon its descent, exhibiting signs of the greatest terror."

On July 17, 1875, a sea monster was reported by one Captain Garton, of the steamship *Norman*. He said that a few days prior to this date, he had seen, just off Plymouth, the head of a gigantic sea monster that was raised at least 10 feet above the ocean. The animal carried on a constant diving action but could clearly be seen as having black and white stripes running the length of its body. He said that the throat was pure white and that the head, which was extremely large, was all black. A pair of saucer-sized black eyes protruded an inch or more from the head. The body was round, like a fish barrel; and the length was more than 100 feet.

A few weeks later, in the same general area, the yacht *Princess* spotted a sea monster that could have been the same one mentioned by Captain Garton. Those aboard the yacht reported that its head looked like that of a turtle or a snake, and when it raised its head 6 to 8 feet out of the water, it held it there for almost 10 seconds at a time. The head, those aboard the yacht reported, seemed to be about 2½ feet in diameter. No estimate of length was given, since the monster never seemed to show itself all in one extended piece.

The newspaper *Straits Times Overland Journal*

A sea monster called a physeter, *capable of swamping a ship through use of its powerful blow holes.* Photo, Navy Library, Washington, D.C.

reported in 1876 that the ship *Nestor*, upon arriving at Shanghai, told of seeing a sea monster. A sworn statement was made to this effect by both the ship's surgeon and the captain. An excerpt from the doctor's report reads in this fashion: "The captain saw an object which had been pointed out by the third officer as a 'shoal'! Surprised at finding a shoal in such a well-known track, I watched the object, and found that it was in motion, keeping up the same speed with the ship, and retaining about the same distance as first seen. The shape of the creature I would compare to that of a gigantic frog. The head, of a pale yellowish colour, was about twenty feet in length, and six feet of the crown was above the water. I tried in vain to make out the eyes and mouth; the mouth may, however, have been below water. The head was immediately connected with the body, without any indication of a neck. The body was about forty-five or fifty feet long, and of an oval shape, perfectly smooth, but there might have been a slight ridge along the spine."

Another account of a sea monster sighting is that reported by the master of the bark *Georgia*, from Rangoon, in May, 1877. He positioned the sighting accurately on his chart and then described the animal in this fashion: "The colour being stated as grey and yellow makes me conclude that the animal had swum for a long time with its body in a straight line, without diving and the part, exposed to the sunbeams, had dried up." He gave its length as 40 to 50 feet and supported his report with the testimonies of other eyewitnesses.

Whales were thought to be capable of flooding passing ships till they sank. Photo, Navy Library, Washington, D.C.

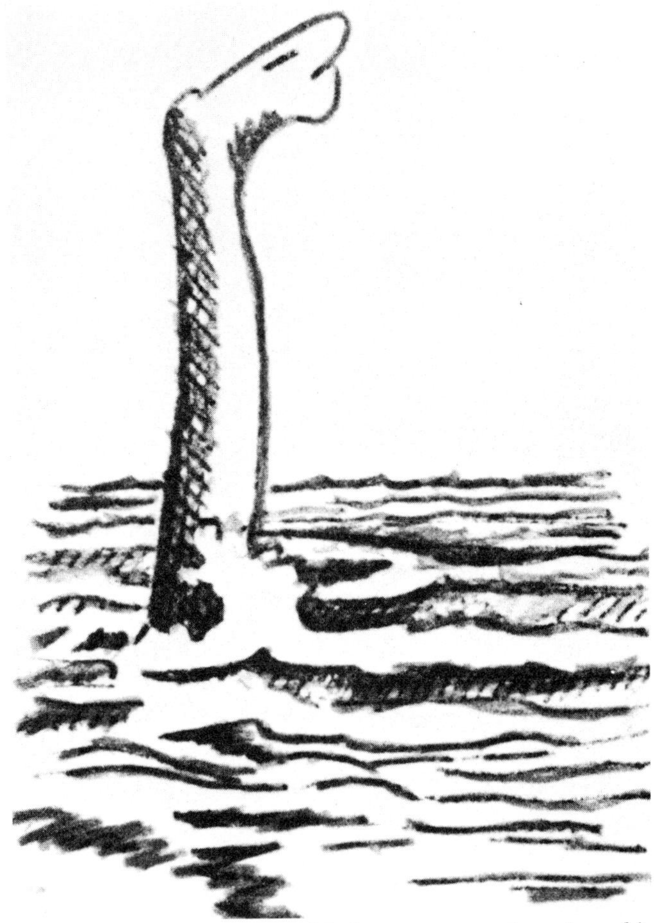

In 1891, a Dr. Farquhar Matheson reported seeing this monster in the Loch Alsh. Photo, Main Navy Library, Washington, D.C.

From Australia, the vessel *Sacramento*, out of New York harbor, reported that on July 30, 1877, in mid-Atlantic, the helmsman called out that there was a sea monster to be seen. The captain confirmed the sighting. He said that at least 40 feet of the animal was visible, that it appeared to be as thick as a flour barrel, and that it was yellowish in color. Its head was flat, and the eyes were quite visible.

The helmsman, who had a closer and longer look than the captain, made a sketch of what he had seen. At the time this drawing was published, according to a spokesman for the National Library of Australia, the helmsman recalled the sea serpent as having a body like that of a large snake, with a length of about 50 or 60 feet. Its head was like an alligator's, and the color was a reddish brown.

The Illustrated London News *of November 20, 1875, carried this illustration, showing a giant sea serpent attacking a whale.* Photo, Navy Library, Washington, D.C.

Another interesting account appeared in 1879. It was written by a Major W. H. J. Senior, of the Bengal Staff Corps, British Royal Army, traveling aboard the steamship *City of Baltimore.* He stated that "I observed a long black object abeam of the ship's stern on the starboard side, at a distance of about three-quarters of a mile, darting rapidly out of the water, and splashing in again with a sound distinctly audible, and advancing nearer and nearer at a rapid pace. In a minute it had advanced to within a half mile, and was distinctly recognizable as the veritable "sea serpent." I shouted out, "Sea Serpent! Sea Serpent! Call the Captain!" Dr. C. Hall, the ship's surgeon, who was reading on deck, jumped up in time to see the monster. . . ." The doctor continued the story from there, and his account was fully documented and witnessed by numerous other passengers aboard the ship.

The year 1879 appears to have been a vintage year for sea monsters. The British ship *Privateer* saw one 100 miles west of Brest, and frequent coverage of sea monster sightings commenced to appear in *The Bulletin,* issued by the United States Fish Commission. An example of this coverage (in volume 4, April 15, 1884) reports a strange creature sighted near Saint Elmo. No more accurate words can be found than those used in the report itself.

"While the boats of Captain Seymour's bark *Hope On* were on the watch for whales off the Pearl Islands [between 40 and 50 miles from Panama] the water broke a short distance away, and Captain Seymour made ready for a whale. But a head like that of a horse rose from the water and then dived. The creature was seen by all the boat's crew. Captain Seymour describes the animal as almost 20 feet in length, with a handsome horse-like head, with two unicorn-shaped horns protruding from it. The creature had four legs or double jointed fins, with a brownish hide, profusely speckled with large black spots, and a tail which appeared to be divided into parts. The creature was seen on two different days, and if whales had not been about at the time, an effort would have been made to catch it. Captain Seymour and his officers agree in considering that the creature is peculiar to the locality, and that it could easily be killed with lance and guns. It is important to notice that officers of the Pacific Mail Company state they have seen the animal on several occasions, but not so closely as did the officers and men of the *Hope On.*"

The report concluded that a race of such marine animals does exist. It is suggested that there are probably several varieties, one of which is characterized by a serpentine neck and a head that is small when compared with the size of the body but large when compared with the thickness of the neck. Volume 5 of the *Bulletin,* dated 1885, also mentions a sighting by Captain W. Hopkins, of the schooner *Mary Ogilvie,* who had just returned from a voyage around Australia. "He saw an immense creature which he took to be a species of octopus. His attention was drawn to it by a perfect cloud of sea-birds, and at first he thought it must be a dead carcass. On approaching it, however, he found it was alive, and sluggishly disporting itself. In shape it was like a violin, but of immense size, with some six feelers about the greater diameter of the violin. It lay almost flat upon the water, was of a dark gray above and lighter gray below, and was continually elevating one of its feelers, apparently twice the thickness of a man's arm, to a height of from 6 to 8 feet. It appeared to be vomiting, and as the birds were evidently feeding that accounted for their presence in such numbers. Its size was so great that, had it grasped the vessel, it could easily have capsized it. The captain, therefore, got out of the way as quickly as possible. . . ."

Captain Weisz, of the German steamship *Katie*, in May, 1882, reported that his vessel, when in the vicinity of the Hebrides, sighted what at first was thought to be wreckage from another craft. On approaching for a closer look all aboard were amazed to find themselves in the presence of a sea monster. "The length of the visible part of this animal," the captain explained, "which had not the least resemblance to the back of a whale, measured, according to our estimation, about 150 feet, the bumps, which were from three to four feet in height, and about six or seven feet distant from each other, were smaller on the tail end than on the head end, which withdrew from our observation. At our arrival at Newcastle, I learned that some days before some fishermen of Lewis had observed the same or similar animal."

In 1891, at Pablo Beach, Florida, a sea monster is said to have suddenly appeared in the midst of a group of bathers and caused a panic. And in 1893, a Dr. Farquhar Matheson, an English surgeon from London, reported seeing a serpent during a boating holiday in the Loch Alsh. "Suddenly I saw something rise out of the Loch in front of us. . . . a long, straight, neck-like thing as tall as my mast. I could not think what it was at first. I fancied it might be something on land, and directed my wife's attention to it." He continued to describe all the circumstances in some detail and then said, "I can think of nothing to which to compare it so well as the head and neck of the giraffe, only the neck was much longer, and the head was not set upon the neck like that of a giraffe; that is, it was not so much at right angles to it as a continuation of it in the same line."

It is not easy to conclude this survey of the years between 1850 and 1900 with the mention of so few sea monster sightings. There were well over 100 reports made known to the public during this time, and it can only be assumed that as many more went unreported.

Bathers at Pablo Beach, Florida, saw this sea monster in 1891. Photo, Naval Photo Center, Washington, D.C.

Numerous sea monsters are shown in this old print. Photo, Naval Photo Center, Washington, D.C.

A Canadian serpent was sighted in Second Narrows, Vancouver, British Columbia. Its length was estimater at 80 feet. Photo, Provincial Archives, Victoria, British Columbia.

7

1900 to 1950

At the beginning of the twentieth century, many strange sea monster stories began to circulate from around the world, such as the one told by Captain Laurence Thomson, of the steamer *Nemesis*, who, in 1900, saw what could only be classified as a monster. Supposedly, it was a 300-foot, rubber-like, tubular sea serpent about 3 feet in diameter. The beast rose from the sea in three "arches" and moved at a great speed in this fashion. The story was printed in *Wide-World* magazine but never gained scientific recognition because of the dubious manner in which the sea serpent is said to have conducted itself.

In 1901, a sea monster story appeared in a Japanese nature newspaper called *Shizen Shimbun*, which told of a horned sea serpent, with large eyes, sharp teeth, and tough hide. This animal was peculiar in that it appeared only on stormy nights and was best seen in flashes of lightning. Although the Diet Library vouched for the authenticity of the newspaper and its story, they could locate no sightings subsequent to this date.

There was also a sighting in 1902, made by the ship *Chillagoe*, under the command of Captain W. Firth. Apparently, he and his crew caught only a brief look at the sea serpent. They described it as about 35 feet long, with four very large fins, and a seal's head of gigantic proportions. The ship is said to have gotten quite close before the sea monster submerged.

Also in 1902, another curious sea monster was reported in the public press. Apparently the story was well received. The creature was sighted from the ship *Fort Salisbury* at about seven bells (3:30 A.M.), when the lookout reported a dark mass, with two masthead lights, moving through a phosphorescent ocean. The officer of the watch, his helmsman, and the man on lookout later agreed that as they closed on the object, they saw that it was a sea monster with two brilliant eyes. They gave the length as being between 500 and 600 feet, with a possible width of about 30 feet. Since all this took place at night, with no evidence that spotlights were used, it leaves a good many questions unanswered.

In 1905, the steam yacht *Valhalla* was cruising off the coast of Brazil on a scientific research mission. She weighed 1,600 tons (which is large for a yacht) and was capable of carrying a considerable spread of canvas. Her recorded speed is said to have been 12 knots under steam, and something close to 16 when sail was crowded aloft in a good breeze. The vessel was owned by the earl of Crawford, who had previously been president of the Royal Astronomical Society. On this occasion, his scientific party included several prominent intellectuals of the period. Two of these brilliant gentlemen were Michael J. Nicoll, F.Z.S., and E. G. B. Meade-Waldo, F.Z.S. This latter English scholar, in as unemotional and objective a manner as possible, reported the sighting in the following fashion:

"On 7th December 1905, at 10:15 A.M. I was on the poop of the *Valhalla* with Mr. Nicoll when he drew my attention to an object about 100 yards from the yacht; he said: 'Is that the fin of a great fish?'

"I looked and immediately saw a large fin or frill sticking out of the water, dark sea-weed brown in colour, somewhat crinkled at the edge. It was apparently about 6 feet in length and projected from 18 inches to 2 feet from the water.

"I got my field glasses on it (a powerful pair of Goerz Trieder), and almost as soon as I had them on the frill, a great head and neck rose out

Drawing made from an eyewitness sketch of the Valhalla *sea monster of 1905.* Photo, Illustrated London News.

of the water in front of the frill; the neck did not touch the frill in the water, but came out of the water in *front* of it, at a distance of certainly not less than 18 inches, probably more. The neck appeared about the thickness of a slight man's body, and from 7 to 8 feet was out of the water; head and neck were all about the same thickness.

"The head had a very turtle-like appearance, as had also the eye. I could see the line of the mouth, but we were sailing pretty fast, and quickly drew away from the object, which was going very slowly. It moved its head and neck from side to side in a peculiar manner: the colour of the head and neck was dark brown above, and whitish below—almost white, I think. When first seen it was about level with the poop of the yacht, and on the starboard side.

"I made it out by the chart to be in about S. lat. 7°4′, long. 34°20′, but I think this is not quite correct. Mr. Nicoll got the correct position from the captain. The depth of the water where we saw it was about 300 fathoms, but quickly went to as much as 1300 fathoms.

"Since I saw the creature I consider on reflection that it was probably considerably larger than it appeared at first, as I proved that objects, the size with [*sic*] which I was well acquainted, appear very much smaller than they really are when seen on the ocean at a similar distance with nothing to compare them with. (Signed) E. G. B. Meade-Waldo."

There is a footnote to all this. "The correct position seems to have been 7°14′ S., 34°25′ W. This puts the ship about 23 miles 106° from C. Branco (near Parahiba, Brazil), and just outside the 100-fathom line. Two miles landward from this position, there is a chartered sounding of 37 fathoms; three miles seaward, one of 1,340 fathoms (Admiralty Chart No. 528)."

Mr. Nicoll described the event in the following manner: "On the 7th December, 1905, when in latitude 7 degrees 14′ S., longitude 34 degrees 25′ W., and about fourteen miles from the coast of Brazil near Para, a creature of most extraordinary form and proportions was sighted by two of us. At the time we were under sail only, and were slowly making our way to Bahia.

"It was about 10 o'clock in the morning, and I was leaning on the rail of the poop deck, when a large fin suddenly appeared close to the ship at a distance of about fifty yards. This fin resembled that of no fish I had previously seen, and

At the turn of the century, sea charts were still being decorated with sea monsters.
Photo, Schiffahrts Museum, Brake, West Germany.

This Scandinavian print depicts a giant whale that was washed ashore. Photo, Thijs Mol, Arnhem, Netherlands.

I pointed it out immediately to Mr. E. G. B. Meade-Waldo, who was on the deck with me at the time, and we watched it together for several minutes.

"As we passed slowly by, a long eel-like neck surmounted by a head, shaped somewhat like that of a turtle, rose out of the water in front of the fin. This creature remained in sight for a few minutes, but we soon drew ahead of it, and it became lost to view, owing to the ripple of the water. Owing to the fact we were under sail at the time, it was not possible to go about and make a closer inspection, and with great regret we had to be content with the view we had of this remarkable monster. (Signed) M. J. Nicoll."

A book entitled *Adventures in Peru*, by C. H. Prodgers, an English businessman, contained an account of a sea serpent sighted in 1905. This took place off the island of Fernando de Noronha, several hundreds of miles north and east of Cape São Roque, Brazil. The author stated that he got a good look at the monster from a distance of about 50 yards and described it as having a head as big as that of a cow, with a body that appeared to be as round as a flour barrel.

During 1906, according to that eminent authority Dr. A. C. Oudemans, Holland was beset by numerous sea serpent reports. The newspaper *Oprechte Haarlemsche Courant* told of a sea serpent that passed a given point twice a day, as though it were either going to or coming from a favorite feeding area. This was confirmed during the same year by a group of workmen on the beach at Zandvoort, who described a similar eel-like creature about 60 feet in length. Also, in 1906, the Dutch were reporting sightings from as far away as the Indian Ocean. A passenger on the steamer *Java* saw the head of a sea monster raised about 6 feet out of the water. The man described the serpentine beast as having brown, smooth skin.

The steamer *Vondel*, in 1907, sighted a sea serpent that in general was similar to most of the others. This too happened in the Indian Ocean. The animal was described as black. In Sogne Fjord, Norway, in 1908, a Dutch tourist is reported to have seen the same sea serpent or one almost the same.

A few months later, a somewhat similar beast was sighted by a ship's master named Merlees. His written report says that "some distance ahead, I saw a black mass which I thought to be a cap-

A Professor J. W. Van Dervoort envisioned this battle between monsters of the ocean before the great flood. Photo, Thijs Mol, Arnhem, Netherlands.

sized boat. On coming closer and examining it with a pair of binoculars, I realized that it had a peculiar shape. This appeared like a supporting frame over which canvas had been pulled taut. The ribs were very evident. When seen from the side at a distance, it appeared indented, for the ridges were quite plain. I had a good view which permitted me to make a fairly accurate estimate as to its size."

Captain Merlees goes on to say that it was about 16 feet long, 5 feet wide, and black in color. According to his description, it also had a large mouth, was without scales, and possessed very small eyes protected by membranes. Today we know of many families of amphibious beasts that could satisfy most thoughtful monster sightings given to date. Among these are several kinds of giant lizards that reportedly inhabit remote interior regions of Africa. Comparatively little is known about these vicious-looking animals except that they originate near the headquarters of the Ubangi River in South Central Africa. While native names and spelling vary widely, the descriptions do not. Two of them are said to resemble an elephant in size, while a third has a snake's head, a long powerful tail, and leaves behind it a track about five feet wide. During the Triassic period, many such reptiles are known to have existed. Consequently, if we make a composite of these lizards, we would come up with an amphibious beast that ranged from 25 to 30 feet in length. It would have smooth skin, colored either dark brown or black, a long serpentine neck, a snake's head, and four short legs.

A Captain Jorgenson, master of the sailship *Felix*, arrived in Canada in July, 1910, on a voyage from Norway and told of sighting a sea monster. This, he said, had happened just north of Ireland, and the beast had a long neck, a small head, and a lumpy body.

This rendition of a Netherlands sea serpent combined details of sex, the sea, and youthful sweetness. Photo, Nederlandsch Historisch Museum, Amsterdam, Netherlands.

The following month, off Scotland, a Mr. W. J. Hutchinson reported that he and a party had gone duck hunting in a small sailboat. As they left the bay and entered the ocean, a school of whales leaping in the air attracted their attention. Some distance behind the whales was seen a long neck, atop which was a head similar to that of a horse. Hutchinson says they were no more than 150 yards away from the beast and so had a good and distinct look. Using the height of the mast as a measuring device, they estimated the neck to be at least 18 feet high.

A Japanese version of a sea serpent that could be seen only by light of an electrical storm. Photo, Diet Library, Tokyo, Japan.

On December 13, 1910, at 1300 (1 P.M.), in mid-Atlantic Mr. Van Erp, the third officer aboard the steamship *Potsdam,* reported catching sight of a sea monster. To him and to others he called as witnesses, the creature appeared to be a gigantic snake close to 130 feet in length. It was about 3 feet in diameter and was moving very fast. From time to time, it reared its head over 8 feet above the surface of the water. The forward part made a great deal of splashing as the serpent swam, as did a split tail.

In October, 1911, the Dutch newspaper *De Zee* reported that the third officer of the steamer *Amsteldijk* saw a sea serpent in the North Atlantic. It first appeared in the form of a great mass of foam. Then, in the center of this, they could distinguish a dark body. The serpent reared its head about 8 feet out of the water, showed approximately 30 feet of body behind the neck, and then splashed back into the depths of the sea.

William Cook, in company with three ladies, was in a small sailboat off the shores of England in 1911. He saw what he at first took to be black rocks but which suddenly came alive. For safety, he took in the sail and rowed for shore. Although Mr. Cook could not discern the head of the sea serpent, he gathered that in total length it was somewhere near 90 feet. Apparently, the animal was engaged in feeding, for after a time of constant thrashing about, it submerged and was seen no more.

Captain Ruser, the master of the passenger ship *Kaiserin Augusta Victoria,* with the pilot still aboard in Scandanavian waters, saw what could only have been a sea serpent. This was in July, 1912. He was very conservative in giving it a length of only 20 feet, but he reported that it was of such bulk, it could not have been a mere water snake.

Not all sightings were made from on board ship. Some, as in the case of a sea serpent seen in the summer of 1914 in a Norwegian bay within

Professor J. W. Van Dervoort published a book in which he described the existence of this giant "nautili." Photo, Thijs Mol, Arnhem, Netherlands.

the district of Nordland, were witnessed by men, women, and children on shoreside locations. The monster swam about the bay for approximately five minutes. It was evidently frightened off by the commotion it caused along the shoreline and was last seen heading out toward open water.

On May 25, 1917, the British merchant steamship *Hilary* was attacked and sunk in the Atlantic Ocean. This was indeed unfortunate for our research into sea serpent sightings because just two days prior to this time, the *Hilary* had come upon a sea serpent. The documentation of this has long been lost; however, Captain Dean, master of the ship, and an officer in the British Royal Navy, later wrote of the matter in these words:

"After a stormy winter we enjoyed some wonderfully fine weather during May 1917. On the morning of, I think, the 22nd the conditions were perfect—brilliant sunshine and not a cat's paw to ruffle the sea. Beyond the horizon to the northward the peaks of Iceland could be seen, the only breaks in our all-round view of sea and sky.

"It was about 9 a.m. and I was sitting at the writing table in my cabin, which was immediately under the bridge, when a report was shouted down to me: 'Object on the starboard quarter.' In three seconds I was on the bridge, asking. 'Was it a periscope? Where is it?'

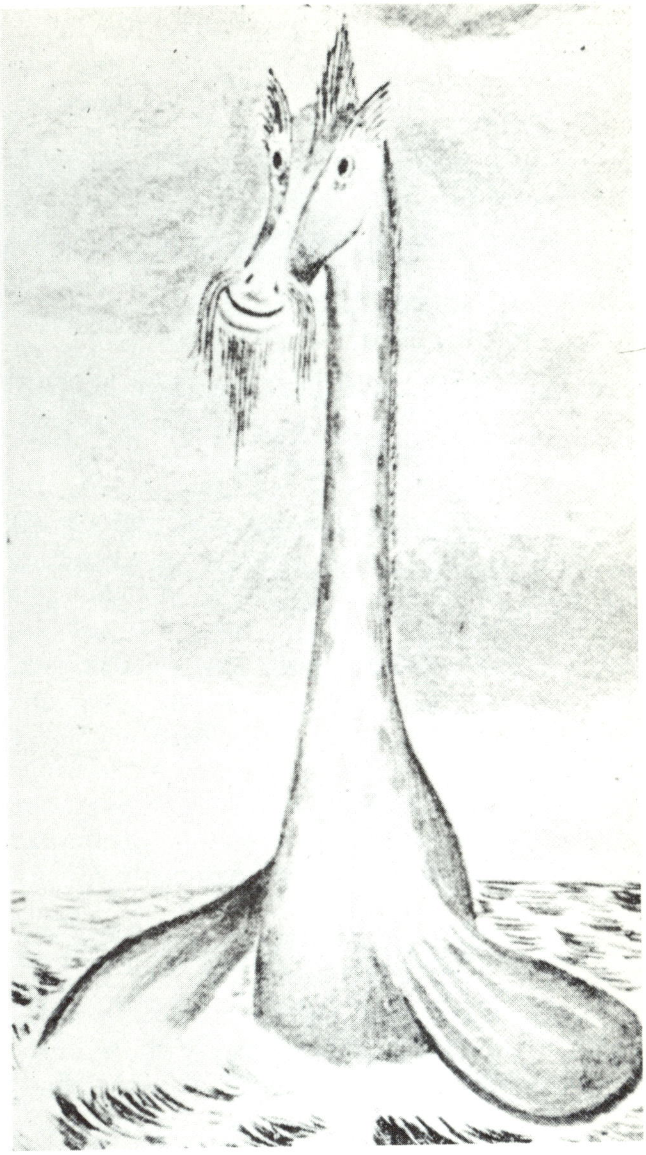

A sea serpent, as seen by the officer of the watch aboard the Corinthian, *August 30, 1913*. Photo, Illustrated London News.

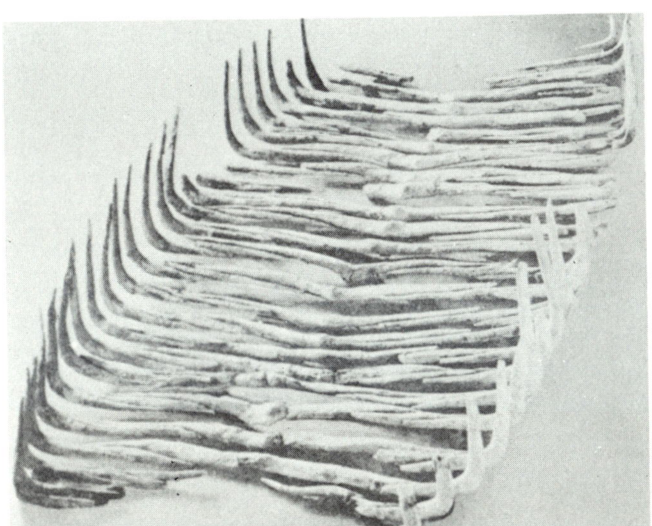

Rib cage of a supposedly extinct giant marine reptile of the genus Placodus. Photo, Senckenberg Museum, Germany.

" 'No, not a periscope,' replied the officer of the watch. 'It looks more like a living thing, but it is not a whale,' and he pointed to an object which at first glance suggested to my mind a tree trunk with only the knobby ends (from which branches and roots had been cut) visible. A careful look through my glasses, however, made it clear that the thing was alive, and that the 'knobby ends' were in fact its head and dorsal fins.

"We never missed a chance in those days to do a bit of anti-submarine practice, and it at once struck me that here was a good target; so I turned to the 1st lieutenant, Lieutenant-Commander Charles M. Wray, R.N.R., and told him to get our three 6-pounder guns' crews up, so that each one in turn should have a run. Like most of the armed merchant cruisers on that patrol, we had, in addition to our heavier guns, two 6-pounders placed on each side, just abaft the bridge, and part of our defense scheme was to have three guns' crews, one of which was always on watch between these two guns, ready to close up instantly starboard or port, according to which side a submarine might be sighted. . . .

"Before taking the liberty of using the stranger as a target, however, I thought it would be a good thing to have a closer look at it, so told Lieutenant Harris, the navigator, to turn the ship round and head towards it. In due course we approached our object end on, and when we were about a cable from it, it quietly moved out of the way, and we passed it about thirty yards off on our starboard side, getting a very good view of it while doing so.

"The head was about the shape of, but somewhat larger than that of, a cow, though with no observable protrusions such as horns or ears, and was black, except for the front of the face, which could be clearly seen to have a strip of whitish flesh, very like a cow has, between its nostrils. As we passed, the head raised itself two or three times, apparently to get a good look at the ship. From the back of the head to the dorsal fin no part of the creature showed above water, but the top edge of the neck was just level with the surface, and its snake-like movements could be clearly seen. (It curved to almost a semi-circle as the creature moved its head round as if to follow us with its eyes.) The dorsal fin appeared like a black triangle, and when the creature was end on, this fin was seen to be very thin and apparently flabby, as the upper part turned over sometimes like the top of a terrier's ear when cocked. The fin was estimated to be about four feet high when in the position highest out of the water.

"I was anxious to get as good an estimate as possible of the length of the neck, i.e., the distance from the head to the fin, so asked the 1st lieutenant, the navigator and the officer of the watch each to note, as I did, the length, and we compared with following results:

"1st Lieutenant—'The Length of one of our boats (28 ft.).'

"Navigator—'Not less than fifteen feet.'

"Officer of Watch—'Length of one of our boats.'

"Myself—'Twenty feet.'"

"From which it may fairly be assumed that the true length of the neck was probably not less than twenty feet; and, assuming that the dorsal fin would be just behind the junction of the neck and body, the total length of the creature would be about 60 feet.

"The creature did not seem to be in the least alarmed by the presence of the ship, but continued to bask on the surface, now sinking down till only the tip of its nose and fin were visible, and anon rising again till the whole head and the fin to a height of four feet were above water.

"The ship was steaming at twelve knots all the time, so the creature was soon 1,000 yards astern, and we then turned round, and I gave the order, when the range was approximately 1200 yards, for the first gun's crew to fire five rounds rapid.

"The gunlayer straddled his object (with the third and fourth rounds), but his best shot was some twenty yards off being a hit, and the creature ignored the performance entirely.

"The second gun's crew were now closed up, the helm was put over, and the range opened again to about 1200 yards, and the five rounds rapid repeated. This time two shots seemed to fall very close to the object, but still quite failed to disturb its equanimity.

"The third gun's crew was now closed up, and on my order, 'Five rounds rapid,' the first shot fell very close to the creature, the second was a clean hit, and produced at once a furious commotion, which reminded me more than anything else of a bather lying on his back in smooth water and kicking out with all his force to splash the water, only of course, the commotion on this occasion was on a vastly greater scale. It continued for perhaps three seconds, and then stopped, and we saw no more of the creature, though we passed close to the spot where it was last seen, so I gave orders for the ship to be put back on her patrol course, and left the bridge, telling the navigator to make a brief notation of the whole circumstance in the log.

"It was soon after 7 a.m. on the second (or maybe the third, I am not certain) day after this that brother Hun got us. A sitting shot we gave him, it being a bright, smooth morning, and the low sun right in our eyes. Our fires were extinguished at once, and when the ship had lost her way and was steadily settling down, it became necessary to take to the boats. In due course our poor old ship went under; the boats made for the land, which was fifty miles away. . . ."

So ends Captain Dean's story, except for one amusing note. It seems that he had been warned that entering a sea monster sighting in a ship's log would most certainly bring disaster. He did not believe in such nonsense, but concluded by saying, "Well, if ever I again found myself at sea in command of a ship, and anything of the sort was sighted, I should leave it alone and *make no entry in the log*."

Now it is interesting to take a look at a sea monster sighting from another angle. How did such an event look when viewed through the periscope of a submariner during World War I? The submarine that sank the *Hilary* cannot be traced. An exhaustive search of German war logs reveals no such event, so it must be assumed that the U-boat involved was later sunk. If such were the case her papers would also be lost. However, in the case of the World War I German submarine *U-28*, records of a sighting have been preserved. George Gunther Freiherr von Forstner tells what happened after they tropedoed the steamer *Iberian.*

"On July 30, 1915, our *U-28* torpedoed in the North Atlantic the British steamer *Iberian* (5223 tons) loaded with valuable freight. The ship, which was about 180 meters long, sank rapidly, stern first, the depth at this point being a few thousands of meters. When the steamer had disappeared for about 25 seconds it exploded at a depth which we could not know, but 1000 meters would be a good safe guess. Shortly thereafter pieces of wreckage, among them a huge marine animal which made violent movements, were ejected out of the water, flying approximately 20 or 30 meters high.

"At this moment there were six of us men on the bridge, myself, the two officers of the watch, the chief engineer, the navigator and the helmsman. We at once centered our attention upon this marvel of the sea. Unfortunately we did not have the time to take a photograph because the animal disappeared in the water after 10 or 15 seconds. It was about 60 feet long, looked like a crocodile, and had four powerful paddle-like limbs and a long pointed head.

"The explanation of this event seems easy to me. The explosion of, or in, the sinking steamer caught the 'undersea crocodile,' as we came to call it, and ejected it out of the water."

Nor did the end of World War I bring an end to sea serpent sightings. Here is an example of what can be found in England's postwar records of the period. A Mr. Mackintosh Bell relates the following experience: "In 1919, after being demobilized following the end of the war, I went as had been my custom for years to Brims Walls, Orkneys, to spend a month's holiday by taking my place as one of the crew in a cod-line fishing boat, belonging to fishermen friends and the first day I was there, I think it was about 5 August, I went afloat with a crew of four at about 9:30 AM for the purpose of first hoisting lobster creels and then for cod fishing. On arriving at the creels, which had been set in a line between Brims Ness and Tor Ness, my friends said, 'We wonder if we will see

In 1919, Mr. Mackintosh Bell had his wife make this sketch of what he had seen off Hoy, in the Orkneys. Photo, Liverpool Museum, Liverpool, England.

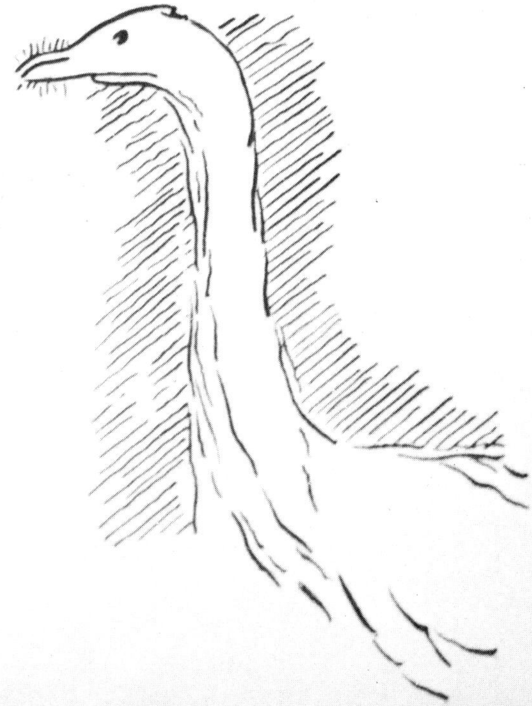

that sea monster which we often come upon, and perhaps you will be able to tell us what sort of a beast it is.'

"We got to the creels, hauled a few, and were moving slowly to another, when my friends said very quietly, 'There he is.'

"I looked, and sure enough about 25–30 yards from the boat a long neck as thick as an elephant's fore leg, all rough-looking like an elephant's hide, was sticking up. On top of this was the head which was much smaller in proportion, but of the same colour. The head was much like that of a dog, coming sharp to the nose. The eyes were black and small, and the whiskers were black. The neck I should say, stuck about 5–6 feet, possibly more, out of the water.

"The beast was very shy, and kept pushing its head up, then pulling it down, but never disappearing out of sight. The body I could not see. Then it disappeared, and I said, 'If it comes again I'll take a snapshot of it.' Sure enough it did come and I took as I thought a snap of it, but on looking at the camera shutter, I found it had not closed owing to its being swollen, so I did not get a photo. I then said, 'I'll shoot it' (with my .303 rifle) but the skipper would not hear of it in case I only succeeded in wounding the animal, and it might therefore attack us.

"It disappeared, but as was its custom swam close alongside the boat about 10 feet below the surface. We all saw it quite clearly, my friends remarking that they had seen it numerous times swimming just the same way after it had presented itself on the surface. My friends said that they had seen it the year before in just about the same place. It was a common occurrence they said.

"That year (1919) was the last of several years in which they saw it annually. It did not show itself again for two or three years, and then it was only seen once. As to its body, it was, as seen below the surface of the water, dark brown, getting slightly lighter as it got to the outer edge, then at the edge appeared to be almost grey. It had two paddles or fins on either side and two at the stern. My friends thought it would weigh 2 or 3 tons, some saying as much as 6 tons. Not only my friends, but others as well, lobster fishing, got many chances to see it.

"I might say that since 1919 all cod and other deep sea fish have deserted the Pentland Firth. I think that the reason for this is that such monsters hide in the rocky caves, which are always covered with deep water."

This commercial fisherman concluded his narration with the remark that "if we get a really hot summer it will be seen again." Although he did not explain this remark further, warm water could be a factor in the travels of any sea monster and might therefore be of importance to our search.

During these same years, scientists were exploring the jungles of South America with an eye toward discovering what the Indians identified as a giant boa. This snakelike animal, so they professed, was a massive aquatic beast. The first substantiating report came from a Catholic priest named Heinz, who traveled the Amazon River during a flood in 1922. He wrote that he "was being ferried to my quarters by canoe, when abruptly I saw something startling mid-stream. I positively recognized it as a water-snake, at a distance of about 100 feet.

"It was coiled into two rings and drifted quietly downstream. I estimated its body to be as thick around as an oil drum and that its length was approximately 80 feet. After we had gone some safe distance away, my Indian boatman stated that the beast would probably have crushed our boat like match wood."

The *Bombay Evening News* of October, 1921, carried a story about a sea monster that read: "The upright neck stood ten feet perpendicular from the surface of the sea, while tapering very slightly towards the head. The diameter of this column-like figure seemed to be about 18 inches along the water. It seemed to have large scales and, in colour, was a light olive green at the back, shading down to a dirty yellow in front. The head appeared somewhat like that of a gigantic tortoise, or turtle, but a body having to correspond to such a head would need be, at least, fifty feet long.

"The viewer, being almost on the waterline himself, could see no sign of any body beneath the surface. The creature remained in a vertical position for about 30 seconds, swaying slightly, and then sank beneath the surface. The most amazing part of the report being that the entire encounter happened not over one mile from shore in a well enclosed bay."

During August, 1923, another report tells of a sea serpent that was sighted several times in the Thames estuary. This report is especially noteworthy because it comes from the hydrographic ship H.M.S. *Kellett* and is signed by the captain, a man named F. D. B. Haselfoot. He states that on two occasions he "observed, rising out of the water, approximately 200 yards from the ship, a long, slender, serpentine neck, which was about six to seven feet long. It remained upright, in each instance, for from four to five seconds. Seeing it with the naked eye only, there did not seem to be anything resembling a positive head."

It should be remembered that at this point in history, we are no longer dealing with sea monster sightings as reported in old history books, but rather with accounts given in modern newspapers, radio broadcasts, and documentary films. As a consequence, there are several approaches to the subject that we should keep in mind. First, there is the negative reaction of the public toward any mention of a sea monster. Second, it is frequently very difficult to dig out objective facts. Young reporters, on being assigned to cover a sea monster sighting, generally do so with tongue in cheek.

Nevertheless, let us turn to the pages of the *Clearwater* (Florida) *Sun*, a newspaper known for its objectivity in such matters. In 1948, they reported an upsurge in monster sightings. Some of the reports are hard to refute, and they seem to center on a "thing" that puts in an appearance every four years.

In March, 1948, the newspaper carried the story of monster "tracks" found on the Clearwater beach. They could not be readily identified and, consequently, caused quite a stir. Letters poured in from around the world. One, from a man named A. Lilius, then in the Montreal office of *Holiday* magazine, stated: "Could you put me in touch with the man who found the monstrous footprints on the beach? When I was in South Africa, I found on the Natal coast several similar mysterious footprints and took clear pictures of them. I wonder how much the prints in Florida are like my pictures."

The story cited above was carried in the March 7 issue. On March 21, we find this article: "County Highway Patrolman J. F. Peacock reported yesterday he had received a call from Indian Rocks [a nearby Florida community] Friday night stating a near panic had been created among residents of that beach area north of the Indian Rocks Hotel by the appearance of a 'sea monster' which chased one individual, identified as John Moore.

"Moore's call dispatched the county officer to the scene where several residents, in a case of jitters, told the officer that some large animal had come from the gulf, stamped about on the beach and then returned to the water. Huge three-toed footprints bore out this story."

Peacock said he studied the footprints carefully and was personally convinced that if the whole thing was a prank, it was one of the most masterful ever perpetrated in Pinellas County. He said frankly he did not see how a man or several men could take the time and equipment to put the large tracks in the sand without leaving the slightest trace.

An early version of a sea monster painted on a porcelain from the Netherlands. Photo, Thijs Mol, Arnhem, the Netherlands.

An 1,800-pound monster called the "Sea-Hag" lived in the Gulf of Georgia, British Columbia. Photo, Provincial Archives, Victoria, British Columbia.

"The officer said that some of the tracks were deep in hard sand that would have taken at least a 1,000-pound weight to make. Moreover, said Peacock, the tracks gave the appearance of being a running monster in some places on the beach, and, 'I don't see how a man could make anything like that,' he declared."

A few days after this story, the *Clearwater Sun* ran this follow-up: "Tales of a sea monster's tracks about Clearwater and the beaches increase," the newspaper said. "Latest explanation of the tracks came from a Clearwater Beach resident, who asserts he had seen unusual three-toed creatures in the silt marshes of Cuba which leave tracks similar to those found on Clearwater Beach and Indian Rocks Beach.

"According to this gentleman, he was in Cuba some time ago and had his attention called to large 'critters' of sloth-like appearance which inhabited the marshes adjacent to Morro Castle in the Havana harbor. He tells the following story:

" 'This critter walks on his hind legs leaving three-toed footprints. . . . It has a head similar to a crocodile, but a much shorter jaw. The critter has large teeth. I am convinced that the tracks found on the Clearwater Beach are not those of a human prankster because where would you find a man with an eight foot stride? And if a prankster were using stilts he would have to have about three times his own weight added to them to get the depth of the footprints found.

" 'Moreover, the tracks on the beach showed where their maker had upset the lifeguard stand in front of Everingham's Pavilion and left evidence that some large rough-skinned animal had rubbed against the wood of the stand. That doesn't seem to be the work of a prankster.' "

A fish-market operator told customers his fishermen reported hundreds of footprints on Dan's Island, south of Clearwater Beach. Tracks were also on Honeymoon Island, north of Clearwater.

"From Tarpon Springs," the *Sun* reported, "comes the unconfirmed report that a sponge boat crew battled over 12 hours with some large denizen of the deep recently when it attacked one of the boat's divers."

Four months later, four fliers, from a flying school near Clearwater, publicly claimed they saw the culprit monster in shallow water off nearby shores. A spokesman for the group said that "it had the shape of a log, was over 15 feet long and moved in a clockwise direction."

This same spokesman, in a lengthy interview with a reporter for the *Sun*, stated, "It was only a few feet off the north end of Clearwater Beach, and was easy to see because of the white, sandy bottom. I'm positive that it was furry and its legs, or flippers, were white with the ends cupped like hands." The white, he said, was in contrast with the black body. "The informant stated that the head was shaped like that of a hog's."

During these mid-century years, there were other sea monster reports going around. Some came from Canada, others from Norway, and a few from the Amazon region of South America. East Africa developed an exceedingly strange sea monster that within a few years became known as the Kilindini sea serpent. It was accorded various and sundry physical shapes during 1949 and 1950. Then a convocation of wiser scientific minds agreed that it had probably been a hippopotamus migrating by water. They noted that such gigantic beasts sometimes swim far out to sea, perhaps by willful intention or perhaps merely by force of tides.

Vancouver Island, in Canada, also had a well-authenticated sea serpent sighting in 1950. A Judge Thomas Brown, former president of the King's Bench of Saskatchewan, saw a monster that reared its head 4 or 5 feet out of the water; behind the head there were a number of coils or loops. He estimated the length of the sea serpent at 35 to 40 feet.

During 1950, the acting postmistress of a small resort called Summerland looked up from her job of canceling letters and saw a sea serpent out on the water. She watched it disport itself for about ten minutes. Her description generally agreed with other descriptions of sightings in that area. These said that the sea monster had a big head, gigantic eyes, a mane resembling seaweed, and a long neck.

In summary, it can be said that during 1948, 1949, and 1950 the number of sea monster sightings were extremely numerous.

"Nessie," as actually photographed by a 12-inch lens, from a distance of 4½ miles, in early morning mist. Photo, Loch Ness Phenomena Investigation Bureau.

8

1950 to 1971

Perhaps the most famous sea monster of all time is the Loch Ness monster. *Loch* means "lake," and Loch Ness is the largest of the Scottish lagoons. There are other lakes and other monsters in Scotland, as for example the Loch Morer monster, which was sighted only 20 miles from its more illustrious counterpart in Loch Ness; but none are quite so well publicized as the latter.

Both these neighboring lakes are deep, ranging from 500 to 1,000 feet from surface water to craggy bottom. The scenery is truly breathtaking, and as a consequence, these are favorite places for tourists to come to boat, fish, and swim. Loch Ness, during the undisturbed moments of early morning or in the peaceful lassitude of a still evening, offers a perfect setting for monster sightings. As might be expected, reports are numerous, many of them from well-educated persons. Clergymen, schoolteachers, scientists, and tourists claim to have seen the Loch Ness monster. Unfortunately, none carried cameras. However, recently volunteers, equipped with electronic gear, long-range cameras, radio communications, and swift boats, have tried to contact the monster of Loch Ness. Great Britain's Joint Air Reconnaissance Center—the identical organization used to pinpoint Hitler's

V-2 rocket-launching sights against London—was called upon to act as the coordinating nerve center. The undertaking had everything going for it except the monster, and that animal refused to cooperate. The project, although thorough and costly, was a failure.

There have been many recent attempts to establish contact with "Nessie." An organization called Loch Ness Phenomena Investigation Bureau, Ltd. (LNPIB), teamed up with the World Book Encyclopedia and sent a twenty-nine-year-old submariner named Dan Scott Taylor, Jr., from Atlanta, Georgia, on the trail of the elusive monster. Taylor, an oceanographer by training, used a small, bright-yellow submarine to chase readings that registered on the highly responsive electronic gear of his vessel. "We know the monster exists," he stated after a six months' venture. "I never got close enough to get a photograph, or punch out a biopsy—two of the things we most wanted—but I think we made believers out of all the scientists on the trip."

Sonar equipment aboard Taylor's submarine and other boats used in the hunt documented earlier findings that large, eellike fish inhabit Loch Ness, and perhaps similar lochs and fjords around the world. Taylor said that once he got within 20 feet of a monster but was unable to chase it down. "They move at a rate of about 12 knots. My sub can make only 2 or 3 knots," he said.

Other sea monster sightings have been compared with the Loch Ness monster. A sea monster story appeared recently in the *Irish Evening Herald*, in Dublin. "The Loch Ness monster," the lengthy article reads, "now has a cousin here in Ireland. A number of Irishmen have reported sighting it.

"Those seeing it say their monster lives in the deeps of Glendarry Lakes near Achill Sound over in County Mayo. This, as we are all aware, is the wild part of our country. The lake is about 4 acres in extent, wooded on two sides and reputed to be very deep. There have been stories for years that it hides some form of prehistoric life.

"Recent events have been frightening a number of local people half out of their wits. The beast has been coming ashore they say. There was Gay Denver, 16-year-old student, mindfully pushing his bike up the rise which overlooks the lake. And there was Ireland's Nessie, humping along and climbing a turf bank into the woods.

" 'It was moving in a sort of jumpy way like a kangaroo,' Gay breathlessly told the *Evening Herald*. 'It had a long head like a sheep and a long neck and tail. The hind legs were bigger than the front ones.' Gay called the monster about 12 feet long, dark in colour and, 'much bigger than a

Some who have seen the Loch Ness monster have said that it resembles this sixteenth-century sea monster, depicted by Olaus Magnus in 1555. Photo, Suomen Kalastusyhdistys, Finland.

horse. I got a terrible fright, jumped on my bike and rode away as fast as I could.'

"John Cooney said he had been passing the lake for more than 30 years and never saw any sign of a monster until just the other day. 'My friend, Michael McNulty, had given me a lift,' he related, 'and, as we came round a bit of a bend, we saw it just ahead in the middle of the road. It was between 8 and 12 feet long, I thought, with a long swan-like neck—only much bigger. The tail was very thick. It was moving at an angle to us and we couldn't see exactly how long it was. And it was weaving and curving. We could see it very clearly indeed.'

"Mr. Cooney went on to say that the monster was dark brown and was shiny and scaly. The eyes were glittering. 'I don't know if it actually looked at us,' he said, 'but it disappeared in an instant into the thick underbrush and could be heard heading back toward the water. We neither of us stopped to make any further inquiries.'

"Cooney, who stressed he was a nondrinker, said he and McNulty were stunned by what they had seen. Over in Scotland, the Loch Ness phenomena investigation unit which keeps tabs on the original Nessie when she appears, found the Irish reports no laughing matter.

"They have always said there might well be other lakes, the deep lakes of Canada and Sweden, just waiting for discovery. Such lakes, they say, were once sea fjords until the land level rose sometime between the end of the Ice Age and a date 5000 years ago. Nessie and her relatives were cut off then from escape back to the seas. Trapped in the lakes, they probably survived because the food and environment were right. At least that's what the monster experts say."

One official of the Loch Ness monster watch, Lieutenant Colonel Carter-Wright, ret., when queried about the Irish sea monster, rationalized the likelihood of such a beast in these words: "There is a branch of the biological and anatomical sciences, called *teratology*, which treats of abnormalities or monstrous growths in the structure of both plants and animals. While most ancient scribes dealt with such deviations from the normal as something of satanic origin, they unquestionably were subjects of scientific investigation similar to that which we are affording the Loch Ness monster. Most just went unattended. As a consequence, it was not until the 18th century that whispered superstitions commenced to yield to physical observations and the character and causation of monstrosities became an important branch of natural history. Our Loch Ness investigations are simply an extension of teratology and as in all scientific research we must be patient, so awfully patient."

And with a wry grin he added, "Even to the extent where people laugh at us."

Almost one year after Dan Scott Taylor, Jr., brought his submarine back to the United States a large London newspaper carried the following story: "A group of 25 oceanographers announced that they will launch an expedition to the Scottish glens to ascertain if Morag, a sea monster, really exists.

"This enterprise is to mark the group's contribution to *European Conservation Year*, and is the first serious attempt to view what villagers say is the huge humpbacked beast that moves through Loch Morer, the deepest lake in Britain, at 1,017 feet. This beautiful lake, surrounded by steep, pine-clad hills, is about 20 miles from Loch Ness, the haven of the other legendary monster.

"Although legends and stories about these monsters are age old, scientific interests were not aroused until recently, at which time a pair of fishermen saw a 'monster' on the loch, or landlocked bay, in western Scotland. The monster rammed their boat. One of the fishermen attempted to beat it off with an oar, but the oar snapped in half when it struck the monster."

If the Loch Ness monster does exist, it still may never be seen because of a man-made factor: *pollution!* A story datelined London, July 27, 1970, was headed, "Pollution Threatens Loch Ness Monster."

"Pollution is either killing the Loch Ness Monster, or has in all probability already done so, a British scientist declared today.

"Douglas Drysdale, of the British College of Naturopathy and Osteopathy said to newsmen that he had analyzed samples from rivers and streams feeding into Loch Ness and found the pollution level so high it must reduce marine life in that lake in northern Scotland.

The monster of Lake Utopia, Charlotte County, New Brunswick, Canada. Photo, The New Brunswick Museum, Saint John, New Brunswick.

"A spokesman for the Loch Ness Phenomena Investigation Bureau, which checks and records reported sightings of the hump-backed creature known affectionately as 'Nessie,' said that if pollution was all that bad, 'why have not every one of our fish died?' "

Only time and electronic gear will reveal the truth or falsehood of these avowed encounters. Yet none of our "almost" successes should be lightly dismissed, for other lakes, in other parts of the world, have long harbored sea monsters. Canada, our neighbor to the north, is a good case in point.

Newspapers in Saint John, New Brunswick, recently carried a story of the eighteenth anniversary of the last reported sighting of their famous Lake Utopia monster. They featured an interview with eyewitness Mrs. Fred McKillop, then approaching her ninetieth year. "It is still fresh in my mind," she told reporters, "and I was never so frightened in all my life."

"It was a clear beautiful day," she recalls, "the men had all gone fishing and had left me to sit with two of my grandchildren. We were all watching the lake and it was beautiful. It was so clear it resembled glass and there wasn't a ripple showing.

"Suddenly, as I watched, the water commenced to boil and churn, and make waves which came in and broke on the shore. Then a huge creature of some sort emerged from the water, at least it showed part of its head and part of its body. It resembled a huge black rock, but it moved and churned all the time. I was alone with the grandchildren at our cabin, and was so terrified that I took the children and ran into the cabin and locked the door.

"After a short time had passed, I realized that whatever it was belonged in the lake and so we were in no danger. It was then I went outside again and watched it.

"It moved back and forth across the lake, boiling and churning the water, and making great waves. I don't recall exactly how long I watched, but it seemed to be a long time.

"I had never before heard of the Lake Utopia monster, and therefore had no idea what it was. When the men returned home I told them about it and they said that must have been what it was. I recall it as a very frightening experience."

The initial report of this lake Utopia monster was given by a Mr. Leith Adams over 100 years ago. He said that lumbermen employed at a sawmill along one far end of the lake observed that the water, usually very placid, was suddenly disturbed by the splashing of some object. Several of the men stated that it was fully 10 feet wide and approximately 30 feet long.

During the following two days, the same sort of

disturbance was seen repeatedly, but in different parts of the lake. The reports produced such a reaction that the shoreside residents set out large hooks baited with salt fish and pork. Adams said that he found many attached to logs near the shore and all around the lake. The *Illustrated News of Canada* published the first report on the monster, in which it was stated that the beast was raising "Old Ned" when in a sporting mood "because it was sending up logs, spruce edgings and other material and causing the water to boil and foam as if a geyser had suddenly broken loose."

Some years later, this same newspaper said that at one time a company formed in St. George, with capital of $200, for the purpose of procuring nets and apparatus for the capture of the monster. Nets were made and set out, and a party of well-known gentlemen, one of whom was an officer in the fisheries department, went from St. John to assist in the capture. "They were all armed to the teeth, but returned without a nibble."

The report given by numerous people who claim to have seen the Utopia monster are contained in the Ganong Collection at the New Brunswick Museum. "Strangely enough," one newspaper concluded in presenting the story of

This Canadian sea monster, nicknamed "K.D." (for "King of the Deep"), caused so much furor that a company was formed to hunt it and other lake monsters. The company failed to make a catch. Photo, Provincial Archives, Victoria, British Columbia.

the beast, "the monster usually appears soon after the winter ice has broken up in the lake." That is why, every year, about that time, people visit Lake Utopia, hoping to see the ugly head of the famed monster breaking through the thin ice and opening the lake for the start of the fishing season.

Meanwhile, back in the ocean, strange things were being reported at the mid-century mark. Life magazine published a story to the effect that a saucer-shaped, animalistic blob, weighing 7 tons, with matted hair, and ivory in color had washed ashore in Australia. When discovered, it was relatively fresh; but by the time word had been gotten to someone knowledgeable, and naturalists had made the trek to the specific location, the "thing" was badly decomposed. Nevertheless, it was established that it did have fish gills and a gullet, although it "lacked both bone structure and eyes." As writers stated at the time, the absolute terror to be found in this "thing" rested in the fact it was "completely unlike *anything*."

"That Monster Back Again" was the headline in the *Sun* of Clearwater, Florida, during June, 1952. Under the heading "Indian Rocks," the story read: "Huge tracks found on the hard-packed sand early Monday morning by Commander R. J. Brown of the U.S. Coast Guard, Tampa, has caused a flurry of conjecture among 'Rock' residents."

The deep tracks, as deep as those of a tractor, were found on the beach in front of Walter M. Reelle's residence. Brown pointed them out to Reelle, who took plaster casts of the imprints. But the thin plaster mixed with the wet sand, and only the outline of one print and a perfect likeness of one claw were made. According to Reelle and other observers, the 40 feet of tracks led from the water's edge to a seawall ashore near the beach side of Relle's residence. The tracks were washed out shortly after Relle had taken the print, which showed the claw on the left front foot as more than 1 inch in width and a definite curve to its 1½-inch length. Reelle said that the tracks showed a series of similar claws across the front and one side of the foot.

As a hunter of numerous giant sea turtles and turtle eggs in Florida, Reelle said that there was nothing to compare with these odd tracks. He described the impressions as being about 1 foot long and 7 or 8 inches wide. Traces in the plaster cast showed what Reelle described as partial impressions of the row of claws.

According to Reelle, the creature was a mammal having four feet. He said that the tracks indicated the legs were about 36 inches apart. However, according to some of the observers, the second set of tracks, apparently those made by the hind feet and

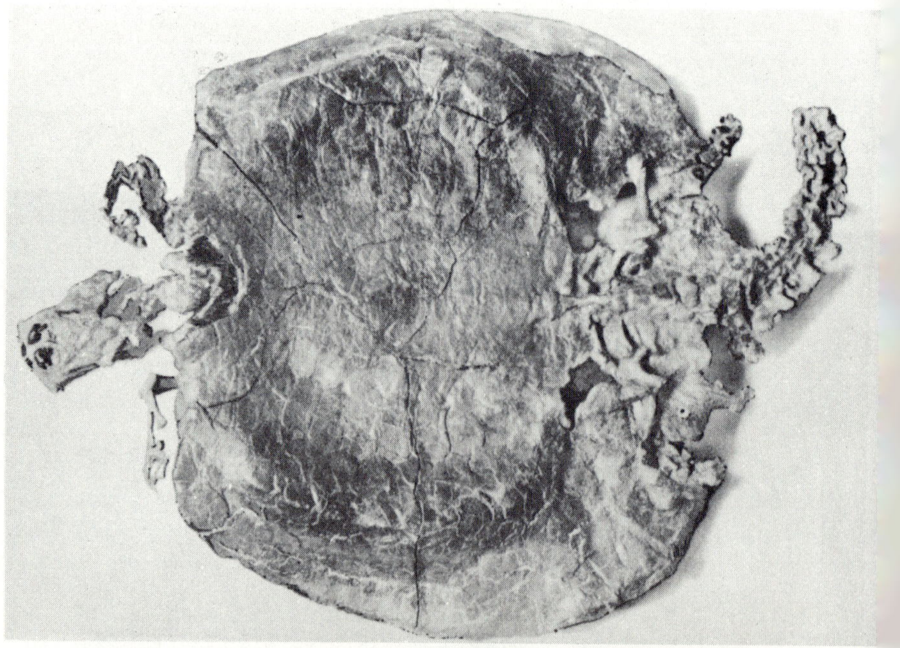

Giant, square-headed turtle, over 200 million years old. Called Henodus chelops, *it possessed powerful, toothless jaws.* Photo, University of Tübingen, Tübingen, Germany.

differing from the other pairs, were only about 8 inches from those made by the front feet. By actual test, a 200-pound man walking alongside the tracks made hardly any impression in the sand; and according to Reelle, the tracks were imprinted in the sand to a depth of 2 inches.

Not all sea monster sightings are well publicized. For example, during 1962, a Mr. Edward McCleary reported a sea serpent tragedy that should have made headlines. Instead, it received scant mention. McCleary told the story of how he and three companions were out skin diving in swirling fog, just off the coast of Florida, when a carnivorous sea serpent struck at them. The green-eyed beast was turtle-headed and long. Before the attack was over, McCleary's three companions were dead.

Some time just prior to the McCleary incident, a group of fishermen on the western coast of Vancouver Island, Canada, made an interesting discovery. They found what appeared to be the remains of a sea monster. One man, the captain of the fishing boat *Dana*, assisted in carrying the remains to a nearby town. Unquestionably, the find was a rarity of sorts, for it possessed the head of a goat, a spinal cord that was close to 50 feet in length, and vertebrae that were as much as 6 inches in diameter.

Officials of the Fisheries and Forestry Department were called in for an opinion and, after lengthy consideration, announced that it might be a "ribbon fish." According to the *Daily Press* of Canada, a Dr. Forester said that "the head, what we could see of it, resembled a ribbon fish, but we are making a further comparison to a small specimen of ribbon fish which we have at the station." He concluded his statement to the press by saying that these remains were not of the whale family and only in a minor degree resembled those of a shark.

The doctor's reference to a "ribbon fish" immediately placed new emphasis on that type of marine life. As far back as 1879, a British physician, Dr. Andrew Wilson, identified this species. Since then it has also become known as a tapefish, or king-of-the-herring. It is a serpentine sort of creature, possesses a rather odd-shaped fin that runs the length of its back, and is known to inhabit waters of great depth. No captured specimen of ribbonfish has exceeded 25 feet in length. Because

Huge marine crocodile, 150 million years old. This specimen has a head over 4 feet long. Photo, University of Leningrad, Leningrad, Soviet Union.

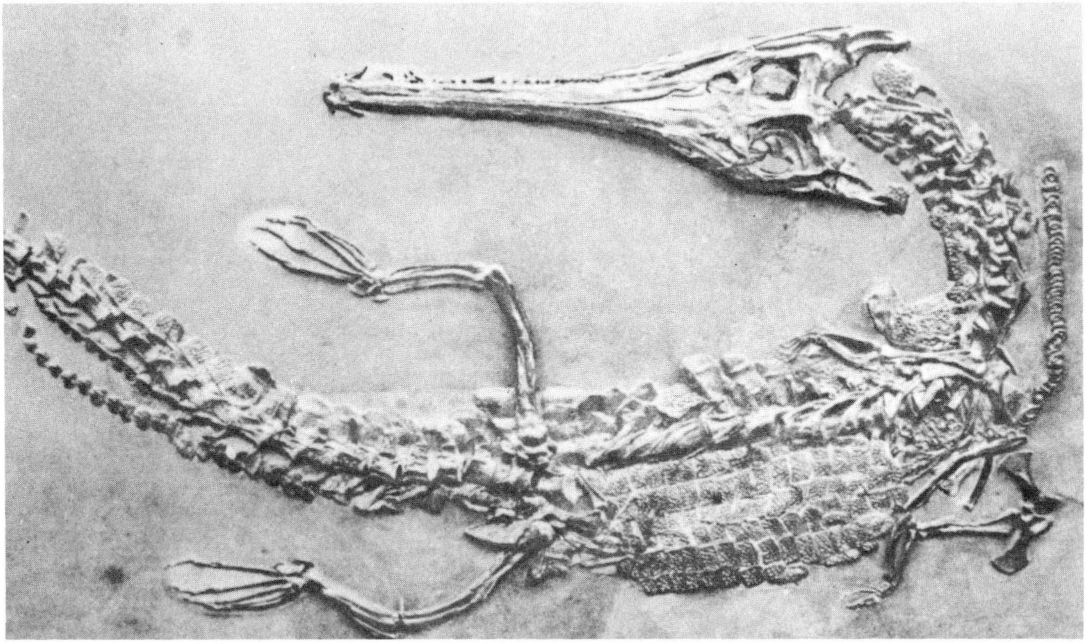

of the odd-shaped fin that resembles a mane, the fish is invariably identified as a young sea serpent. However, we must refer to the Naval Oceanographic Office statement that "fish have the ability to grow rapidly whenever their environment is favorable." Perhaps the environment was favorable, and a monstrous serpent did develop.

California has its current sea monster sightings. In fact, so many are seen, at such frequent intervals, that they are seldom reported. One 200-year-old story is about a sea monster seen in the vicinity of Santa Catalina. Initially, this was reported by uneducated fishermen who might have distorted the facts. But shortly after World War I, more reliable observers recorded what they saw. One such person was the secretary of the American Tuna Club, an organization known as the forerunner of all big game fishing clubs in this country. Official papers show that the secretary recorded reports of sightings of the animal on several occasions. It was described as having a long, massive neck, with great pie-plate eyes. It was covered all over with a sort of mane of thick, matted black hair. Although the observer, with witnesses aboard his power boat, had more than one look at various times, so little of the beast's entire bulk was visible that they never satisfactorily estimated its length.

To confirm the California sea monster, it is first necessary to establish the credibility of its observer. One reliable witness was Captain George C. Farnsworth, president of the American Tuna Club and at the time considered to be one of California's top-ranking Catalina fishing guides. He is the sport fisherman who developed the technique of using a kite, fashioned to a trolled fishing line, to give his bait a lively, lifelike movement. This method of sport fishing is mentioned because numerous commercial fishermen who work nets or longline for their catches have made the comment that the monster most often seems to appear when the kite technique is being used somewhere in the vicinity. Perhaps the large eyes, which have been described in many ways, by numerous people, are attracted by the kite itself.

Briefly, such a rig is made by attaching a sizable kite, capable of floating several hundred feet above the surface of the water, to the leader and towed across wind, so that the kite itself and the bait are swung to one side of the boat. This takes the bait out of the wake and places it where it can more easily be seen. The sportsman, by lifting on his rod, can make the bait skitter along the surface, leap, plunge, and splash. Large fishes such as bluefin tuna and massive sharks readily strike at such bait. There is no evidence, however, that the Catalina sea monster has ever struck at it.

Captain Farnsworth said the eyes were at least 12 inches in diameter and were close-set. He also described it as having a tangled sort of mane that appeared to be about 2 feet in length. These eyes have warranted such descriptive words as "fiery," "dead flat," "vacant," "piercing," and "sharp." Unfortunately, in no single case has the direction of the light flashing on these eyes ever been recorded or described in detail. A noon sun would give a far different effect than the light at five in the evening.

Many sightings were discussed among the hard-working commercial fishermen that ply their trade in waters nearby to San Clemente and Santa Catalina. Captain Mario Lapana, skipper and owner of the fishing boat *Barbara C.*, said that his 100-foot wooden vessel is a typical drum-rigged trawler. This type of fishing boat uses a power-driven drum to bring its net in over the stern and therefore offers a crewman an excellent view of what is going on around them.

"We were recovering our net by reeling in on our drum," Lapana said, "when one of the hands yells, 'lookit there.'

"I quit working for a look and my eyes was first struck by a sport boat hauling a kite through the air. Pete, that was the hand on the winch, he yells, 'No cap'n, not there—over to the port side.'

"Well, I looked and sure enough there was a head—no mor'n twenty fathoms distance—which was maybe fashioned like the skull of a cow. It was sitting on a neck as thick as a good-sized cargo boom. But it was the eyes that held me; they were large, but sort of dead-like. The drum was still reeling in a good catch at a right smart clip and I had to turn my eyes back to the deck.

"When I looked up again, no more than a quarter-minute later, this here animal was gone."

Captain Lapana is a rugged-faced seaman, with skin weathered to the color of dark mahogany after more than fifty years on fishing boats. He is an alert man, despite his years, well fortified with muscles, and would stand to gain nothing by consistently spinning such a yarn. When pressed for details, he is naturally hesitant.

"In all my years of working drift nets, purse seines, otter trawls, or whatever kind of gear was on deck to draw in fish, I ain't never come across the likes of that. It is hard to compare what I seen with anything else in the sea, mainly because it wasn't like anything else. But that animal was big indeed."

It often takes prying and getting acquainted before most fishermen will tell their stories of sea monsters. Owner-operator Captain Bible O'Rourke, of the 70-foot *Cindy*, out of Point Judith, Long Island, is a typical case in this respect. During his school years, he worked his way through college as a crewman on commercial fishing boats. After working hard for a B.S. degree in business administration, he soon realized that he preferred the heave and pitch of a sturdy deck to the confines of a shoreside office. He returned to fishing in 1946, and has been hard at it ever since. Cautiously at first, he began to tell of his experience.

"I was working on a shrimper out of a place called New Harbor, Maine. There was just two of us—me and the skipper. There wasn't much of a sea running—maybe sea state 2 on the Beaufort scale—but it was dark and I was on the after deck, with the skipper in the wheelhouse. We were making our first drag, when our diesel began to labor. The skipper yelled that the twines were snagged and threw the engine into reverse.

"We wanted to work fast in order to keep our net from getting all ripped to hell and gone, which was well over 100 fathoms out, so the skipper flashed a big marine searchlight used to flood the open stern deck.

"That's when I got the shock of my life. Standing as I was, hard by the port rail, I looked directly into blazing green eyes that popped out of the black night like a pair of lighted basketballs. These eyes belonged to a snake-type creature with a body about 24 inches in diameter. There wasn't much time to study what was 10 or 20 fathom off toward our stern; the animal blew softly and gently submerged, revealing that it had run afoul of our net.

" 'What'n hell was that?' the skipper yelled through the rear port in the wheelhouse, 'A killer whale?'

" 'No, my God no, it was a—,'then I choked up. How in the Lord's name can a man report that he has just snagged a .sea monster without holding himself up to public ridicule?"

Now let us return to our California sightings. During June, 1953, the crew of a fishing boat were amazed to have a sea monster surface adjacent to their vessel, look them over, and then quietly submerge without making another appearance. They reported the matter to the Coast Guard by radio and in a later interview confirmed what many other commercial fishermen already knew—the sea monster was frighteningly big and had a serpentine neck and a triangular head. The captain of the fishing boat, a rugged individual named Randazzo, in acting as spokesman said, "It had protruding eyes, each about one foot in diameter."

On the same coast, although considerably farther north, a lumber boat operated by Captain Lou Baggon is on record as having sighted a similar large-eyed beast. In this instance, however, both the captain and his wife, who frequently sailed with him, said that the animal had snakelike loops.

During July, 1970, another story, which originated in Los Angeles, was carried by many newspapers across the country, including the *Newark* (New Jersey) *Evening News*. It said that crewmen on the freighter-passenger ship *President Grant* knew that there was something wrong with their ship but could not find the cause of their trouble. The engineer officer did notice that they had developed severe vibrations at the bow. Further, he noted that the wave normally formed at the bow had developed a decidedly different shape. On making port, the captain was amazed to learn that his 14,000-ton ship had skewered a massive beast onto the underside of its bow. After much examination, it was decided that the remains were those of a large marine animal of uncertain species.

In a similar encounter, the Grace Line's passenger ship *Santa Clara* reported by radio that she had struck a 45-foot monster, with eellike head and a body approximately 3 feet thick. The event was witnessed by the chief mate, Mr. William Humphreys; the third officer, Mr. John Axelson; and the ship's navigation officer, Mr. John Rigney. When the *Santa Clara* docked, reporters interviewed Mr. Axelson.

"On noon of 30 December, these three men had gathered on the ship's bridge to shoot the sun for navigational purposes. Mr. Axelson abruptly noted that a snake-like head had come up out of the sea off the ship's starboard bow. He called to the others and pointed out the sea monster and all watched in amazement as the beast first came close by the bridge and then fell astern. They described the head of the serpent as being two-and-a-half feet wide, about five feet long, and approximately two feet thick. The tubular-shaped body of the serpent was approximately 3 feet in diameter and as the animal came abeam of the ship, it was noticed that the water was stained red. There emerged about 35 feet of body, which had a skin that was without scales, appeared dark brown, and gave no evidence of hair or fins. It was believed, because of the color of the water, that the sea monster had been hit by the keel of our ship. This assumption even now seems valid, as the animal twisted and thrashed in a manner suggesting it was severely hurt. There appeared no section of the serpent-like body on the opposite side of the ship, which might have suggested how much of the torso had been cut off."

This carved panel from Nigeria depicts the mythical story of the great sea serpent. Photo, British Museum of Natural History, London, England.

Here was a sea monster sighting made by three experienced and trained mariners so certain of what they had seen that the incident was duly entered in the ship's log. Any false entry into such a permanent record book is considered to be a criminal act and may be punishable up to loss of license and/or imprisonment. It is hardly likely that three officers, each with a promising career ahead of him, would suddenly dream up a fantasy with which to decorate their logbook.

However, as logical as all this might appear, these men were nevertheless scoffed at and their sighting ridiculed by shore-bound adjudicators who proclaimed the animal to be an oarfish. An oarfish is a relatively rare fish, about which little is known; it grows to approximately 20 feet, and is frequently thought of as a young sea serpent.

There are those people, of course, who would argue that an oarfish can grow to a length of 35 or 40 feet. The Seaorama, Clearwater, Florida, has several oarfish on exhibit, and the attendant staunchly claims that they grow to massive lengths.

Strangely enough, delayed sea monster reports are to be found in Australian records during the late 1900s. A marine animal or sea serpent was seen after 1950, but the report of it took some time to reach the public. Ron Spencer (or Ron Secar; later reports vary) claimed that on half a dozen occasions a sea serpent had surfaced near his small boat in Reception Bay, close to Brisbane. He stressed that the head was like that of a bull, being several feet long and as many wide but without horns.

On the southern coast of Australia, two women, one of them a nurse, saw a sea serpent in the harbor at Darwin. Both asserted that it could not have been less than 100 feet in length, that it moved with great speed, and that it possessed a back lined with lumps. By 1955, a series of such sightings had been recorded, all with a common factor: speed. Whatever creature was on hand, it moved fast and was at times noisy. So many reports aroused the interest of the Australian navy. That organization immediately suspected a Russian submarine of prowling their territorial waters and launched an exhaustive investigation. Nothing came of the matter. The whole thing was finally laid to rest by a longtime commercial fisherman who told several writers that the sea monster was an old friend of his: a giant ray, about 40 feet in length, capable of speeds up to 50 miles an hour.

In England, about the same year, a book called *Gower Journal* told of that author's having seen the head of a monster appear out of the surf near Middlesex. It resembled the head of a horse but was larger. Shortly after the book was circulated, a John Hadley, of Wembley, England, reported seeing a sea monster on the opposite side of the British Isles. In 1953, some fishermen from Scotland reported coming across a sea monster with a camel's head, a giraffe's neck, and the body of a gigantic serpent.

During October, 1954, the captain of a boat called *Sea-Fern,* while sailing off Newport, Rhode Island, told of seeing a monster that would have weighed about 20 tons. "Its head was generally round and appeared to have a sort of horn."

So it is that sea monster sightings go on and on: 1955, 1960, and now the 1970s. People see the monsters, report them, and hope that someone will take their sightings to heart.

This Egyptian stone carving shows natives appeasing a sea serpent. Photo, British Museum of Natural History, London, England.

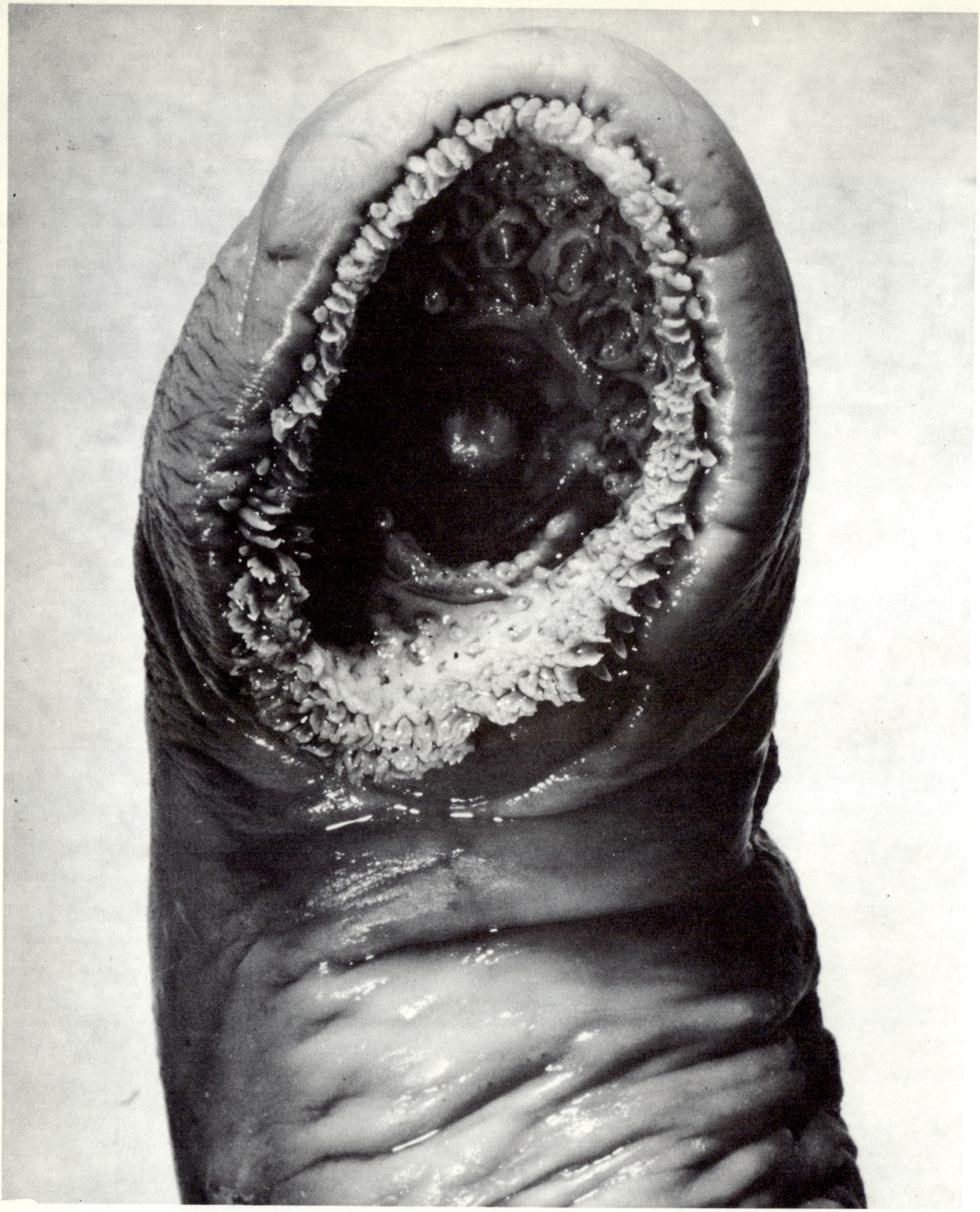

This is the formidable mouth of a sea lamprey. There are twenty-four known species living in North American, Eurasian, and sub-Antarctic waters. Photo, United States Naval Oceanographic Office, Suitland, Maryland.

9

Conclusions

Throughout the history of sea monster sightings, numerous people have attempted to rationalize or explain such appearances. As a result of this continued practice over the centuries, interpretations have come to fall into certain categories. An excellent commentary on this is to be found in the *Bulletin of the United States Fish Commission*, volume 5, dated 1885, which contains this rationalization by a Mr. Richard A. Proctor: "All sorts of explanations of sea serpent stories have been advanced. Because one captain has mistaken a lot of floating sea wrack, half a mile away, for a sea monster, therefore the story of a sea creature seen swiftly advancing against wind and sea, at a distance of less than 200 yards, meant nothing more than misunderstood seaweed. Another mistakes a flight of birds in the distance, or a shoal of porpoises, or even a range of hills beyond the horizon, for some sea serpentine monster, and forthwith other accounts, however manifestly inconsistent with such explanations, are regarded as explained away. Then, worst of all, some idiot invents a sea serpent to beguile his time and find occupation for his shallow pate; and so soon as the story is shown to be only a story, men of sense and standing, as incapable of the idiocy of inventing sea monsters as I am of inventing a planet, are supposed to have amused their leisure by sending grave reports of nonexistent sea monsters to men under whom they (the seamen, not the sea monsters), held office, or by taking oath before magistrates that they have seen sea creatures which they had invented, and by parallel absurdities."

This chapter will note the most popular versions of all the "logical" explanations advanced over the years and analyze just how logical or illogical each might be.

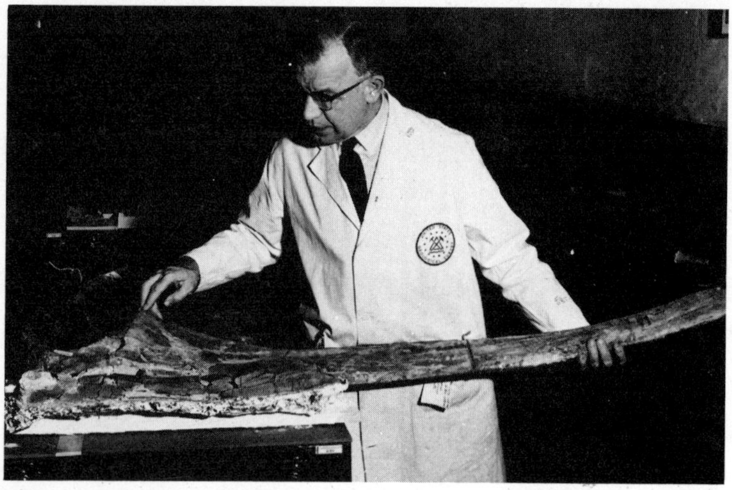

Dr. Frank C. Whitmore, a paleontologist with the United States Geological Survey, holds the jawbone of a 20-million-year-old whale found at Calvert Cliffs, Maryland. Photo, United States Department of the Interior, Geological Survey, Washington, D.C.

Teeth and molars retain distinctive features better than most parts of the body. This is an Ice Age mammoth's lower molar tooth, which was dredged from the sea floor of the Atlantic continental shelf. In all likelihood, it is more than 200 million years old. Photo, United States Department of the Interior, Geological Survey.

PORPOISES SWIMMING IN A LINE

This justification is as ancient as the initial report of the sea monster himself. It is easy for people to visualize a school of porpoises playing games and, in so doing, forming a queue that resembles a gigantic serpent moving through the sea. As far back as 1803, a sea monster sighting was rationalized as "a number of porpoises, following each other in a train." These words were contained in a letter written by the Reverend William Jenks to the Honorable Judge Davis, which later appeared in the *American Journal of Science and Arts,* volume 2, Boston 1819.

This is not an isolated case. Even in Europe, birthplace of the sea monster report, we find that for every sighting there also appeared a "logical" explanation. A German named Rathke, who had devoted a portion of his life to the investigation of sea monster sightings, in 1841 published voluminous findings in the *Archiv für Naturgeschichte.* He wrote:

"Truly, I know that some believe, that what has been taken for a so-called sea serpent, to be nothing else but a row of porpoises, swimming in a line. But all those persons, by whom the above-mentioned evidences are given were too familiar with the sea, and have too often observed porpoises

Porpoises swimming in a line, if viewed from a distance, on a rough sea, could possibly be mistaken for a sea monster. Drawing by the author.

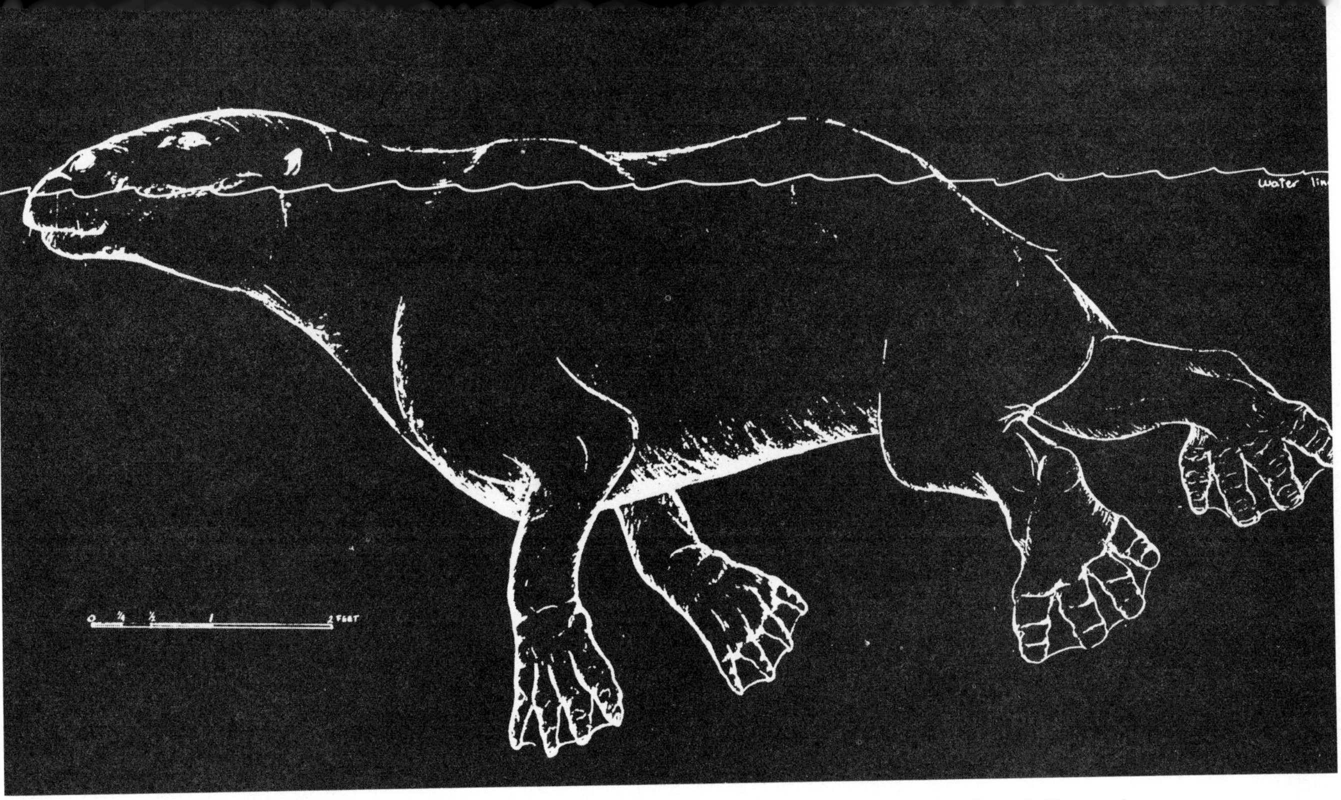

This drawing shows a paleoparadoxia in swimming posture. Scientists believe that this animal lived about 14 million years ago. Photo, United States Department of the Interior, Geological Survey, Washington, D.C.

together, to be deceived by a row of such animals swimming on the surface of the water. If this, however, had been the case, all the observations related to me of the sea serpent's holding its head above the surface, and about the size of it, must have been mere fiction, and this I cannot admit. According to all this, it evidently cannot be doubted, that there is a long serpentine animal in the Sea of Norway, which may grow to a considerable length."

The best early refutation of the porpoises-in-a-line theory is to be found in Dr. A. C. Oudemans's book *The Great Sea Serpent*, which he published himself in 1892.

"I need not say that porpoises swimming in a line, do so very irregularly. They are in the habit of continually throwing up their bodies half above the surface of the water, and so their backfin is clearly visible, but no where the sea serpent is said to have on each coil a backfin. Sometimes one porpoise only is visible, a moment afterwards, three, eight, or more, but never the whole row is seen at once, while the undulations of the sea serpent are constantly visible above the surface, moving with the greatest regularity. Everyone will feel that this explanation is not satisfactory; it does not even explain a single observation. Besides, how to explain the swan-like neck, so often seen by reference to porpoises?"

GIANT SQUID (OR CALAMARY)

There are many stories of weird sea monsters that have suddenly appeared to attack ships, snatch away boaters, and even come ashore to devour individual fishermen. Unquestionably a certain portion of each yarn is based upon fact, for as Dr. Oudemans notes: "It must be remembered, as I have elsewhere said, that until the year of 1873, notwithstanding the adventure of the *Alecton* in 1861 (Note: A French corvette of this name came across a giant squid swimming on the surface and tried unsuccessfully to haul it aboard), a cuttle (squid) measuring in total length fifty or sixty feet was generally looked upon as equally mythical with the great sea serpent. Both were popularly scoffed at, and to express belief in either was to incur ridicule. But in the year above mentioned, specimens of even greater dimensions than those quoted were met with on the coasts of Newfoundland, and portions of them were deposited in

This 1892 photograph was used to show that parts of "genuine sea monsters" then being sold on the open market in large American cities were in fact remnants of whales that abounded on Bering Island. Photo, National Archives, Washington, D.C.

museums, to silence the incredulous and interest zoologists."

Oudemans also mentions a giant squid 80 feet long. Measurements of scars carried by captured whales indicate that squid have probably attained a length of up to 120 feet. However, a giant squid cannot be misconstrued as resembling a sea serpent. Nevertheless, to clarify the matter, we should define those species covered by the label "giant squid." Here we include the entire class of Cephalopoda, which takes in the cuttlefish, nautilus, squid, octopus, kraken, calamary, and any other marine animal that has eight or ten long tentacles, well-developed eyes, and jaws fortified with a sharp beak and that moves by expelling water through a siphon.

Scientists accept the existence of giant cephalopods. Nevertheless, it is extremely difficult to see how these creatures could be mistaken for the sea monster washed ashore in 1930 on Alaska's Glacier Island. That fur-covered "thing" was close to 30 feet in length and had a large head, with a beak at least 3 feet long. Nor can cephalopods be likened to the massive snakelike animal, with very sharp teeth, that was washed ashore in Newfoundland in 1932. Even the fur-clad monster washed ashore in Scotland in 1944 could in no way be passed off as a member of the squid family.

As a result, although we agree that giant squids and octopus exist, they cannot be considered sea monsters.

A giant squid, octopus, or any other sizable cephalopod, if suddenly confronted on the open sea, becomes a dangerous sea monster in fact. Drawing by the author.

RIBBONFISH (OR OARFISH, KING-OF-THE-HERRING)

It is only within comparatively recent times that this species (*Regalescus glesne*) has come to the attention of scientists. Although Dr. Andrew Wilson wrote of this fish in his *Leisure Time Studies*, published in London during 1879, little was known of the species. However, in recent years, numerous of these fish have been brought ashore. Residents of Bradenton, Florida, can attest to this. On March 15, 1966, a Mrs. Rathburn found a 5-foot oarfish washed up on Long Boat Key. It was only about 1 inch thick but a full 12 inches wide, and it certainly could be considered a "ribbonfish."

This was the eighth of the same species found washed ashore in the same general area within recent years. There is apparently no record of ribbonfish exceeding 10 feet in length. However, a book published in England during 1960, *Under the Sea*, by Maurice Burton, states that they grow to "20 feet or more." This cannot be confirmed.

Again we must deflate any belief that ribbonfish, as we know them today, meet the criteria established for even a small sea monster. They have no neck-supported head capable of being reared out of the water, no pie-plate eyes, no teeth or fangs, and no appreciable number of appendages capable of grasping. Most important of all, they fail to measure up to size requirements. Oarfish must be dismissed as being nothing more than an ordinary species of fish.

BOA (AND OTHER GIANT SNAKES)

Without a doubt, a snake swimming in the ocean can be considered a sea serpent. If the snake is of sufficient length, it could very likely be dubbed a sea monster. The question logically follows: How long do the longest snakes grow? This is indeed a moot question. The National Zoological Park in Washington, D.C., lists the following snakes and their probable lengths: boa constrictor (South American), about 20 feet; African python (*Python sebae*), about 35 feet; anaconda (*Eunectes murinus*), about 40 feet.

These are not intended as the ultimate in lengths, but they establish what is currently accepted as normal for such species. As has previously been stated, rumors have persisted for centuries about a South American serpent in the Amazon River that attains gigantic proportions. Like so many old fishermen's stories these yarns very likely contain a grain of truth. How much fact and how little fancy still remain to be settled. One reasonably authenticated case was reported by a member of the Royal Geographical Society of London, in 1907. A party member named Major Percy Fawcett is on record as stating, in writing, that he shot an anaconda snake that showed 45 feet above the water and another 17 feet within the water, giving it a total length of 62 feet. However, such measurements are, at best, guesswork.

In India, the natives relate tales of enormous pythons that have been found far at sea, floating on trees and other flotsam. The *Zoological Journal*, in December, 1827, told of a large boa that was

A 40-foot anaconda snake or a slightly smaller boa, if found at sea, would in truth be a "sea serpent." Drawing by the author.

known to have drifted close to 200 miles coiled around a cedar log. However, all this poses many unanswered questions. How far can one of these giant-sized snakes swim? What is their behavior when they are overtaken at sea? We may never know the answer to these questions. So the assumption must remain that a sea serpent—or at least a sea-borne serpent—does exist. Perhaps, then, this is our only true sea monster. But this is rather hard to believe because many accounts of sightings mention fins, flippers, a flat tail, and other characteristics that are not found in true snakes.

Sea snakes (*Pelamis platurus*) cannot be entirely discounted in any discussion such as this. They are highly poisonous, average about 4 feet in length, with some attaining a length of over 10 feet. This, unquestionably, is large for a snake, but not large enough to be considered a sea serpent. These sea snakes are found principally in the Indian Ocean and the tropical Pacific. All species are reported to be found in sheltered waters along coastal areas, but they have been sighted several hundreds of miles or more at sea. This is not surprising, for they are a true marine species and are capable of remaining submerged for many hours at a time.

ANCIENT ANIMALS (AND EELLIKE CREATURES)

Numerous people claim that what they have seen or what other sightings describe is a remnant from the Paleozoic era. This is certainly true of the coelacanth, the fish decreed extinct, only to be rediscovered in 1938.

Another classic example of survival is the horseshoe crab (*Xiphosura polyphemus*, or *Limulus polyphemus*). The largest of these ungainly animals is only about 2 feet in length. It has a somewhat crescentic shell and a long swordlike tail that is not injurious to man. If this seemingly helpless species of animal life could survive from the Devonian period of some 400 million years ago, why then, so the argument runs, should there not be a strong possibility of survival of giant reptiles of the Permian period, which was a mere 300 million years ago.

Another substantial example offered on behalf of the survival theory is the shark. Mean, aggressive, and completely savage, it is quite a contrast to the harmless horseshoe crab. This bloodthirsty type of undersea animal has even managed to adapt itself to fresh water without sacrificing any of its dangerous tendencies. The Lake Nicaragua shark (*Carcharinus nicaraguensis*) provides a vivid illustration of survival under extreme change in environment. It is known to be a man-eater and has succeeded in adjusting to life in a South American freshwater inland lake.

Conger and moray eels commonly reach a length of 6 to 10 feet. They are a reticent kind of fish, given to coiling themselves in rock formations and striking at an intended victim in much the same manner as a snake. Although people are inclined to speak of these eels in the singular, there are at least 120 varieties of moray, plus numerous species of conger. Suppose, through some quirk of nature, there was a deepwater variety that liked to secrete itself in the trenches and canyons known to exist at 25,000 feet below the surface. Man has seldom reached such depths either to deny or to confirm a hypothesis of this order.

Eels have certain inherent characteristics that lend themselves well to such a hypothesis. They have teeth and are mean when aroused. Their skin is generally without scales, or at most, possess only vestigial embedded scales. Many of them feed on squid, cuttlefish, and octopus, just as whales are known to do. Knowing this, we can justly assume that the giant squid known to exist at very great depths are attacked by giant eels. Certainly the whale, which must surface for air, demonstrates the fact that a dark, remote world of fabulous giants still exists within the deep channels, reefs, and provinces that dot the ocean floor. Man has yet to probe this territory, and until such time as he manages this colossal task, we can only make assumptions.

"September 4, 1854 [*Illustrated London News*]—It is reported by the British brig *Albeona*, arrived at Liverpool, that on the 4th of Sept., last, about five in the afternoon, in lat. 38 S., long 13 E., while the ship was under a light wind and in smooth water, a sea monster of great size and singular appearance was descried. Attention was first directed to it by the broken action of the water, which was otherwise smooth all around.

The animal was discovered protruding its head above water to the length of about 30 feet, at an angle of 60 degrees to the horizon. His head was about 12 feet long and was marked by a white stripe or streak down each side. At about six feet from the termination of the streaks, which were presumed to be its jaws, there was a protuberance on its back like a small water cask. The creature kept its mouth shut, but its eyes were plainly visible. At the point of contact with the water the body seemed about as much as the ship's long-boat round. The general colour of the body was black, but under the jaw was a quantity of loose skin, like a pouch, of a lighter colour than the rest of the animal. While under observation he dipped under water three times, remaining submerged about a minute each time. From the broken action of the water at different points, it seemed as if protuberances, similar to that on the back existed on various parts of the body. From the best conjecture that could be made, it was computed at 180 feet in length over all."

Now let us look at the slight possibility that there is a member of the genus *Basilosaurus* still living. This marine animal was declared extinct and first described by a scholar named Harlan in the year 1824; later the name was changed to that of zeuglodon by a Professor Richard Owen. This marine species lived in the Tertiary period, and almost all the characteristics of the skeleton are said to relate to pinnipeds, with a few directly traceable to cetaceans. All this leaves an investigative mind uncertain about the order to which the animal really belongs. A Professor D'Arcy W. Thompson, as reported in the studies from the Museum of Zoology in University College, Dundee, stated that the length of the largest skeletons measured 75 feet. The teeth and molars, he claimed, were nearly exact replicas of those of seals. The nostrils were situated at the tip of the nose.

As it stands, no one seems to be certain that the skeleton of one of these Basilosaurus, or zeuglodon, ever did exist. Years of research have only proved there might have been a very complete skeleton. But then again, that which was seen at one time may in reality have been the bony structure of some other aquatic animal such as a paleoparadoxia, which lived about 14 million years ago. A paleontologist of the United States Geological Survey has assembled some very fine fossils of this latter beast.

GIGANTIC SEA TURTLE

Examples of this very large species of marine turtle have been recorded in areas of the South Pacific, and some fine skeletal remains have been recovered in Australia. Relatively current reports say that in Queensland there is a massive turtle known as a Moka moka. One report by a young woman named Lovell, a resident of that area, was recorded in 1890. She said that the sea turtle had either teeth or serrated jaw bones. She went on to state that "what I saw of it was about 27 or 28 feet, but I think it must be 30 feet in all. Whilst its head was out of the water it kept its mouth open, and, as I could not see any nostrils, I fancy it breathes through its mouth. The jaws are about 18 inches in length; the head and neck greenish white, with large white spots on the neck, and a band of white round a very black eye and round upper and lower jaws."

Sea turtles (*Dermochelys coriacea*) have been caught numerous times; however, little is known about them. Consequently, we can only guess how large a sea turtle has ever grown. Nevertheless, it must be noted that there are some sea monster characteristics which a sea turtle does not have. For one thing, the neck is too short to allow the animal to rear its massive head out of the water, an action that has been reported in so many sightings. For another thing, no known species of turtle is capable of swimming with great speed while on the surface.

FLOATING TREES

To anyone who has spent a substantial number of hours afloat, it is logical that fallen trees or masses of floating underbrush can suggest strange and ominous shapes. Recently, the commander of the United States Naval Oceanographic Office, Captain F. L. Slattery, USN, was on a hydrographic inspection trip in Chesapeake Bay, Maryland. One of the members of his inspection team pointed out what appeared to be a bright-eyed long-necked, ill-shaped, small-sized sea monster. The

This massive sea turtle was captured by the Institute of Zoology in Genoa, Italy. Photo, Underwater Naturalist Magazine.

black silhouetted head moved up and down rhythmically as the animal apparently swam against a gentle current; the beast's jaw loosely flapped from what appeared to be the result of strenuous effort. As the small boat carrying the inspection party closed in on the monster, it became quite clear that the water-soaked branches of a tree formed the torso and that light shining through an opening created an eye. The jaw was a flapping branch, and the swimming movement was the result of surface motion on the water. This conglomeration, when silhouetted against the surface of the bright water, with a strong sun well beyond the object, took on form, motion, and reality.

"From a distance," one seaman said, "it sure looked like a live animal."

Keeping in mind that such illusions can and do happen, let us check back through the pages of history. In 1860, the magazine *Zoologist* carried the following article by Mr. Arthur Adams:

"An incident occurred aboard the vessel of which I am surgeon, which, I think, deserves to be recorded as an illustration of optical delusion that might have become a source of error, and given rise to yet another sea serpent. We were sailing among the islands of the Miatan group, at the entrance of the Gulf of Pe-Chili. There was little wind, and the gentle ripples covered the surface of the sea. I was sipping my congo at the open port of the ward room on the main deck, admiring the setting sun, and watching the rounded outline of the blue mountains and distant islands against the sky, and the number of sea birds "wheeling rockwards to their nest," when my eyes rested on a long dark object making its way steadily through the water. After observing it for some time in silence I was sorely puzzled and could make nothing of it. It was neither a seal nor a diver nor a fishing cormorant, for with their forms I was familiar; so I went on deck and consulted other eyes than mine. Sundry glasses were brought to bear on the suspicious object, and the general scrutiny seemed to decide that it was a large snake, about ten feet long (or much longer according to some), working its way vigorously against the tide by lateral undulations of the body. So strong was the conviction that the course of the ship was altered, and a boat got ready for lowering. With a couple of loaded revolvers, some boathooks and a fathom or so lead-line, I made ready for the encounter, intending to range up alongside, shoot the reptile in the head, make him fast by a clove-hitch, and tow him aboard in triumph! By this time, however, a closer and more critical inspection had taken place, and the supposed sea monster had turned himself into a long dark root, gnarled and twisted, of a tree, secured to the moorings of a fishing net, with the strong tide passing it rapidly, and thus giving it an apparent life-like movement and serpentine aspect."

Almost any type of floating object can take on animalistic qualities if seen under the combined influences of tide, wind, and light. In *Nature Magazine*, September 12, 1878, a Mr. E. H. Pringle presented the following story:

"One morning in October 1869, I was standing amid a small group of passengers on the deck of the ill-fated P. & O. SS *Rangoon*, then steaming up the Straits of Malacca to Singapore. We were just within sight, so far as I remember, of Sumatra. One of the party suddenly pointed out an object on the port bow, perhaps half a mile off, and drew from us the simultaneous exclamation of, "The Sea Serpent!" And there it was, to the naked eye, a genuine serpent, speeding through the sea, with its head raised on a slender curved neck, now almost buried in the water, and anon reared just above the surface. There was the mane, and there were the well-known undulating coils stretching yards behind.

"But for an opera glass, probably all our party on board the *Rangoon* would have been personal witness to the existence of a great sea serpent, but, alas for romance; one glance through the glasses and the reptile was resolved into a bamboo, root upwards, anchored in some manner to the bottom—a 'snag' in fact. Swayed up and down by the rapid current, a series of waves undulated beyond it, bearing in their crests dark coloured weeds or grass that had been caught by the bamboo stem.

"Ignorance of the shallowness of the straits so far from land, and of the swiftness of the current, no doubt led us to our first hasty conclusion, but the story . . . shows how prone the human mind is to accept the marvelous, and how careful we should be in forming judgments even on the evidence of our senses."

SEAL (MANATEE, SEA LION, SEA ELEPHANT)

The seal and its related species were the victims of mistaken identity in 1890 when a sighted sea monster plagued Alaska's Bering Island. However, an Arctic expedition took numerous photographs of the seal herds of that area. They brought back a number of prints, which proved conclusively that although a seal might be *mistaken* for a sea monster, it could never rightly be considered as one of the many long-necked sea monsters reported.

A LINE OF FLYING BIRDS

This is an explanation offered for the sudden appearance of a sea serpent, but it is hardly a practical one. Certain birds *do* fly in a long, undulating formation close to the surface of the water. But in vast numbers, this is a rarity, and it is even rarer that the birds should be able to hold

A line of flying birds is hardly likely to form the proper silhouette of a sea serpent and hold the configuration for any appreciable length of time. Drawing by the author.

a continuous pattern. The idea, though, was first presented many years ago. In *Nature*, dated September 5, 1878, a Mr. Joseph Drew wrote the following:

"On Monday, August 5, a number of geologist crossed in the Folkstone boat to Boulogne, to study the interesting formations of that neighborhood, and, when about three or four miles from the French coast, one of these gentlemen suddenly exclaimed, 'Look at that extraordinary object passing across the bow of the steamer about a mile or a mile and a half in advance of us!' On turning in this direction there was seen an immense serpent apparently about a furlong in length, rushing furiously along at the rate of fifteen or twenty miles an hour; it was blackish in front and paler behind; its elongated body was fairly on the surface of the water and it progressed with an undulating or quivering motion, *mirum erat spectaculum sane.*"

"Of course many speculations were started to account for this extraordinary phenomenon, but they quickly changed and settled into the fixed idea that the object before them could be nothing less than the great sea serpent himself; for—

"Prone on the flood, extended long and large,
"Lay floating, many a rood, in bulk as hugh
"As whom the fables name of monstrous size,
"Leviathan, which God of all his works
"Created hughest, that swam the ocean stream."

"The writer fortunately had with him one of Baker's best opera glasses, and, after a few moments' use of this little instrument, the wonder was satisfactorily resolved. The first half of the monster was dark and glistening and the remainder of fainter hue, gradually fading towards the tail. The glasses did not determine the matter until the extreme end was reached, and then it was seen to consist of a mass of birds in rapid motion; those that were strong on the wing were able to keep up well with the leaders, and to make the head appear thicker and darker by their numbers, whilst those that had not such power of flight were compelled to settle into places nearer and nearer the tail. Doubtless these birds were shags (*Pelecanus cristatus*) returning to their homes for the night from the distant waters in which they had been fishing, during the day; perchance it may be wrong to assert positively as to the variety of birds, but in as much as the writer has often seen shags on the Cornish coast in smaller numbers returning

In 1892, the Albatross Expedition *took this photograph on Bering Island to show how sea lions, seals, and walrus could be mistakenly identified as sea monsters.* Photo, National Archives, Washington, D.C.

in single or double file to their roosting places, and since it is stated in works of natural history that they have been noticed occasionally flying in this peculiar manner to the number of a thousand or more, it does not appear an unwarranted liberty in supposing that they really were *pelecani cristati.*"

SEAWEED FLOATING ON THE SURFACE

Here is a prime example of how seaweed can be misconstrued to be a sea serpent. A passenger aboard the ship *Brazilian*, in 1849, told the following story:

"On the morning of the 24th the ship was becalmed in latitude 26 degrees south, longitude 8 degrees east, being about forty miles from the place in which Captain M'Quhae, R.N., is said to have seen the Great Sea Serpent. About 8 o'clock on that morning, whilst the captain was surveying the calm, heavy, rippleless swell of the sea through his telescope, the ship at the same time heading N.N.W., he perceived something right abeam, about half a mile to the westward, stretched along the water to the length of about twenty-five or thirty feet, and perceptibly moving from the ship with a steady, sinuous motion. The head, which seemed to be lifted several feet above the water, had something resembling a mane, running down to the floating portion, and within about six feet of the tail it forked out into a sort of double fin. Having read at Colombo the account of the monster said to have been seen by Captain M'Quhae in nearly the latitude, Mr. Harriman was led to suppose that he had fallen in with the same animal, or one of the genus; he immediately called his chief officer, Mr. Long, with several of the passengers, who, after surveying the object for some time, came to the unanimous conclusion that it must be the sea serpent seen by Captain M'Quhae. As the *Brazilian* was making no headway, Mr. Harriman, determining to bring all doubts to an issue, had a boat lowered down, and taking two hands on board, together with Mr. Boyd, of Peterhead, near Aberdeen, one of the passengers who acted as steersman under the direction of the captain, they approached the monster, Captain Harriman standing on the bow of the boat armed with a harpoon, to commence the onslaught. The combat, however, was not attended with the danger those on board apprehended; for on coming close to the object it was found to be nothing more than an immense piece of sea weed, evidently, detached from a coral reef, and drifting with the current, which sets constantly to the westward in this latitude, and which, together with the swell left by the substance of the gale, gave it a sinuous, snake-like motion."

In conclusion, let us cite a remark made by one of the most famous of all historians of the sighting of sea monsters, Mr. Henry Lee; some time prior to 1883 he wrote:

> It is most desirable that every supposed appearance of the *Great Sea Serpent* shall be faithfully noted and described; and I hope that no truthful observer will be deterred from reporting such occurrence by fear of the disbelief of naturalists, or the ridicule of witlings.

This print shows a ship's crew investigating floating seaweed that looks like a sea serpent.

PART TWO

OTHER DANGEROUS MARINE LIFE

A sandbar shark (Carcharhinus milberti). Photo, Marineland of Florida.

SECTION ONE

Killers and Their Close Relatives

I

Sharks

There are nearly 400 varieties of sharks, ranging in size from a few inches (scyliorhinids) to 60 or more feet (the giant whale sharks). Neither the exceptionally small nor the exceedingly large sharks are dangerous to man. But many of the hundreds of types in between *are* extremely vicious and dangerous.

The history of the shark goes back to the Devonian period, some 400 million years ago. One of the most complete primitive shark fossils from this age was discovered in Ohio. And from this, it has been possible to trace the development and population growth of the shark.

The ancient Hebrews recorded shark attacks, as did the Egyptians and Romans. Personal observations at times differed concerning the physical characteristics of sharks, but all agreed that this marine beast was a savage killer.

For example, here is an extract from a book published in London during 1657, entitled *A True and Exact History of the Island of Barbados.*

"There is a fish called a Sharke, which he as is a common enemy to saylers and all others that venture, in Calmes, to commit their naked bodies to the sea (for he often bites off Legs, sometimes Armes, and now and then swallows the whole body, if the Fish be great): So when the Saylers take them, they use them accordingly. Sometimes by putting out their eyes, and throwing them over board; sometimes by mangling and cutting their bodies, finns, and tayles, making them a prey to others, who were mercilesse Tyrants themselves; And in this kind of justice they are very Accurate."

However, before relating examples of some of the more interesting general behavior patterns as well as the behavior of individual sharks, let us look at the general configuration of this fishlike vertebrate. With the exception of a few species, sharks possess a torpedolike body, have especially well-developed jaws and teeth, and a skeleton that is made up of cartilage instead of bone. Sharks, although generally referred to as "fish," are in truth members of a special class called Selachii. The name is derived from the Greek word meaning *shark*. There are other classes of aquatic animal life belonging to this group, such as rays and chimaeras, and when spoken of in the collective sense are called Selachians. The hide of these animals does not have scales, but, instead, is covered by minute close-packed skin-teeth called dermal denticles. Unlike bony fish, sharks, with a few exceptions, fertilize their eggs with a definite union of male and female. The male possesses a pair of organs, commonly known as clappers, positioned just inside the pelvic fins. The female shark possesses two matching openings, housed within another single opening, the cloacal opening, through which the live young will emerge after a gestation period of 6 to 9 months. Most of these live young are born in small numbers, in comparison with other marine life; from 50 to 100 for sharks against several million eggs for true fish.

On each side of a shark's head, there are from five to seven gill openings, which do not have the covers found in most fish; and these openings lead from the wall of the gullet to the surface of the body. All these gill openings, or at least the greater portion of them, are on the sides of the body.

In most cases (with only a few notable exceptions), sharks are meat eaters and appear to become excited by the smell of blood. They are ravenous eaters that fare on fellow sharks, turtles, smaller fish, seals, squid, and other living marine species. However, an inventory of the contents taken from the stomach of a captured specimen suggests that they are almost indiscriminate in their selection of food. For example, a large whitetipped shark (*Carcharhinus longimanus*) brought to the public docks at Miami, Florida, yielded the following stuffing: four old beer cans, one large monkfish, numerous small squid, one modest-sized eagle ray, the lower portion of a moray eel, a glass bottle, one badly worn woman's shoe, a collection of unidentifiable bones (three of which appeared to be human bones), a leather glove, a shark hook fitted with wire leader, and a great mound of large shrimp. Unquestionably, they are scavengers and can be enticed by chumming with garbage, raw meat, decomposed fish, and other odoriferous refuse. Both the basking shark and the whale shark are notable exceptions to the rule. These two massive sharks generally subsist on small shrimp, tiny fish, and plankton. A number of sharks possess specially developed crushing teeth that enable them to feed on clams, oysters, other shellfish, crabs, and lobsters.

Shark attacks on man have occurred mostly at night and upon injured swimmers, especially if the wounded have white or very light skin. These attacks take place mostly offshore in deep salt water. But sharks do inhabit fresh and brackish waters and have not lost either their size or their viciousness in the process of adaptation. The Lake Nicaragua shark (*Carcharinus nicaraguensis*) of South America is a freshwater species that grows to about 10 feet in length and is on record as having attacked many humans. Although sharks generally inhabit warmer waters, they are found in all seas and are known to migrate from one to another. They are mostly night feeders, which explains why attacks on humans occur more frequently during late evening or complete darkness.

An inventory of shark attacks reveals that the worst violators, in order of frequency, are: first, mackerel shark; second, requiem shark; third, sand shark; fourth, hammerhead shark; and fifth, many sharks of unknown species. Numerous studies have been made of shark behavior patterns, and all of them show that sharks are nonconformists. They will respond to an established set of circumstances in a certain way for a long period of time. Sud-

This 1,400-pound tiger shark was hunted and killed through use of the tiny Perry Cubmarine pictured at right. Photo, Perry Cubmarine.

A spiny dogfish shark. Drawing, United States Naval Oceanographic Office, Suitland, Maryland.

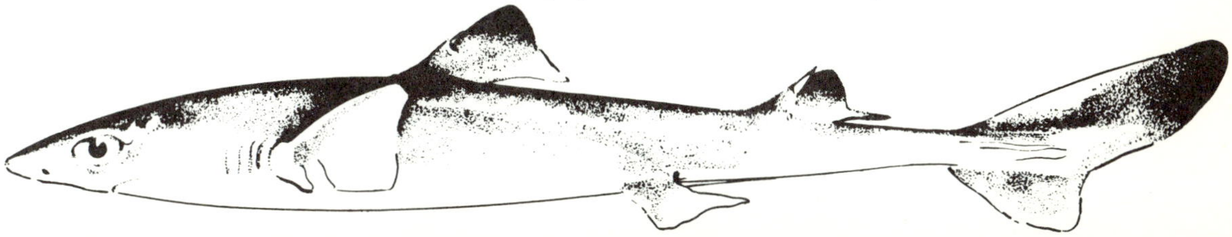

denly they will change, as if fighting the very set of circumstances to which they were responding favorably. It is interesting to note that in one series of experiments, almost half of a shark's brain could be removed without changing his feeding habits or other behavior to any noticeable degree.

In some instances, for no apparent reason, sharks will suddenly explode into massive rage. During such fits, they have been known to chase a man onto land, or leap into a boat after their tormentor, or as happened in Hawaii, hurl themselves upon a dock to get at a fisherman. On a beach in Puerto Rico, a foolhardy bather fired a spear gun at a pair of great white sharks (*Carcharodon carcharias*) that had chased swimmers out of the water. The dart hit what later was adjudged to be the male shark which then went after the bather. The bather quickly leaped ashore. This did not deter the frenzied 15-foot shark. It propelled itself onto the beach and lay there for several moments snapping its jaws, then flopped clumsily back into the surf, and was gone. Bystanders were so stunned, they simply watched as the brute escaped in the surf.

Some fishermen have only words of praise for sharks. In Cuba one commercial shark fisherman used to proclaim that these beasts are "all good except for their bite." He should know, for upon catching a large shark, he would skin it for its hide, boil it to obtain about five gallons of high-grade oil, eat a good portion of the meat, and pickle a part of the flesh for sale at a nearby food market. Whatever was left over he made into fertilizer, and that, too, he sold to local farmers. This fisherman was also an authority on the bite of a shark, for he walked with an obvious limp, having lost his entire left foot to the sharp teeth of a whitetipped variety.

Other commercial fishermen—such as those who fish for herring, menhaden, shad, and cod—curse the shark as being destructive to gear and ruinous to the schools of fish they seek to locate.

"On one occasion," Shampoo Jones, captain of a modest-sized fishing smack said, "I'd invested every penny available to me in longlining for tuna. That season we had an invasion of shark. Every hook came up with nothing more than half a fish,

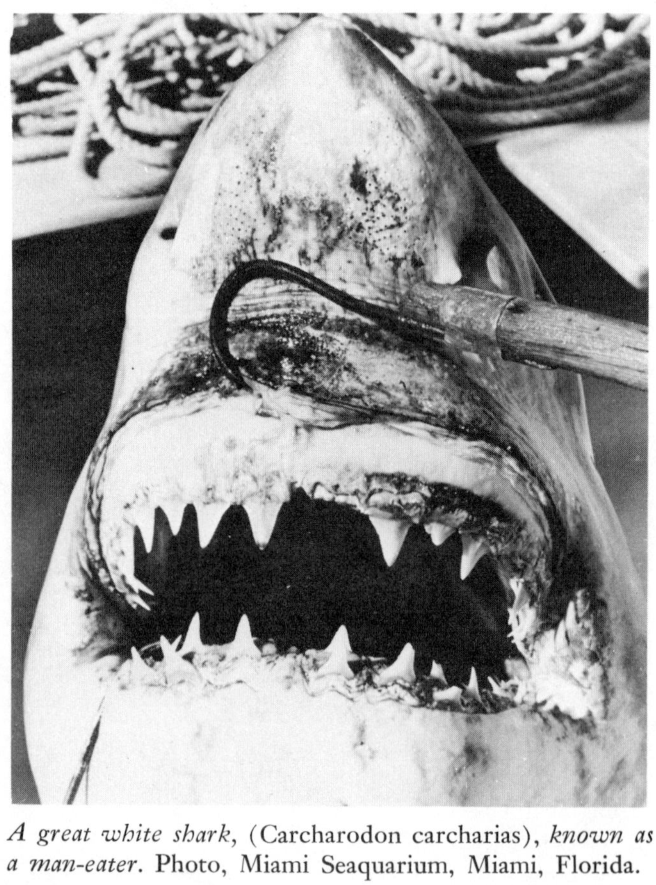

A great white shark, (Carcharodon carcharias), *known as a man-eater.* Photo, Miami Seaquarium, Miami, Florida.

The eye of a large tiger shark. Photo, Miami Seaquarium, Miami, Florida.

or a whole shark. And sharks weren't bringing nothing on the market that year. So I went broke, flat broke, and I ain't finished paying for it yet."

The reverse is true in the South Pacific, where the natives worshipped Kamo-Hoa-Li, said to be a monstrous shark God. Numerous Polynesian groups also used shark wrestling as a test of virility. During certain periods, in celebration of religious holidays, sharks would be enticed into a shallow pool. There a contestant would endeavor to get astride the giant beast by diving off a canoe and then hooking his fingers into the gill slits. Once this could be accomplished, the shark would eventually become inert, sink to the bottom, and literally smother to death. The contestant would then pull his prize ashore, slice off a fin, and lay this token of bravery at the feet of his favorite maiden.

A great deal of research has been done on the shark, yet today we are learning more about sharks than ever before. Some facts that are just now coming to light appear to refute established theories about sharks. For example, it had long been thought that sharks would drown if they were not constantly on the move because water must continually flow against their gill openings to supply them with needed oxygen. If the flow of water ceases, no oxygen supply circulates through the gill slits and the animal suffocates. Now a shark has been discovered to which this theory does not apply. There is a small island off the coast of Japan, called Takara Jima Island, where large sharks come each year to sleep, starting in about March. Everyday at noon, the male of this species sticks its head into an opening in the coral reef and takes a siesta. Should all available holes be occupied by sharks taking snoozes, a latecomer simply rolls over and sleeps on its back.

It has been reported that divers have found these sharks to be completely inert during this nap period. They can be easily captured and handled with relative safety. Obviously, in such a condition, no water flows over or through their gill slits.

Another so-called truism about sharks is that they will not coexist with porpoises. We know this is not true in captivity, for many of both varieties swim together in large tanks without being combative. Now, however, it has been established that sharks and porpoises will even feed together in their natural habitat. This does not mean that sharks will not eat porpoises; they will eat anything. On the other hand, porpoises are far from being helpless. They are fast, powerful, and can butt with the power of a battering ram. It is hardly likely, though, that they would be any match for a large, hungry mako shark. After all, sharks have even been known to kill and eat elephants if they catch the pachyderms bathing in shallow ocean water. They do this by first hamstringing the elephant, and then, once the great animal has been immobilized, they literally chew off its feet until it drops.

Other interesting information about sharks is being compiled. For example, when sharks die, they die from the tail forward. Therefore, in handling what appears to be a dead shark, beware of the mouth. Another noteworthy fact is that sharks become meaner and more aggressive in water that is of low salinity. Also, sharks have been found to be more sensitive to low-frequency sounds but are not likely to be driven off by sounds of any nature, regardless of tonal qualities.

Commercial shark fishing has often been tried but with little success. One reason is that sharks, unlike many bony fish, do not have swim bladders or lungs. Consequently, they are what is known as "heavy fish": they quickly sink when shot or stunned by dynamite. This leaves the fisherman with only two proven methods of shark fishing: one is by hooks and lines; the second is with nets. If nets are used, they must be exceptionally strong. The commercial value of a shark increases in direct ratio to its size. A strong net of large dimension is a costly item, not generally within reach of an average fisherman. Worse yet, if the fisherman is successful and manages to ensnare several large sharks on his first try, the chances are the gear will have to be brought ashore for overhaul because massive sharks often rip the nets to shreds. A giant tiger shark, deemed a fine catch, could weigh about 1,500 pounds and be about 14 feet in length. Hoisting such a beast aboard would be a monumental undertaking.

A question invariably arises: How large can a shark grow? Unquestionably, the largest of the family is the whale shark (*Rhincodon typus*), which sometimes grows to a length of 50 feet or more and weighs many tons. This shark was first discovered in South African waters in 1828. It is a plankton eater, has very small teeth, and emits a constant croaking sound that is thought to be a form of sonar which the shark uses to navigate.

One of the more interesting sharks, about which many stories are told, is the remora. A comparatively small fish, averaging about 3 feet in length, the remora shark (*Echeneis naucrates*) has a flat, suckerlike disk on the top of its head. With this powerful vacuum cup, it adheres to a larger fish and rides without effort, even enjoying morsels of food that are overlooked or strewn about by its host. In the South Pacific, the remora is used for catching fish and turtles. Natives fasten a ring, to which is attached a long length of line, through the shark's tail. The remora is kept in a bucket of water until a likely host is sighted, at which time it is dumped into the sea. Invariably, it fastens its

A boatload of spiny dogfish sharks. Photo, United States Department of the Interior, Fish and Wildlife Service, Washington, D.C.

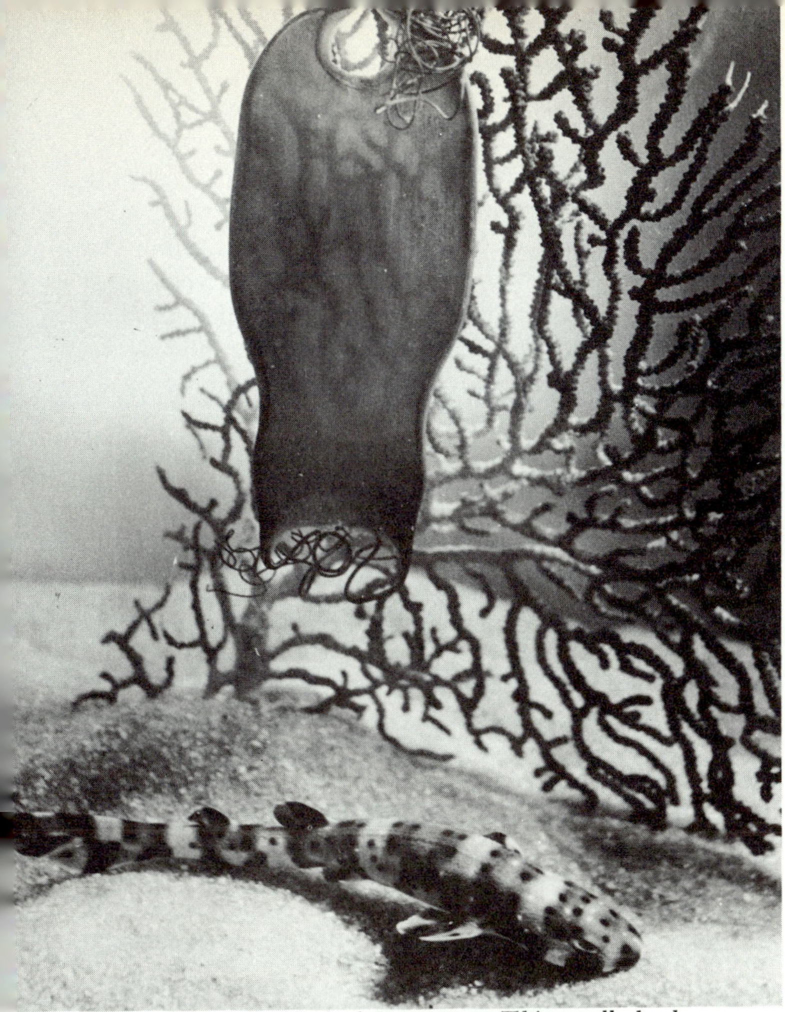

Some sharks are hatched from eggs. This small shark pup has just emerged from overhead "mermaid's purse," where it developed for 7½ months. Photo, Marineland of the Pacific.

A tiger shark. Photo, U.S. Navy.

An unpredictable man-eater swims past an alert diver. Photo, Miami Seaquarium, Miami, Florida.

unbreakable grip on the target fish or turtle. The fisherman then pulls in a lucrative catch.

In mythology, there are stories about how remora sharks, each affixed with a line, were allowed to adhere to an enemy ship that was then pulled ashore to its destruction.

Here is a summary of what the United States Navy has found in its research on sharks. In this respect, the Office of the Assistant Secretary of the Navy for Research and Development, headed by Dr. Robert A. Frosch, has stated that for those venturesome skin-divers who find themselves in the position of directly engaging an overly attentive shark, there are several effective weapons available. The most basic is the shark billy, a short club used primarily to fend off this animal at close range. For offensive use, there is the bang stick, or power-head. This device consists of a cartridge-firing chamber on the end of a probe several feet long, that can be discharged by backward movement into a fixed firing pin; as the muzzle end of the bang stick is driven against the top of the shark's head, it detonates. The force of the discharge, when applied to the region of the brain, is sufficient both to immobilize the shark and to cause its death. However, the user should be aware that the explosive discharge of the weapon can attract other sharks to the area and that an inaccurate shot can produce a very unhappy shark in an extremely excited state.

"A more sophisticated approach is the syringe-type pharmacological weapon proposed a few years ago by a group of Australian divers," it is stated in the office of Naval Research *Reviews*. "A barbed hypodermic needle is rammed home by means of a spear or lance, and an incapacitating dose of drug is delivered to the shark's body by means of a spring-driven syringe."

With respect to syringe-type weapons, the same publication says that current research is concerned primarily with the development of drug loads suitable over wide ranges of temperature for sharks that differ greatly in size. This method, they note, also has its drawbacks. The drugs tested so far require up to several minutes to cause death, but during the first several seconds after the drug is administered, the shark may become highly agitated.

Naval Research and Development scientists say that the dangers inherent in handling lethal anti-shark devices cannot be overemphasized. The hazard posed by any shark-attack deterrent must always be weighed against the actual likelihood of attack. In general, the deterrent must always be readily available and dependable if it is to be of value; this means that it must be capable of enduring long periods of safe storage, must require minimal maintenance, and must still be available for instant use.

The United States Navy also says that in evaluating anti-shark devices, it is important to test them against as many members of a species as possible. Otherwise, the conclusions reached may not be reliable, and there is nothing less comforting than an unproved shark deterrent. A good example of such a danger occurred when one anti-shark device was tested against twelve tiger sharks. Eleven of the sharks swam back and forth through the device at will, but the twelfth refused to do so. If only that twelfth shark had been used in the tests, completely erroneous results would have been obtained.

It is difficult and expensive to test adequately any anti-shark device or technique. One difficulty is the small number of dangerous sharks that can be kept alive in captivity long enough to be used

The mouth of a 575-pound tiger shark. Photo, United States Naval Photographic Center, Washington, D.C.

There are about four hundred varieties of sharks; they range from exceedingly small to exceedingly large. Photo, Marineland of Florida.

This drawing of a man-eating hammerhead shark was made over a hundred years ago as a warning to American fishermen. Photo, United States Fish and Wildlife Service, Washington, D.C.

When it has been hooked, a large dead shark is a problem. It must be hoisted quickly before it can sink. Photo, Bureau of Commercial Fisheries, Washington, D.C.

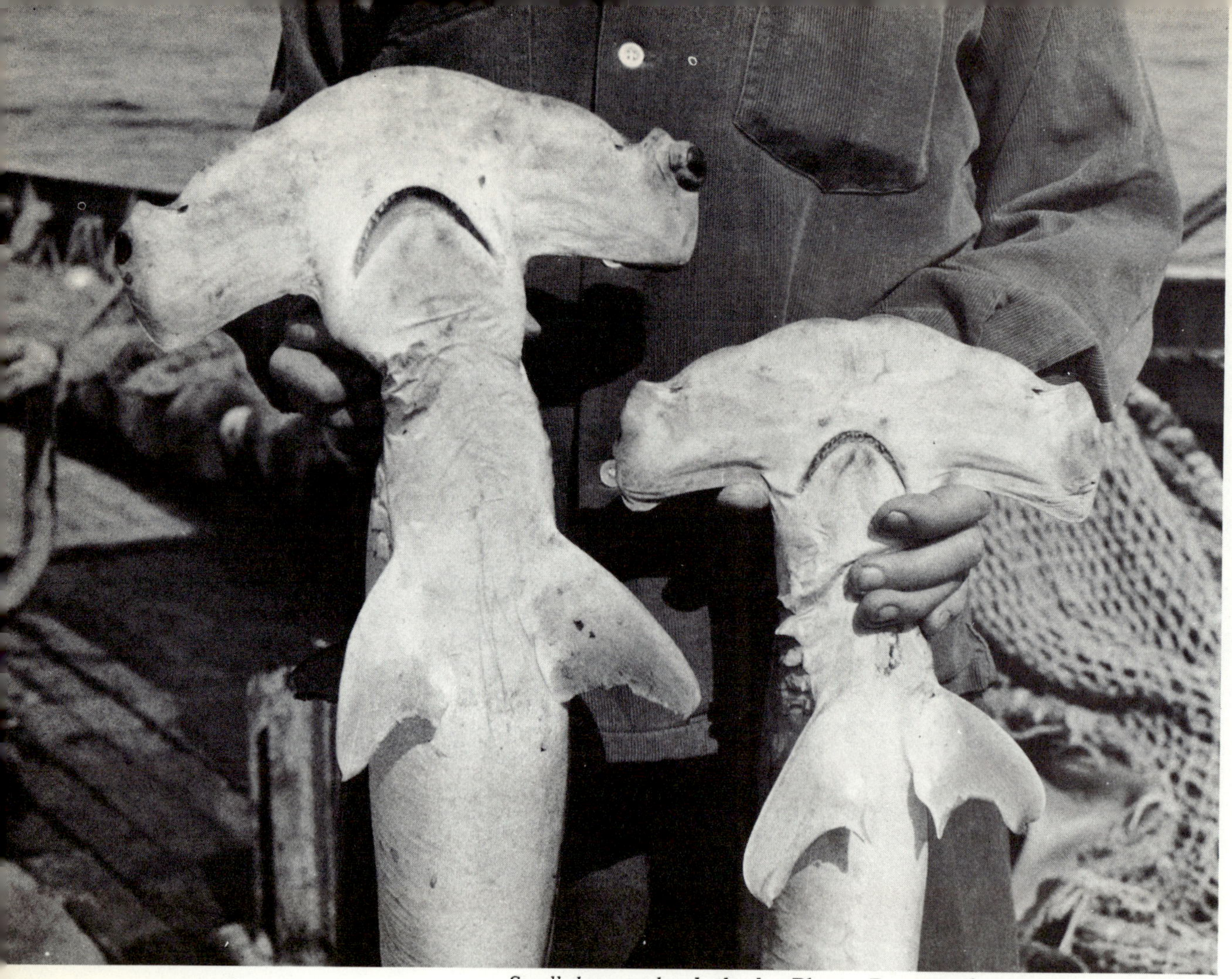

Small hammerhead sharks. Photo, Bureau of Commercial Fisheries, Washington, D.C.

for reliable testing. In fact, none of the larger sharks, such as the great white and the mako, have been kept alive in captivity for more than a few days. Although sharks are difficult to kill quickly, they are easily injured during capture and often die within a few days. As a consequence, healthy captive sharks are valuable, and the larger ones may be priced from $200 to $500. Feeding and maintenance costs are also high. Under these circumstances, the testing of chemical and mechanical devices designed to kill a shark could be very expensive.

But it is not sufficient to conduct experiments on sharks in captivity. If conclusions on the effectiveness of the deterrent are to be proved valid, tests must also be made on sharks in their natural surroundings. The main problem in this regard is the maintenance of adequate facilities for longtime observation, with due regard for human safety, in areas where there are numerous sharks.

The mouth of this white shark reveals tough, sharp teeth capable of snapping a man's arm in two. Photo, Bureau of Commercial Fisheries, Woods Hole, Massachusetts.

In the final analysis, shark deterrents can only be based on shark behavior, and that is completely unpredictable. In order to illustrate this statement, let us examine a few reports recorded by professional divers.

Norm Lewis, an experienced deepwater salvage man, said, "About sharks you can't ever tell. For instance, I was once on the bottom at fifty feet and very much preoccupied with testing out a mixed gas system using a semi-closed-circuit backpack recirculator. It was a new concept about which I was very uncertain and my nerves were up tight. Suddenly I became aware of something very unfriendly happening. A fifteen-foot mako shark had gotten to swimming around and around me. Each time around he got a little closer and it looked to me like he was building up to make a strike. It crossed my mind that something had to be done to shake him out of his deadly pattern. So one time around I leaped at his underbelly. The shock of hitting my target made both of us break off the attack. That big mako turned tail and fled."

Jimmy Hill, a diver for Siebe, Gorman & Company, tells of a shark encounter in this fashion: "At first I thought it was a big skate. But it wasn't. It was a 6-foot 'tiger.' He glided round towards me, and then seemed to change his mind, and settled on the bottom. I gripped my spear and got him in the side. I could do nothing with him on the bottom, but he shot up, and there I could handle him. He was swimming on a level with my head, snapping around at the spear. I played him like that for a quarter-an-hour, with George (another diver) trying to get his harpoon in. Finally the shark made an extra effort and got clear. He broke the barb clean off the harpoon, which was of ½-inch mild steel. Then he turned and came at me. He missed me, and after that he seemed to have had enough. He turned round and swam away."

A Torres Strait pearl diver, a native called "Treacle," was the victim of a severe attack; a shark grabbed him by the head. The scars on his face, neck, and arms were evidence of the mauling he suffered. How he managed to free himself nobody knows. In fact, nobody knows why either of these sharks broke off their attack as they did. Ample evidence is available to prove that many sharks are not so easily frustrated.

Commercial shark fishermen consider this man-eater a valuable catch. Photo, Bureau of Commercial Fisheries.

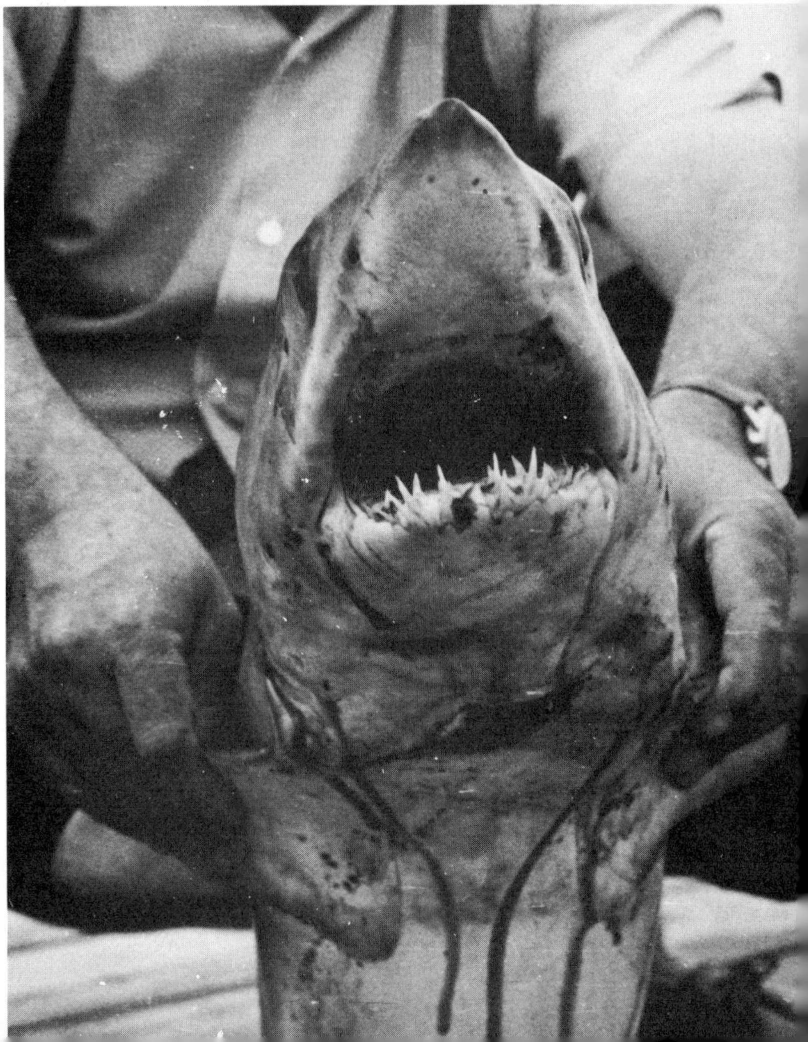

According to scientific studies, a shark dies from the tail forward. The jaws remain dangerous to the last. Photo, Bureau of Commercial Fisheries, Washington, D.C.

A frightening sight to a diver. This whitetip shark was estimated to be over 10 feet in length. Photo, Bureau of Commercial Fisheries, Washington, D.C.

It seems that whenever a sizable shark is caught, someone asks that intriguing question: How old is the beast? There is no certain method by which a shark's age can be judged. With a fish, the rings in the scales (like the rings in a tree) indicate yearly growth. But sharks do not have scales.

One marine biologist stated that it is quite conceivable that a shark, barring injuries, might live forever. His reasoning for this assumption was based on the fact that in some very large sharks, he found a regeneration of cells, a process that ceases when the final aging process (senility) sets in. With most animals, age can be computed from a chemical analysis of the bones. But sharks do not have bones. The framework of a shark, being entirely cartilaginous, disintegrates along with body flesh, leaving nothing but teeth.

Throughout this brief discussion of sharks, we have indicated that they are unpredictable and will eat anything they happen upon. For example, the

A large bull shark swims over a diver (note hand at lower left). Photo, Miami Seaquarium, Miami, Florida.

A 35-foot whale shark caught off Fire Island, New York. Photo, Library of Congress, Washington, D.C.

following story, which is in the archives of the Institute of Jamaica (Kingston, West Indies), is fully documented:

"During the eighteenth century an American privateer was being pursued by a British man-o-war. This was in Caribbean waters, and the Yankee skipper, finding himself hemmed in, threw his ship's papers overboard. The British vessel made its capture and took the suspected privateer in to Port Royal, Jamaica, where the Yankee skipper was placed on trial for his life. However, the court was faced with having to dismiss the case for lack of evidence. Before this could happen another British ship arrived in port and produced the Yankee privateer's scuttled papers. It seems that when off the coast of Haiti, this second British ship had captured a large shark, which when opened, revealed the lost papers. These documents were taken in to court and used to convict the entire crew of the privateer, along with its captain. The original papers in this case have been preserved and are in the Institute of Jamaica, Kingston."

Healthy sharks, such as this 10-foot specimen, can cost up to $500. Photo, Marineland of the Pacific.

Underside of giant manta ray reveals gill openings and three remora sharks going for a free ride. Photo, Wometco Miami Seaquarium, Miami, Florida.

2

Giant Rays and Sawfish

Selachians may be divided into two categories: the entire range of sharks and the rays, skates, and sawfish. There appears to be one known link between these selachians: the monkfish, also called the anglerfish or goosefish. It is a very ugly and otherwise unattractive fish. Because of its appearance and its dangerously wide mouth, the goosefish, when caught, was usually thrown back as a trash fish. But it has recently been discovered that this fish makes good eating. Goosefish grow to about 4 feet in length and attain a weight of approximately 50 pounds. They have a skin that is a mottled brown on top, tapering to a dirty white on the underside of the belly. Commercial fishermen occasionally bring them up off the bottom with their nets when working an area from the Carolinas on north to Newfoundland.

The name *anglerfish* has been given to this variety because the first dorsal spine is elongated in such a manner that it forms a very practical fishing pole. This fishing pole bends forward and dangles an enticing tip somewhere in the vicinity of the broad mouth. When a curious, or perhaps hungry, predator comes swimming after the bait, the goosefish opens its cavernous mouth and sucks the aggressor into its belly. The suction power of these anglerfish is tremendous. Their stomachs have often been found to contain fish as large as themselves and even, upon more than one occasion, ducks and other seabirds.

In one instance a beer can was found in the belly of an anglerfish caught a considerable distance offshore in the Atlantic Ocean. This might have come about because anything touching the fishing appendage appears to trigger the mouth into opening, thereby setting the suction process into action. Although it is known that the fish can regurgitate, it might be that the metal container became lodged in the gullet.

Goosefish are voracious eaters. They are found from tidemark to depths slightly over 1,500 feet. Unlike their close relative the sharks, these goosefish do not give birth to living young; they lay eggs in a band of mucuslike substance that eventually floats to the surface. One such strip may be 25 feet or more in length and may contain an estimated 2 million eggs.

At the opposite pole of the selachians is the vast suborder containing perhaps four hundred species. The largest ray is the giant devil ray, also known as a manta. Like all rays, it is somewhat disk-shaped; has cartilaginous vertebrae; is covered with skin made of tiny toothlike, enameled projections; and is completely without scales. A ray, like a

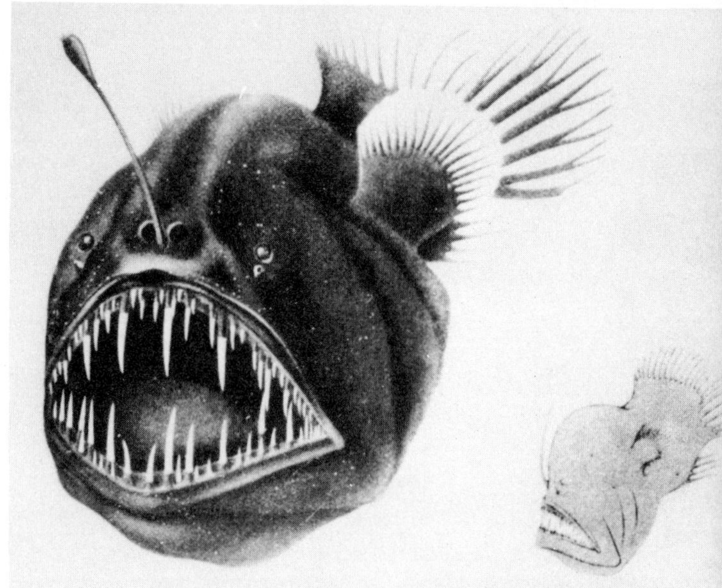

The name anglerfish *is applied to fish having a very practical fishing pole in the forward part of the head.* Photo, United States Fish and Wildlife Service, Washington, D.C.

True rays have a head, trunk, and enlarged pectoral fins united into a flattened disk and ending in a long, stout tail. Photo, Marineland of Florida, Miami, Florida.

Goosefish are the only known connecting link between sharks and rays, skates, and sawfish. They are also called monkfish and anglerfish. Photo, Bureau of Commercial Fisheries, Woods Hole, Massachusetts.

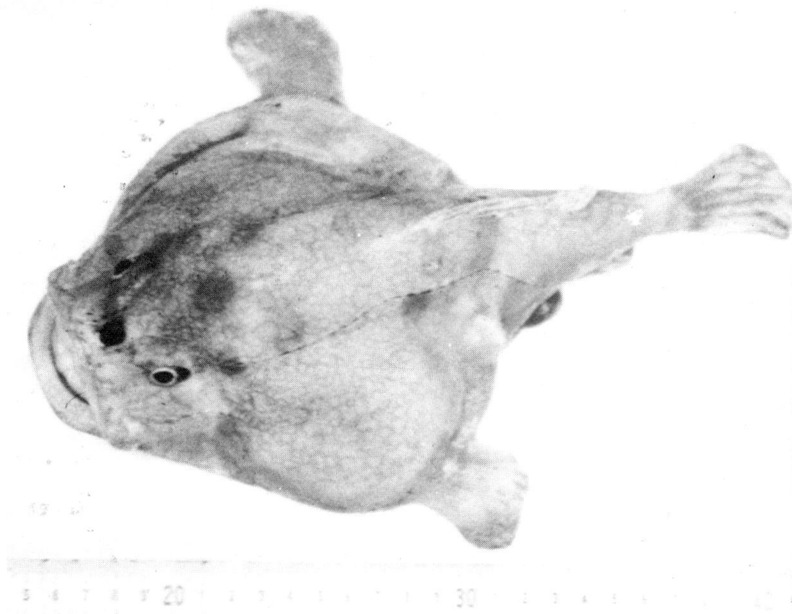

The odd configuration of a captured goosefish. Photo, Bureau of Commercial Fisheries.

shark, has five pairs of gill slits, which are on the lower surface of the body. Also like the shark, the male ray possesses a pair of sexual organs called *claspers*, and fertilization occurs within the female. The young of all species except skates are born alive. The skate's eggs are encased in a tough capsule that has streaming threads at either end. These streamers catch in bottom seaweed, and the eggs are allowed to develop without first being washed ashore. Skates and rays are found all over the world, except in those seas known to be cold. The devil ray prefers the warm waters of the Gulf Stream and is found in abundance in the Gulf of Mexico and adjacent waters.

Because of their enormous size, many weird stories have been concocted about devil rays. They grow to a width of 20 feet and may weigh 3,500 pounds or more. When they leap out of the water and smash back with a thunderous splash, they create a startling sight. An entire school participating in an activity such as this is a frightening spectacle, calculated to inspire many a strange tale. Why rays engage in these leaping activities from time to time is not known. It is supposed that they leap to free themselves from barnacles or as part of a courtship ritual.

Devil rays are not considered good eating and, unlike most other ray fish, do not subsist on the bottom of the sea. Rather, they cruise gracefully near the surface, with rhythmic undulations of their great wings. Devil rays have enormous mouths, on either side of which are large scooplike appendages. In order to gather food, rays open their gigantic mouths and swim through schools of tiny crustaceans, such as minute crabs, shrimp, and other creatures that populate the deep scattering layer.

Devil ray is a complete misnomer. However, they will indeed put up a tremendous fight if harpooned. In the old days of the hard-hat diver, especially before the introduction of underwater telephones and the development of scuba, men working on the bottom lived in mortal fear of the

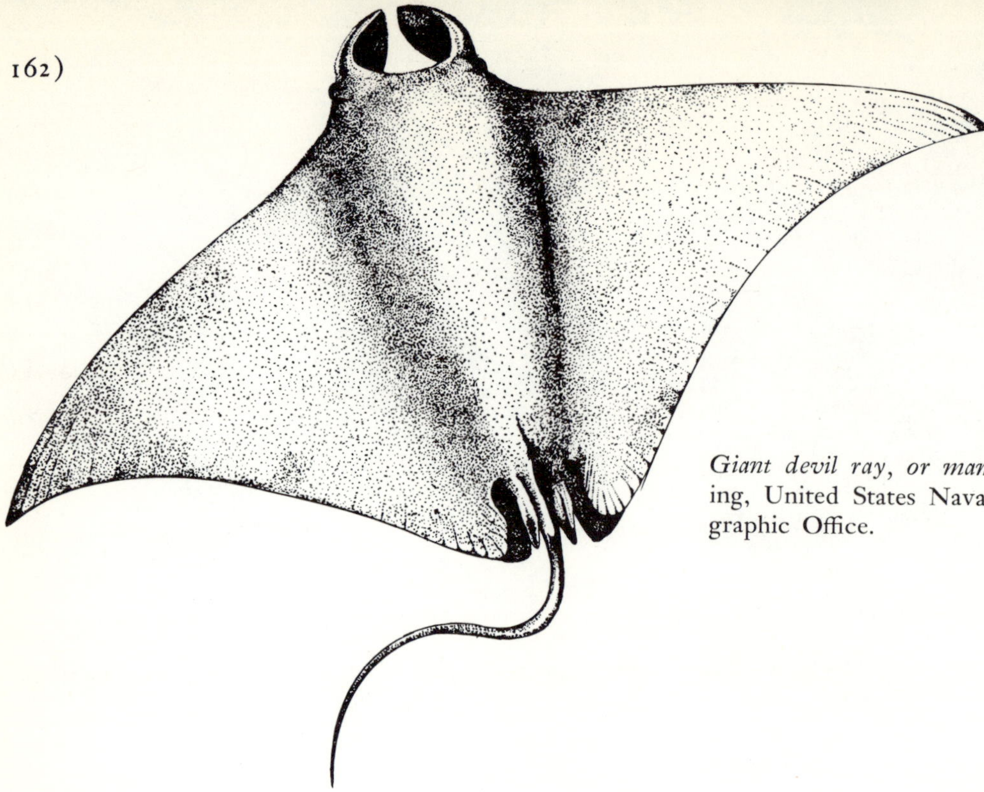

Giant devil ray, or manta. Drawing, United States Naval Oceanographic Office.

manta. Old-timers who have worked the bottom of the Gulf of Mexico tell harrowing stories of how these great beasts, swimming through a work area, would become entangled in a diver's lifelines and take off in mad flight. This irresistible force, whipping through the water at high speed, was certain to tear the gear clean off a diver's head. Not many survived such an experience. Tom Caras, now nearly ninety years of age, still remembers the horror of such moments.

"There was two of us working at offloading a cargo of bananas from a sunk ship that'd gone

A head-on view of a manta ray. Photo, Wometco Miami Seaquarium, Miami, Florida.

A large spotted eagle ray. Photo, Wometco Miami Seaquarium, Miami, Florida.

down no more than a day before. It was shallow water, only about ten fathoms at most, or else them bananas would have been ruined. I had just signalled, by tugging on my life line, that I was on the bottom and about to go to work. My partner had already been down ten minutes and was hard at it.

"In shallow water like we was in, it is clear and any sunlight comes through real good. Suddenly a shadow closes in overhead—a deep, dark shadow meaning something big is passing through the water between me and that bright sun. Then it's gone.

"Me and my partner pressed our copper hats together, in this way we could hear each other talk real good, for metal under water carries sound just fine. 'A devil ray,' he tells me. 'A big one's been hanging around.'

"Being just a young squirt at this diving business, I don't have any real idea of what a devil ray is like. But I don't like the sound of that name, so I takes out my long diver's knife and stand there feeling kind of silly.

"Again that shadow, then *blam*, I'm whipped off my feet and go swooshing through the water end over end. My knife went flying and my head banged around inside that copper pot fit to knock my brains loose. Then, just as sudden, I'm sitting on the bottom holding the chopped ends of both my air hose and my life line. Now there is really a shocking moment, to find yourself on the bottom along with both ends of your air hose.

"My partner was a cool one; he shuts off my air escape valve and frees his own air hose from its coupling. We both live off the one hose until they can break out some new gear and bend some of it to my intake. Later, we found out the rest of the story. Our tender topside had seen this here devil ray make his first pass, so he grabs a fire axe and stands by. When he seen my gear fouled by that ray the second time around, he swings that axe hard, and cuts me clear with one solid blow athwart both of them lines at one time."

Years ago, Mexican fishermen used to entertain tourists with a superthriller of a spectacle. They would take a party of visitors out in a fishing

Small manta ray taken by Florida fishermen. Photo, Library of Congress, Washington, D.C.

vessel and, after sighting a devil ray floating unconcernedly across the surface of the water, would launch a tiny cockleshell carrying two men, a harpoon, and numerous fathoms of tough line. The cockleshell would get directly on top of the manta and square in between the two great wings; at that point a harpoon would be thrown. Instantly, the two men would drop flat and secure their line. The manta would take off with a great splash, while at the same time thrashing its two massive wings inward. There was actually little danger of the boat being crushed because it was low and flat, and the wings of a manta cannot bend inward to any appreciable degree. However, if a man were caught upright in the boat, unquestionably he would have had his head knocked off. The scene that ensued was spectacular. But seldom, if ever, was the great beast tamed; for after a period of time considered sufficient to entertain the viewers, the manta would be cut loose.

Sawfish are special members of the group Selachii. They are true ray that in many respects resemble the shark, but they act like neither the ray nor the shark. They prefer warm, shallow water and inhabit rivers and streams, where they kill small fish by slashing them severely with their great, wide, dangerously toothed saw. Common sawfish grow to lengths of about 20 feet and weigh close to 1,000 pounds. They are found in the warm, shallow waters throughout the world. Sawfish have no commercial value and therefore are not held in high regard by commercial fishermen, in whose nets they easily become enmeshed. Extracting a 15- or 20-foot sawfish from a close-meshed net can be dangerous, expensive in terms of repairs, and extremely frustrating. Once alongside a fishing vessel, an entangled sawfish flails its weapon back and forth, ripping the net, damaging wooden-hulled fishing boats, chopping other netted fish into useless morsels; and in the end, they have to be cut out, often at expense of a sizable portion of net.

Like the shark, which it resembles in appearance, the sawfish bears live young, which average about 2 feet in overall length. Before birth, the weapon of a young sawfish is covered with a protective coating. This is quickly dispersed at birth. It is interesting to note that young sawfish need no pattern of attack. Instinct provides them with an understanding of how to flail their deadly saw. As they grow, it is astounding to see this wondrously odd creature move amongst a school of fish and, within seconds, cut themselves out a sumptuous meal. Frequently, a rather large fish will become impaled on the

A sawfish newly born in captivity. Photo, Aquatarium, Saint Petersburg, Florida.

Sawfish are true rays that resemble sharks. Drawing by the author.

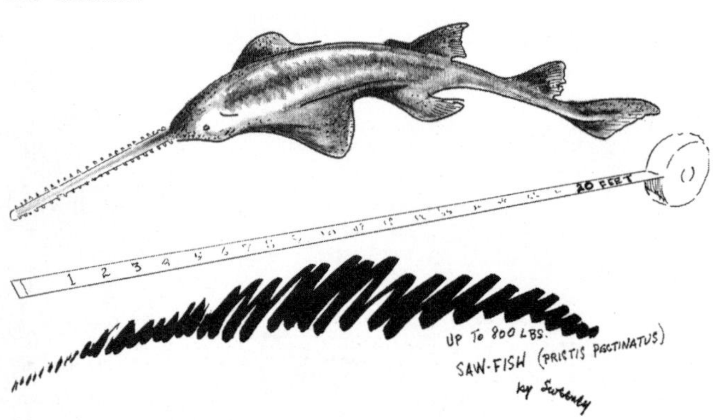

jagged teeth of the saw. On occasions such as this, the sawfish will return to the bottom, scrape the fish off the saw, and enjoy the remnants as a meal.

These fish are notoriously sloppy and wasteful in their eating habits. Anyone studying an aquarium of mixed fishes will note that those on the top, which either strike at food or attempt to chew it down, leave a large train of pieces filtering their way to the bottom. This same feeding characteristic is to be found in many of the ocean's larger fishes, notably the shark. Actually, there is a very good reason for this. The entire cycle of marine life must be well fed, and those in the layers between the surface and the bottom also require fresh food. As these bits of newly killed fish sink slowly to the floor of the ocean, many are devoured en route. The morsels that reach the floor of the ocean are food for crustacea such as lobsters, shrimps, and crabs, and for other species that live on the bottom. Furthermore, many fixed polyps, anemones, and sea urchins are carnivorous; a regular sprinkling of fresh meat is always a welcome event.

Obviously, nature, in its infinite wisdom, wastes

A sawfish in its natural habitat. Photo, Rex Gary Schmidt, United States Fish and Wildlife Service, Washington, D.C.

A smalltooth sawfish (Pristis pectinata). Photo, Wometco Miami Seaquarium, Miami, Florida.

nothing. When a tiger shark rips off the rear end of an Atlantic bonito but fails to devour the forepart, this meat does not go to waste. As the great chunk drifts lower and lower, it attracts an ever-changing group of hungry predators, each pulling, biting, gnawing, and scattering tidbits of food. With luck, a chunk worthy of the attention of a sawfish will land on the ocean bottom and be devoured without a single wave of that deadly saw having been required.

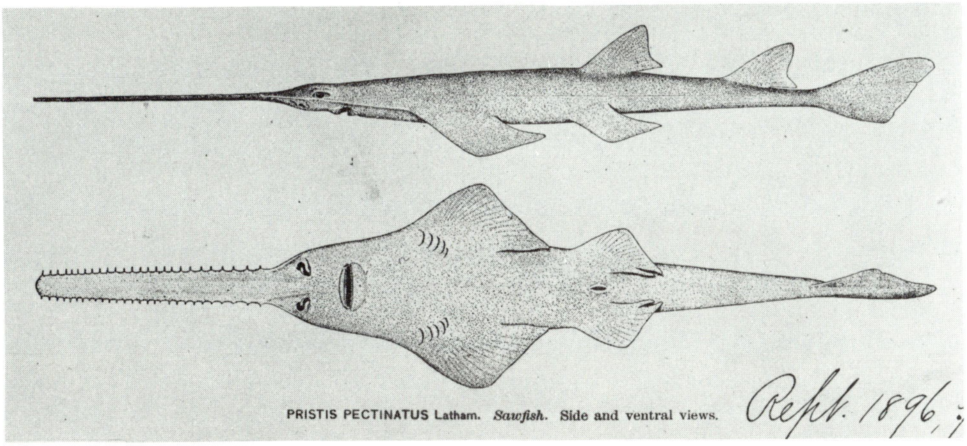

This drawing was published in 1896 as a warning to fishermen that sawfish are dangerous. Drawing, United States Fish and Wildlife Service, Washington, D.C.

Giant groupers have been known to swallow a diver in one gulp. Photo, Bob Noble, Marineland of the Pacific.

3

Giant Groupers and Great Barracuda

A huge fish called *giant grouper* (family Serranidae), jewfish, Australian black rock cod, or sea bass, has become an increasing menace to commercial deep-sea divers. Until recently, the best textbooks on marine fishes disregarded fantastic stories told by commercial divers and listed this species as weighing less than 700 pounds, with a total length of less than 8 feet. Furthermore, these same textbooks say that groupers are sluggish, not likely to devour a man, but that they present a hazard to underwater swimmers because of their massive weight.

Saturation diving, which is a technique allowing deep-sea divers to live on the bottom at great depths, is bringing us new information about the size, physical characteristics, and behavior patterns of the giant grouper. He can, as has now been well substantiated, develop into a supergiant grouper. So it now appears that those old fish stories, like old folk tales, contain a considerable amount of truth. This should make the editors of the *Sydney* (Australia) *Herald* feel better because over the years they have factually reported incidents wherein giant groupers have attacked deep-sea divers or literally swallowed them whole—copper helmet, rubber suit, and lead shoes. The newspaper

A giant grouper caught in Florida waters. Photo, Library of Congress, Washington, D.C.

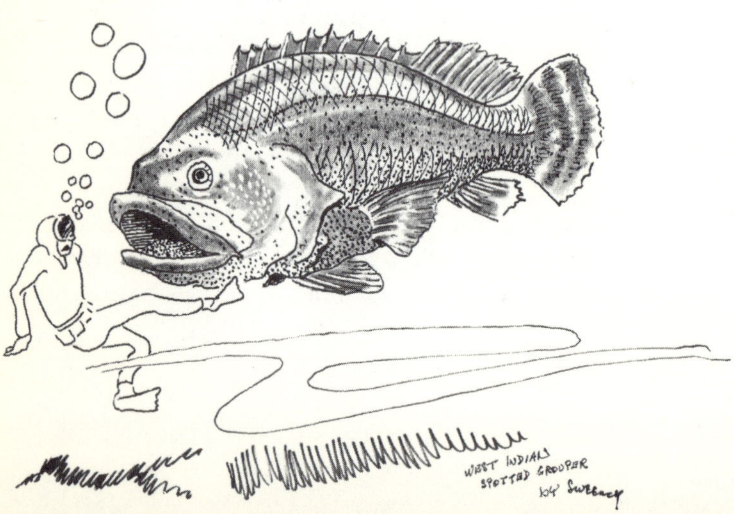

ignored the doubters and even went as far as to suggest that the mysterious disappearance of pearl divers, who were working in regulation canvas suits, could be attributed to their having been gulped whole, thus leaving no traces of themselves. The November 30, 1943, issue of the paper carried a brief story about groupers attacking salvage divers, apparently attracted by light on their copper helmets. One grouper, the article states, "swallowed an entire helmet and swam off with the diver."

Until recently, such yarns were scoffed at. Deep-sea divers who work on oil rigs in the Gulf of Mexico can confirm the truth of such stories and at the same time give new information about the size and characteristics of this beast.

"We used to give a grouper a punch in the nose and they would move off," said Flencer Burton, a master diver who does bell diving and goes down on a helium-oxygen mixture in order to gain greater depths, "but the deeper we go, the bigger,

Divers have found that the deeper they go, the bigger groupers appear to become. Photo, U. Erich Friese, Taronga Zoo, Sydney, Australia.

more aggressive, and better organized they appear to get."

After eighteen years of scuba, salvage, underwater television, marine construction, and deep-sea diving, Burton finds himself in a quandary. "The other day, for the first time in my life," he said, "I refused to leave the submersible decompression chamber on account of a fish. Sharks never succeeded in chasing me off, but these groupers have me stopped. One time last month one followed me through the open bottom hatch of a bell and stuck its nose up and into the helium atmosphere."

According to Burton, the groupers that had been annoying him were in the 2,000-pound class. "One of them could easily have swallowed a Volkswagen, and you just don't go messing around with a fish that big unless you sure as hell know what his reactions are going to be."

Two other divers readily confirmed this estimate of weight and size. One of them is an undisputed authority on grouper ingurgitation capabilities. He and his partner were being annoyed by an especially aggressive giant grouper. "Every time he'd open his mouth," said diver J. T. L. Ponds, "it would cause a massive current. That's how they eat, just point their jaws at a passing fish and suck in, and they never seem to miss.

"So we were working on the mud skirts of an oil rig, which is where they seem to have hollowed out places for themselves. This large economy-size kept sliding in close to me. I even hit him with a big wrench I had in my hand, but that didn't so much as make him blink. Maybe he figured it to be a love pat.

"Suddenly I seen Jake's [his diving partner] feet sticking out of this here fish's mouth. I'm telling you, it was something like a bombshell to see this. So I grabbed his life lines and started to pull hard. Nothing happened. That fish just remained there without so much as blinking one of those big eyes

A brace of giant groupers (Percomorphi), *here labeled "jewfish," caught in California waters.* Photo, Library of Congress, Washington, D.C.

This drawing of a blackfish was distributed to fishermen over one hundred years ago as a warning of danger. Photo, United States Fish and Wildlife Service, Washington, D.C.

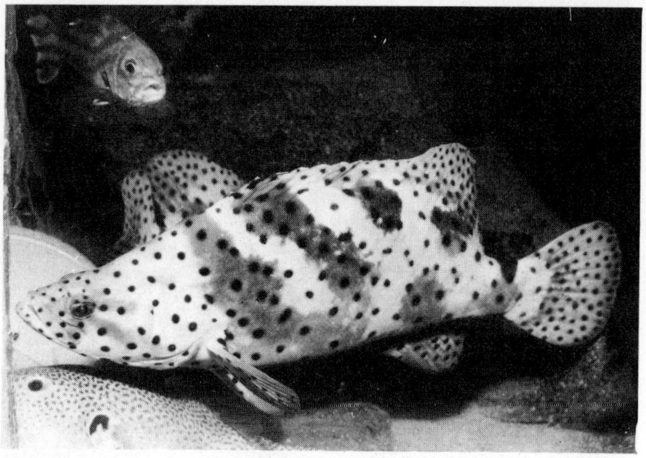

This is a young spotted grouper, sometimes called a panther-fish. Photo, U. Erich Friese, Taronga Zoo, Sydney, Australia.

of his. That gave me a sudden inspiration. I hit him square in the eye with one of my fists. He just spit Jake out."

Numerous interviews with deep-sea divers confirm the fact that the deeper they go to work, the bigger the groupers get. The consistent presence of men working underwater seems to be the factor that slowly attracts these big fish. Several divers agree that the larger groupers seem to stand guard over the smaller ones. "They almost act like they've been rounded up against an invader," one diver said. "Then is when you have to move carefully, for the big fellows—the sort of outriders—commence to inch closer."

Little is known about the personality or behavior of the giant grouper. At shallower levels, they always appear to be rather friendly old fellows holed up in caves. They seem to live by themselves at these levels and only venture forth after divers have been working in their vicinity for some time. There seems to be a definite territorial boundary that separates one grouper's domain from another's. Each seems to have staked out a certain claim that all others apparently respect. Divers say they have never seen one grouper cruising through another's domain.

An odd fact is that few reports can be found relating to the presence of very small jewfish or groupers. Yet the young and small of stature must be around and must be possessed of the same persistent curiosity that makes their elders so dangerous to the working diver.

These giant fish have not been considered good eating in the past for some inexplicable reason, but now they are being used successfully as food. Their meat is deemed quite tender and rather tasty.

Great barracuda (*Sphyraena barracuda*) are formidable killers, with a modus operandi so swift and so certain of success that they are seldom seen. Thus their attacks always evoke the cry of "Shark —shark!"

Several years ago, a tourist who was standing knee-deep in the almost placid surf at Nassau had a leg bitten off just below the knee by some sort of marine animal. The police investigated and asked numerous questions of other vacationers who had been in the vicinity.

"Q. Was a shark fin previously seen in the water?"

"A. No, but only a shark could bite so hard as to sever a shinbone."

"Q. Did anyone see the fish that struck?"

"A. No, it was the vacationer's scream that first attracted attention. Then someone started yelling, 'Shark—shark—clear the water.' "

"Q. Have you heard of sharks previously being reported at this beach?"

"A. Never."

Since there appeared to be no evidence to the contrary, the police simply listed the event as a shark attack, although they knew of course that it was more likely to have been a barracuda, which abound in such waters. Again, in 1953, at Juno Beach, Florida, a young wader had her left leg bitten between the knee and the ankle. This time the attacker was positively identified as a barracuda. In 1962, along the shores of Mexico, in water scarcely a foot deep, a man was badly bitten and his foot almost completely torn off. Again the attacker was spotted and recognized as a barracuda. Unquestionably, barracuda will attack and severely wound a man, even when that person is standing in shallow water. It is impossible for sharks that are of sufficient size to sever a man's leg, to attack in such shallow water without first being clearly visible. On the other hand, the long, lithe body of a barracuda lends itself very well to a hit-and-run technique.

Generally speaking, this pelagic fish (that which feeds and exists above the floor of the ocean) may be divided into two species: the northern barracuda and the great barracuda. The northern type is common off the eastern coast of the United States, from Cape Cod to Panama. It generally does not exceed 12 inches in length. The great barracuda is an 8- to 10-foot power plant, with large, sharp canine teeth. It is found in the warm waters of the Caribbean. However, generalities such as these can be misleading because there are numerous species of both large and small barracuda and each one is different. Some are more aggressive than others; certain species cannot be frightened off as easily as others. This is a highly dangerous fish with an unpredictable nature. Temperament appears to change with the advent of certain phases of the life cycle such as spawning and with abrupt or extreme variations in water temperatures.

Even in the smaller species of barracuda, there is apparently a strong propensity for killing. In the clear water adjacent to the shores of Florida, barracuda have been seen darting at smaller fish, snapping at them with their piercing fangs. After holding the dying fish in their mouths for a moment or two, they spit it out and dart at another passing fish. This is not ferociousness engendered by hunger; it is simply inherent viciousness, the urge to kill.

Imagine such an urge to kill in an 8-foot, torpedo-shaped body, noted for its speed and infallibility of aim, and you will understand that man has a major problem to reckon with. Any person who has had the thrill of watching a great barracuda strike will never forget it. They move from one location to another sometimes 30 or more feet away, in something less than the proverbial blink of an eye. One moment they can be seen in one place at a dead standstill, and in another second they will be in a new location many feet away with a hapless fish in their powerful jaws. No start-up can be observed. They do not build their speed to a striking force; they simply leap from dead stop to full ahead. It is a breathtaking spectacle.

Giant barracuda, 8 feet long, are known to be plentiful. Photo, Nat Fain, Marineland of Florida.

Large barracuda, perhaps not all the same species, but certainly ingrained with identical pugnaciousness, are found from Hawaii to the Red Sea and from Brazil to northern Florida. They will generally strike at anything light in color that flutters through the water. Sailors on merchant ships catch them for food simply by tying a strip of white

There are twenty known species of barracuda; about five grow to giant size. Photo, Marine Studios, Marineland, Florida.

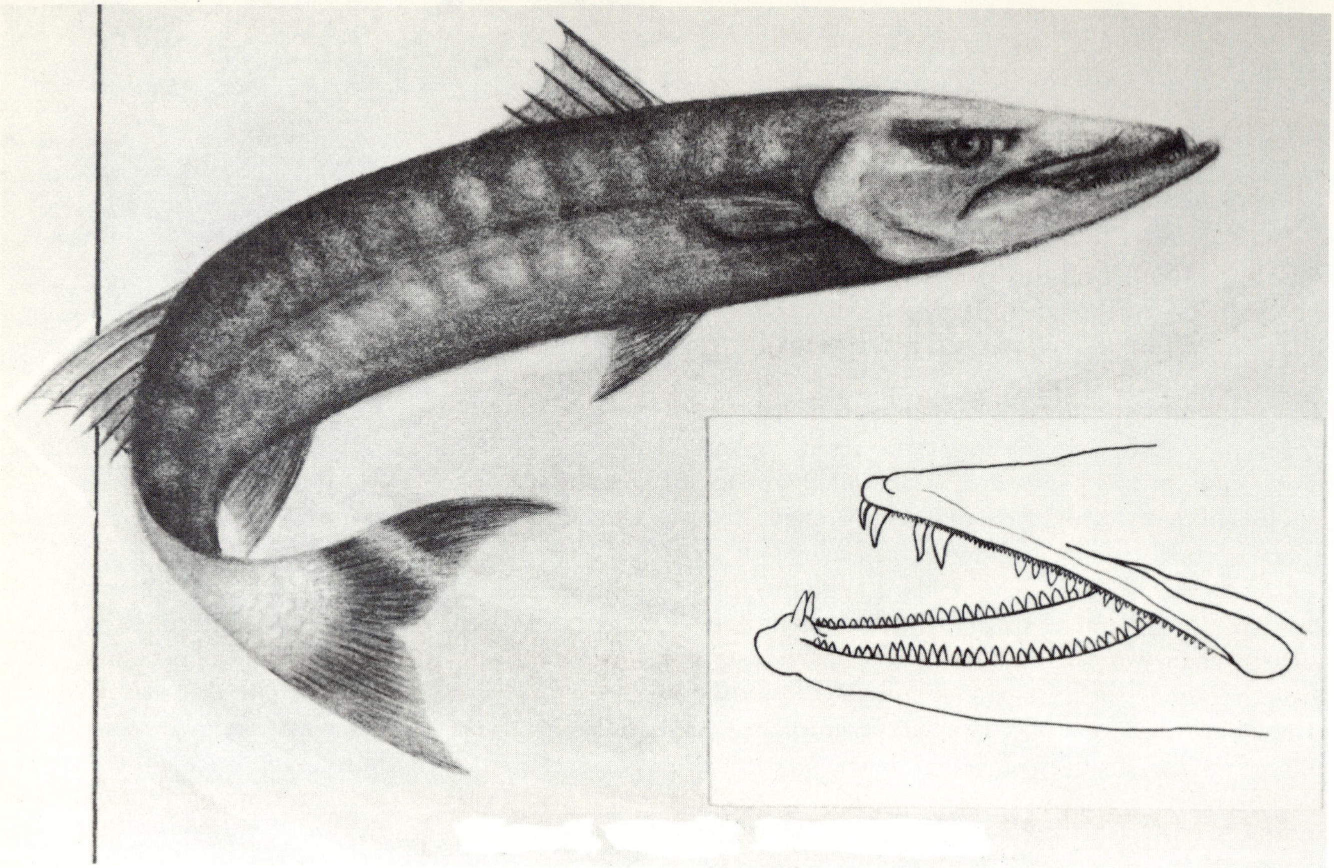

The mouth of a barracuda is large, equipped with powerful teeth, and has an extended lower jaw. Drawing, United States Naval Oceanographic Office, Suitland, Md.

cloth about 6 inches long to a hook and letting this makeshift lure drag behind the ship. They make excellent eating, but a few species are known to be poisonous.

Here is an incident that will give you an idea of the speed with which barracuda can strike. It occurred in the Windward Islands and indicates that these fish can squarely hit a small target in the water in less than a second. A seaman with a harbor-patrol unit had just painted the heavy oars on his dinghy a gleaming white. He used the glistening oars for the first time in an offshore race. With all the force and power in his body, he stroked the boat toward victory. Suddenly, a sledge-hammer blow ripped one oar out of the seaman's tight grip as the blade flashed underwater. Unquestionably, it was a barracuda that dealt the blow. On the end of the retrieved oar dangled the 4-foot-long offender, its piercing fangs stuck in the thick wood. The sailor later reported that the barracuda made an excellent meal.

Young barracuda are schooling fish, and sometimes the sea literally swarms with them. When young, they are not known to attack man; underwater swimmers have frequently found themselves amidst such a school, and none have reported being attacked. It is the lone predator, often named "the most formidable bony fish in the sea," that is likely to attack. A full-grown great barracuda, weighing 100 pounds or more and traveling at an incredible speed, can lacerate a man's side and still keep moving.

Although the patterns of barracuda and shark attacks appear to resemble one another, the teeth marks left by the two are considerably different. Shark bites show the rounded-mouth effect; a barracuda leaves straight teeth lines. Also, sharks sometimes attack in packs, each going round and round to strike again and again. A great barracuda seems to strike only once and then disappears. Group forays on swimmers have not been recorded. Once it reaches maturity, it proves to be a loner.

There are certain positive precautions any swimmer, especially a scuba fisherman, should take to avoid a barracuda attack: Do not annoy, or attempt to handle a barracuda, no matter how small it might appear to be underwater. In the presence of barracuda, back off as quietly as possible and, if feasible, get out of the water entirely. Finally, keep in mind that there is no hard-and-fast formula for dealing with barracuda; they are dangerous, unpredictable, and completely savage when triggered into action.

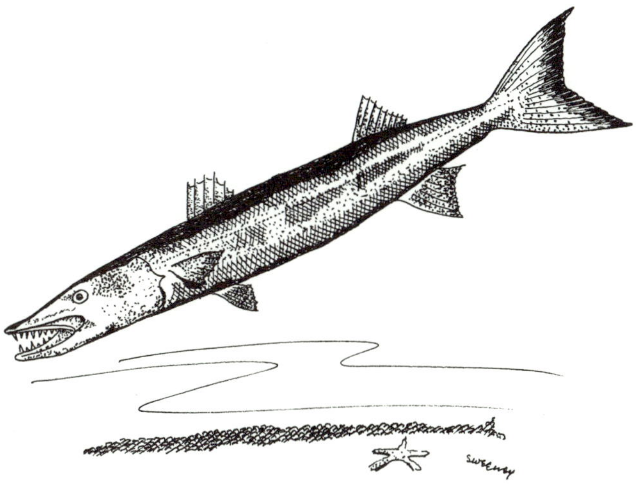

Coloration varies in barracuda. Generally, it is gray or dark green above and chalk white below, with black spots on the sides. Drawing by the author.

An Atlantic Ocean barracuda. Photo, Rex Gary Schmidt, United States Fish and Wildlife Service, Washington, D.C.

Barracuda are pikelike fish and are voracious eaters. Photo, Bureau of Commercial Fisheries, Washington, D.C.

Named Moby-Dick, *this baby albino killer whale is the world's only white killer whale.* Photo, Sealand of the Pacific, Victoria, British Columbia.

4
Killer Whales

Killer whales (*Orcinus orca*), sometimes called grampus whales, are not whales (neither are pilot whales, white whales, and *false* killer whales). All four are members of the family Delphinidae, which includes the cetaceans known as porpoises and dolphins. All these so-called whales either lack true whale characteristics or possess very positive porpoise characteristics.

The killer whale has neither whalebone nor a high dome or beak, which many true whales have. Delphinidae have only four or five pairs of ribs; whereas true whales have eight pairs. With the exception of the narwhal and the white whale, all Delphinidae have the first of their two neck vertebrae fused into one piece.

Size, in the killer whale, is somewhat confusing. A beautifully formed male, capable of striking with power, will grow to about 30 feet. His female counterpart will be no more than 15 feet long at maturity. Actually, man possesses little or no knowledge about many critical aspects of the life of the killer whale. It is thought that male and female killer whales pair off toward the end of our calendar year, that the gestation period is about twelve months, and that the offspring are 7 feet in length at birth.

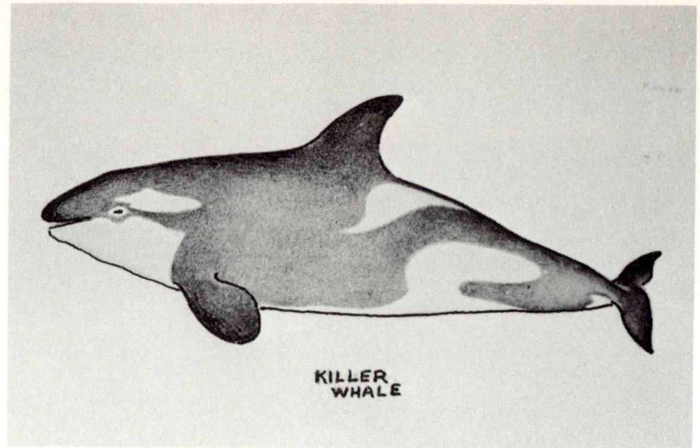

Killer whales are said to grow to about 30 feet in length. Drawing, Naval Research Laboratory, Washington, D.C.

Killer whales are found all over the world. They inhabit the Greenland Sea and the Bering Sea in the far north, and the Scotia Sea and Ross Sea in the far south. Because of an abundant supply of food in colder waters, killer whales are more densely distributed in those areas.

The killer whale is the dominant figure in a broad food triangle. (A *food triangle* is that process wherein a great volume of small marine life becomes the food source for fewer scores of larger marine animals, who in turn are devoured by a narrowing group of still-larger beasts, and so on.) Because a cetacean such as the killer whale is at the peak of the triangle and, therefore, is the terminus of a specified process, it is called *dominant*. The dominant force is not preyed upon regularly by other forces in the triangle. Hence, the killer whale is at the top of a food triangle in which great masses of plankton are devoured by tiny crustaceans, which in turn fall prey to small fish, which are then eaten by bigger fish and porpoises, only to be devoured by killer whales.

The killer whale has no peer; it fears nothing in its domain and has no qualms about attacking any other beast it makes contact with, even the true whale. Would it kill a man? Probably yes.

In the National Archives, Washington, D.C., there is a report by Captain Robert Falcon Scott, whose exploration vessel the *Terra Nova* was caught in the Antarctic ice, where he witnessed a remarkable killer whale attack on January 5, 1911. The photographer for the expedition, Mr. H. G. Ponting, had gone out on the surrounding ice to take pictures. A small pack of killer whales was seen attempting to knock him and several sled dogs into the water. They made this attempt by smashing their backs against the underside of the ice with tremendous force. The ice eventually exploded into fragments, but by then the party had scrambled to more solid ice. This famous and well-documented story has given rise to numerous exaggerated yarns about the attacking prowess of killer whales, but authenticated incidents are rare.

We consulted two famous Eskimo hunters about

Pictures taken through viewing port of an aquarium disprove the theory that killer whales will not live with other marine animals. Here, two former enemies, the killer whale and the porpoise, are shown living side by side. Photo, Vancouver Public Aquarium, Vancouver, British Columbia.

The mouth of a killer whale is formidable. There are ten to twelve strong, conical, pointed teeth on each side of the upper and lower jaws. Photo, Sea-Arama, Galveston, Texas.

Killer, or grampus, whales are not true whales; they are members of the same family as the porpoise. Drawing, United States Oceanographic Office, Suitland, Md.

KILLER WHALE (<u>ORCINUS</u> <u>ORCA</u>)
LENGTH – UP TO 30 FEET

the killer whale. These two venerable and proficient outdoorsmen, Johnnie Kakaruk and Billie Oquilluk, were both men of considerable experience. At mention of the killer whale, the two men began to talk excitedly.

"Eskimos paddle small boats called kayaks, which are small boats made of animal hide. Kayaks are fast, but can easily be knocked over, or made to turn turtle. Either in the presence of large sharks, or killer whales, they are not safe boats. Whenever such an animal is sighted, Eskimos get out of the water quickly. In case of shark, the attack is made with some caution; that is, a shark is more apt to circle a kayak before striking, or attempting to overturn the craft. Not so a pack of killer whales. They suddenly appear and immediately attack. So Eskimos try double hard to reach shore when killer whales are seen. The big sadness in all this is that young Eskimos show their smartness by learning, at a very early age, to handle kayak. They like to travel far and long on open sea. If caught by killer whale in open, they disappear."

However, neither of these two hardy Eskimo hunters had ever seen a killer whale attack a man or even a kayak. Nevertheless, they seemed surprised that anyone would doubt what to them was an accepted fact.

Orcinus orca is a beautifully streamlined animal that is fast, almost unbelievably strong, voracious, and highly maneuverable. Its flippers are unlike those of the dolphin in that they are rounded, not pointed. Most conspicuous are the enormous tail flukes and dorsal fin found in older males. The length of such appendages may be as much as one-fifth of the total body length. Because of the elongated dorsal fin, the killer whale has been mistakenly identified as a swordfish. The great blade cleaving the water looks like a sword. Because of this confusion with the true swordfish, there are many ancient stories about great whales being attacked by swarms of swordfish.

The body markings of the killer whale are one of the most identifiable aspects of the species. Generally, the belly is white, and the back is black. There appears to be no shading of these two colors, no gray areas, simply black and white. To the rear

Sixty-five blackfish deliberately stranded themselves on a South Carolina beach. Photo, William P. Baldwin, United States Fish and Wildlife Service, Washington, D.C.

of either eye is a round area of white that extends backward and a bit upward. To the rear of the dorsal fin is another white area that is rather saddle-shaped. The chin is generally white. The flippers are black on both sides and join the body where it, too, is black.

The killer whale is well armed, with from ten to twelve large, conical teeth that are strong and exceedingly hard. The upper and lower teeth interlock, affording the animal a tremendous bite.

The killer whale feeds on dolphins, seals, penguins, porpoises, and true whales. Examinations of stomach contents have revealed astonishing amounts of such food. The stomach of one 20-foot killer whale held the remains of twelve porpoises, fourteen seals, and six seabirds. This suggests that much of their aggressiveness is motivated by the driving pangs of hunger.

Killer whales seem to have a mass suicidal tendency, as some true whales do. There are innumerable recorded instances of entire schools of killer whales having dashed themselves upon a beach and died. In Britain, a close log of "killer whale strandings" has been kept since shortly after the turn of the present century. Conservationists have towed many of the still-living animals back into deep water, only to have them return to the beach. It has been suggested that in some way nature has given these animals a self-destructing tendency as a safeguard against overpopulation.

One of the most remarkable performances staged by killer whales is their attack and destruction of large true whales, such as a very substantial sperm whale. To accomplish this monumental task, killer whales move in and isolate a single animal from the herd. After completely surrounding the massive animal, the killer whales then harass their intended victim by constantly darting at it from every angle. Finally, one or more of the killer whales will venture to grab the sperm whale by the tail; this slows the tiring beast down even more and lays it open to further attacks. As exhaustion overcomes the sperm whale, its mouth will open and its tongue will hang out. It is this portion of a sperm whale that killer whales seek for food. They strike en masse and devour the entire head, leaving the less-desirable portions of the carcass for other predators, such as sharks and barracuda.

The conical, pointed teeth in each side of the upper and lower jaws identify this member of the killer whale-porpoise family as a blackfish (Globicephala brachyptera). *It stranded itself on the beach at Bull's Island, South Carolina.* Photo, William P. Baldwin, United States Fish and Wildlife Service, Washington, D.C.

Aboard a commercial fishing boat, the first thing to do with a captured swordfish is remove its sword. Photo, Bartlett Bureau of Commercial Fisheries.

5

Swordfish and Sailfish

Swordfish and sailfish are two different species of marine life. They differ both in physical characteristics and in behavior. Yet most people, including big game fishermen, consider these fish to be in the same category. This fallacy should be corrected. The swordfish (*Xiphias gladius*) is a member of the family Xiphiidae. It has been recorded by British authorities as being over 20 feet in length and weighing about 1,000 pounds. American records show that this species of fish (there is only one) runs to 16 feet in length, with a maximum weight of 1,100 pounds. These figures indicate that there is a good deal of uncertainty about this particular deep-sea inhabitant.

However, there are certain clearly established facts about the swordfish. He is one of the fastest fish in the sea, having been clocked at speeds in excess of 60 miles an hour. Also, he could well be classified as one of the most fearless of living marine animals. There are well-substantiated reports on record of swordfish having attacked divers, small boats, ships, and even submarines. No better-authenticated case can be cited than that of the swordfish which pierced the research submersible *Alvin*. At the time, the small three-man oceanographic vessel was being used by scientists from the Woods Hole Oceanographic Institution in Massachusetts.

A 1,000-pound blue marlin caught in Florida waters. Photo, Library of Congress, Washington, D.C.

There is the further well-substantiated incident of the research ship *Gerda,* which belonged to the Marine Laboratory of the University of Miami. On February 14 and 15, 1965, west of the island of Bimini, the ship was struck by a swordfish. Later, after the vessel had been dry-docked, a marlin spear was found broken off in the wood of the hull. The sword had pierced through 2½-inch planking, an oak frame 5 inches thick, and a 1½-inch-thick pine ceiling, for a total thickness of 9 inches.

These cases are not exceptions. The Coast Guard has records showing that small outboard motorboats have also been holed by swordfish. According to the Italian newspaper *Il Resto del Carlino,* a fisherman who had been squatting at the gunwale of his wooden boat died from a rectal wound inflicted by a swordfish that struck upward at the bottom of his vessel. The animal's sword tore up his large intestine, causing severe hemorrhaging. Shortly thereafter, fishermen began to use the iron lids of old potbellied stoves to sit upon.

In 1886, Captain F. D. Langsford, master of the fishing schooner *Venus,* out of Lanesville, Massachusetts, was pierced by a swordfish spear and died shortly thereafter. The International Oceanographic Institution, which has headquarters in Miami, Florida, has stated that swordfish attacks are not confined to men and boats. A 266-pound swordfish taken with rod and reel off Cuttyhunk, Massachusetts, had a 125-pound blue shark impaled on its sword.

In a London museum, there is a remarkable display that shows a 13½-inch-thick timber pierced by the blade of a swordfish. In another section of the same museum, a piece of ship's timber 22 inches thick is shown transfixed by a spear that is 5 inches in circumference.

In 1962, a Japanese commercial fishing ship weighing 40 tons, the *Genyo Maru,* radioed a Mayday (the international signal of distress) after being holed by a large swordfish.

Many people are curious to know why such a fish will attack something as large as a ship. There are several likely explanations. First, a dissection of the backbone reveals that the swordfish is capable of hitting an object at tremendous speeds and still being able to survive the massive impact. This means that in striking, they probably have a braking problem, for their rate of progress is far greater than that of a modern torpedo. Hence, if they miss their primary target, they may inadvertently hit another before being able to come to a complete halt. On the other hand, they possess such poor vision that they may hit at the hull of a ship, thinking it to be the underside of a whale. The fact remains, however, that they will purposely attack man.

Swordfish are generally found offshore, in deeper waters, but occasionally they come in close to feed and may be seen, often in pairs, lolling in an easy manner close to the surface. On the other hand, swordfish have struck at lines trolling as deep as 200 fathoms (1,200 feet). This could have happened because swordfish feed a good deal on squid, which live close to the bottom. They generally use their swords to hit smaller prey with a stunning blow, but they will not hesitate to use these swords as daggers if hard pressed by an enemy.

The general diet of a swordfish is made up of other fish, such as bonitos, flying fish, mackerel, and sardines. Swordfish have been observed striking with lightning speed amidst a shoal of smaller fish, lashing unmercifully with their swords and then settling back to eat the injured fish at a leisurely pace.

There is strong evidence that they are mortal enemies of the shark, and vice versa. Many swordfish remains have been found within the belly of a shark; yet, as already indicated, ample proof exists that swordfish spear large sharks.

Swordfish are known for their characteristic high leaps into the air when they have been hooked. They put up an enormous battle that can last for hours, and sometimes they themselves turn hunter in the process. Many a sport-fishing boat has been sunk by the crashing force of a weighty swordfish.

In New England and Canadian waters, fishermen still go after "baron broadbill" (as they call a swordfish) with a harpoon, a length of stout line, and a keg. Since swordfish like to sun themselves on the surface of a placid sea, commercial fishing boats are able to sneak up on them and sink harpoons in them. This is the beginning of a "Nantucket sleigh ride," in which the small boat hangs on to the line and is towed through the water. Eventually, the keg is dropped overboard, and the swordfish is left to fight against the drag of this as it sounds. When harpooned, a swordfish is inclined to shoot to the bottom and bury its

These remarkably accurate drawings were made in 1875. Top, *an adult swordfish.* Bottom, *a young swordfish.* Drawing, Bureau of Sport Fisheries and Wildlife, Washington, D.C.

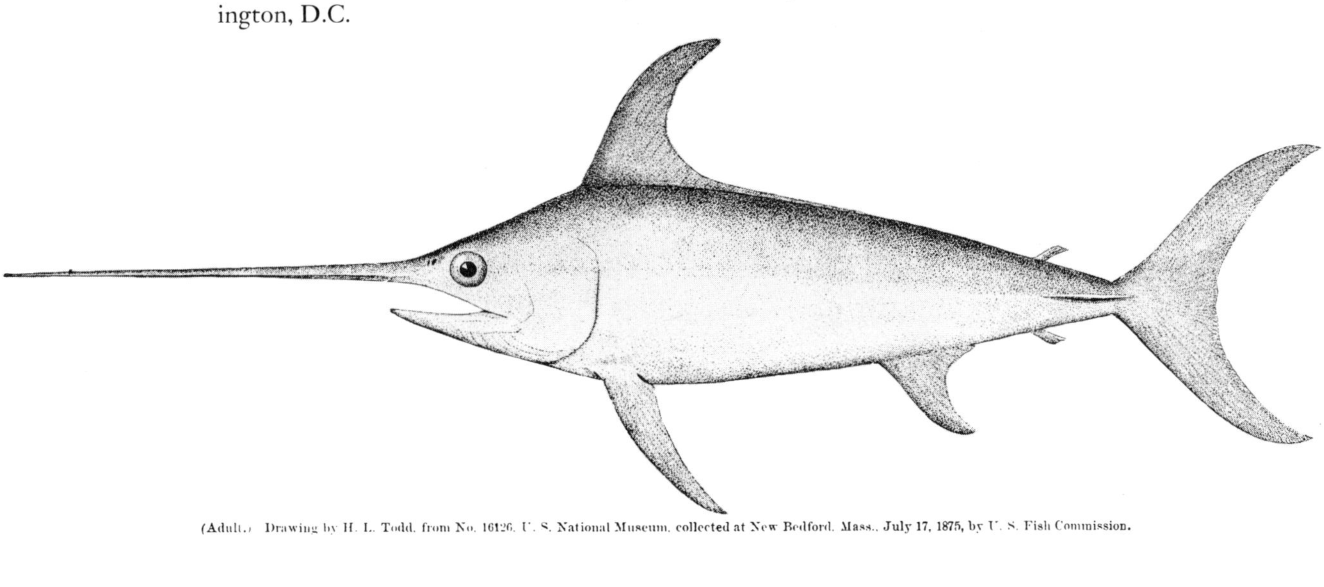

(Adult.) Drawing by H. L. Todd, from No. 16126, U. S. National Museum, collected at New Bedford, Mass., July 17, 1875, by U. S. Fish Commission.

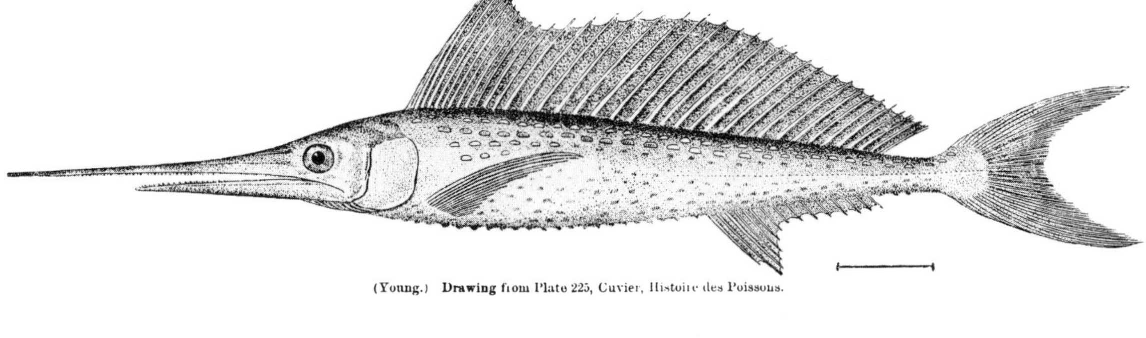

(Young.) **Drawing** from Plate 225, Cuvier, Histoire des Poissons.

A spearfish (Tetrapturus mitsukurii) *is capable of attaining speeds better than 50 miles an hour.* Drawing by the author.

sword, head, and a goodly portion of its body in the mud and sand. The relentless pull of the submerged keg eventually exhausts the great fish, and it is then hand-hauled aboard ship.

Sharks sometimes interfere in the process and will shred a swordfish to bits in a few minutes. "Blood attracts savage sharks," Captain Randy Heiter, of the sloop-rigged *Seahorse*, said, "and there've been days when we had to give up fishing, because shark'd strike as fast as we could sink an iron."

Swordfish vary in color from gunmetal to bronze, with a silvery underlay that becomes almost white at the bottom. Both the leathery fin and the extremely hard sword are blue black. The eye of a swordfish is distinctive, for it is about 3 inches in diameter, is set in a large bony cup, and is a vivid cobalt blue with a narrow rim of pure blue.

During summer months, swordfish work their way north in the wake of the tuna. In the early part of July, they reach their northernmost point, Cape Breton Island, and stay in this vicinity until mid-September.

Sailfish—and here we mean the entire family Istiophoridae, which includes marlin—are all sizable beasts that are capable of great speed and a tremendous punch. They are exquisite to behold, having a perfectly streamlined body completely covered by scales firmly set in the skin, a pointed cutwater that permits them to slip through water with a minimum of friction, and a pair of keels, at the junction of the tail fins, that assure perfect balance while maneuvering at high speeds. They have pelvic fins that can be folded upward into grooves when the fish is thrusting ahead under full power. It should therefore come as no surprise to anyone to learn that the sailfish has been rated by many authorities as the fastest fish in the sea, with a top acceleration of close to 70 miles an hour.

There are two varieties of marlin, the blue and the white, each differing from the other in several respects. Obviously, color is a distinctive factor; the white marlin has a white belly. Another difference is size; the blue measures up to 26 feet or more in length and weighs about 1,000 pounds; the white runs to only about 9 feet and weighs 160 pounds.

Sailfish are said to rival swordfish for speed. This is difficult either to prove or to disprove, for it is almost impossible to measure the maximum speed of any fish. Sailfish, because of their bodily configuration, adopt a unique position when moving at full thrust. They fold their dorsal fins into grooves in their backs; the short anal fins and the long, thin pelvic fins are also withdrawn into body grooves. This leaves the sailfish with a completely streamlined torso capable of sliding through the water with little or no resistance.

There is nothing more beautiful than the sight of sailfish, on a calm sunny day, at the surface, with their dorsal fins fully extended and glistening

This spearfish (family Istiophoridae) was taken in the Gulf Stream off the mid-Atlantic coast. All members of this family are generally large. They are found close to the surface, and they live in warm waters. Photo, United States Fish and Wildlife Service, Washington, D.C.

in the light. Fishermen in the Dominican Republic swear that the dorsal fin, when fully erect, is used by these fish as a sail to propel themselves over the surface. José Vaccaro, a fisherman out of Santo Domingo, swears that he has seen schools of them, scattered across a placid sea, moving as slowly as a widely spread fleet of drifting fishing boats.

Sailfish are valued as game fish because when hooked with line and tackle, they stage an acrobatic show that can be equaled by few other fish. Also, a sailfish makes a colorful and impressive trophy when mounted. In Florida waters alone, about 3,000 are caught each season by avid sportsmen from around the world. In earlier times, it was thought that sailfish, like swordfish, were entirely migratory. The Marine Laboratory of the

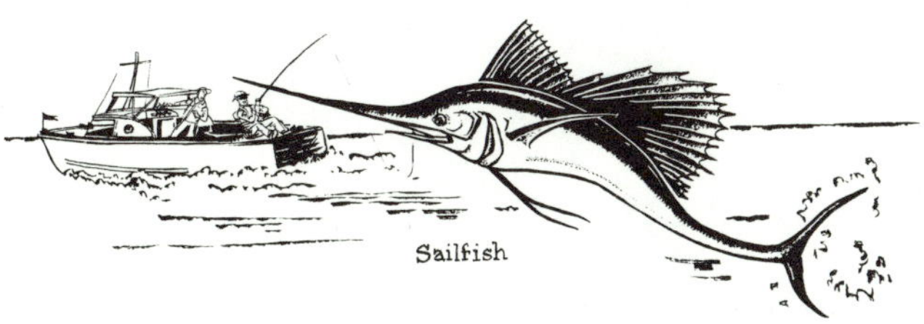

The true sailfish is distinguishable from other members of the family Istiophoridae by its single high dorsal fin. Drawing, Bureau of Sport Fisheries, Washington, D.C.

University of Miami says that this is not so; sailfish do not move out en masse, but are, to some degree, an all-year game fish.

In Florida, there is an annual Silver Sailfish Derby that was at one time known for the regular participation of Ernest Hemingway and that has contributed a great deal to the study and conservation of these fish. Gold buttons are presented to those who bring in a sailfish over 8 feet long alongside their boat and then release it on the spot.

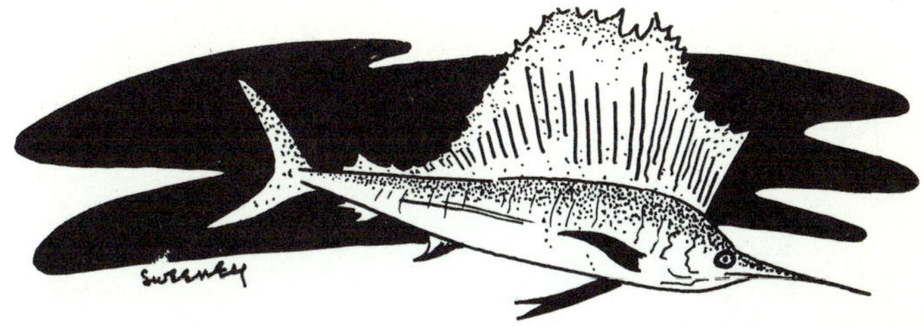

Atlantic sailfish (Istiophorus americanus) *are said to have been clocked at 60 miles per hour.* Drawing by the author.

This allows the fish to recover, grow to a greater length, and again be involved in a contest with some determined angler.

Although sailfish are best known for their ability to tail-walk when hooked (some have been known to cover 30 or more feet), their natural feeding habits are also of keen interest. They most often feed on small fish, commonly called *pilchards*, which they shepherd into tight schools by using their massive raised sails. When the school has been sufficiently impacted, the hungry sailfish darts through the tight group, slashing its bill from side to side. As stunned pilchards drop beneath the school, the sailfish will swim lower and eat the fish. It is said that sailfish develop a feeding frenzy during these pilchard raids and, in their excitement, have been known to leap into passing boats.

Tagging of sailfish by sportsmen, in conjunction with a University of Miami program, has been very successful; and much has been learned about the habits of these fish. Inventories of the stomach contents of captured specimens have revealed a high percentage of octopods and squids, as well as sea snails and sea robins, all of which live at the bottoms of reefs. It has been generally agreed, however, that sailfish will eat just about any live fish it can master with its bill or swallow after only one or two snaps of its powerful jaws.

The green moray has an ugly temper, but it allows a tiny neon goby to swim into its mouth and pick food particles out of its teeth. Photo, Miami Seaquarium, Miami, Florida.

6

Large Eels

This chapter is one of the most important in the book because it is a link between the sea monsters and sea serpents of the past and the marine animals known to exist in the present day. Eels may very well be considered the contemporary thong that draws the two parts of this book into a unified whole. Unfortunately, science knows abysmally little about the behavior of marine eels; therefore much is left to conjecture.

In the first chapter of this book, we noted that, in 1930, Dr. Anton Bruun had retrieved the larva of an eel 6 feet in length. We also explained how such a grub could mature into a specimen 100 or more feet long. Hence, a *sea serpent in reality!*

We know that such fast-moving surroundings as those offered by the continental shelves lend themselves to the spawning and migration of eels, a subject that has baffled man from as far back as the age of Aristotle. Aristotle noted in 350 B.C. that eels went to sea in order to spawn. It was long believed that horsehairs thrown into water were the reproductive source of eels. This theory enjoyed an extended lifetime because it was almost impossible to identify the reproductive organs, eggs, or larvae of the eel. In fact, even today, almost nothing is known about the mating and fertilization processes of some of the eels found in the great undersea swamp at the edge of the continental shelf.

It was not until 1904 that Johannes Schmidt, aboard the Danish research ship *Thor*, brought eel larvae, in their advanced state of metamorphosis, to the surface in a deep-sea trawl. Schmidt devoted the next several years to further intensive research on eels. Finally, he discovered and proved to his own satisfaction that grown eels gather in the Sargasso Sea, at a depth of about 100 fathoms (600 feet), lay their eggs, and die.

Eels are unlike the majority of other deep-sea fishes in that they are without a pelvic fin and their skin is either devoid of scales or possesses only slight traces of what might be scales. Moray and the conger are the best-known marine eels because of their size; they grow to 10 feet in length. The conger (*Conger oceanicus*) is found in waters that range from only a few feet deep up to about 500 feet deep. Like so many others of the eel family, it is thought that this species migrates to an area within the West Indies in order to spawn. After spawning, the adult dies, and the young pass through a leptocephalous state before finding their way back to that zone from which the parent came. When mature, the conger feeds on smaller fish, octopus, shrimp, and shellfish.

The moray eel is something else again. It has a genealogy all its own. Ancient Roman poets such as Horace (Quintus Horatius Flaccus, 65–8 B.C.) wrote that these animals were kept in specially constructed reservoirs on the outskirts of the city and fed the remains of dead slaves. Because of this kind of background, moray eels have won for themselves the fear and respect of fishermen all over the world. In actual fact, the term *moray* pertains to nearly 150 varieties of eels. The 10-foot-long green moray (*Muraena helena*) of the Mediterranean is perhaps the best known. It has deadly teeth, massive jaw muscles, and a very mean look. The sight of a green moray gliding across the surface of a coral outcropping is awesome.

Deep-sea divers tell numerous exciting stories about conger and moray eels. One of the most unusual was related by an English salvage diver a few years back. Mr. W. S. Johnson, an experienced deep-sea salvage diver, at one time an employee of Siebe, Gorman & Company, manufacturers of underwater gear, tells of his experience in this manner:

"I had a curious experience many years ago. I was repairing an old seawall when, to my horror, the head of a great fish suddenly darted out of a hole in the masonary, and tried to attack me. It missed me by only a few inches. Investigation showed that the fish was a huge conger which by some means had become trapped in the worn masonry and was unable to get much more than its head through the hole, for although it was observed many times afterwards to seize passing fish, its body was never seen. A full-grown conger, weighing hundreds of pounds, and measuring 8 or 9 feet in length, is an ugly customer to tackle even when hauled aboard. Once it gets hold of a man's arm or leg, the only thing to do is cut off its head, and even then its jaws will have to be pried open."

The green moray actually has gray skin; a mucous covering gives it the color of green, which is sometimes replaced by gray or brown. Morays have anal, caudal, and dorsal fins that are almost indistinct, being almost completely under the thick, leathery, scaleless skin. They do not have pelvic or pectoral fins, and the coloration in most species can range from green, black, or any other solid color to blotched, spotted, or striped. Distinctive colors generally indicate the specific areas from which moray eels stem. Black or green morays (*L. funebris*) are common in the Pacific near the tropical coasts of the Americas; the common spotted morays (*Gymnothorax moringa*) are to be found throughout the West Indies and in and around Brazil; and the large painted eels (*Enchelynassa bleekeri*) inhabit the South Sea area. Other species are just as specific in color and locale.

Moray eels are generally found coiled in and about offshore coral reefs. From this position, they seem best able to strike at anything passing within a distance of the length of their bodies. They are voracious eaters; most have a diet of live fish. However, many different species live on crustaceans. For this purpose, the moray has a mouth heavily studded with powerful teeth operated by massive jaw muscles. In some species, the teeth are

About 120 species of moray, or painted, eels exist in tropical and subtropical seas. Shown here is the green moray, which may attain a length of about 10 feet. Photo, Marineland of Florida.

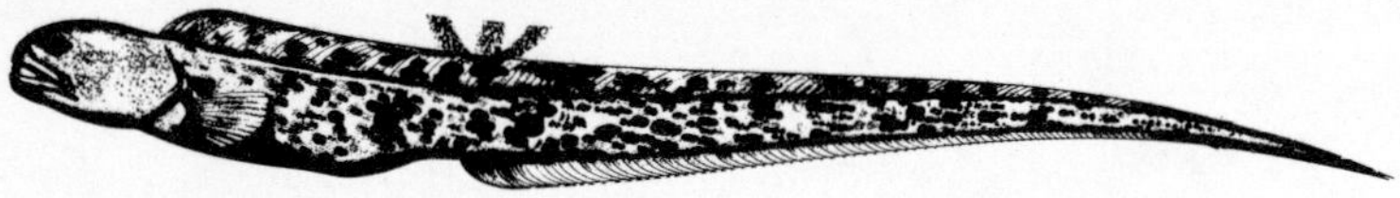

so thick within the jaws that the mouth cannot be properly closed, which tends to emphasize this beast's vicious appearance.

Although it has never assumed proportions comparable to frolicking with porpoises as a popular underwater pastime, the taming of moray eels can be a pleasant and rewarding endeavor. It is said that they make docile underwater pets, for they are actually very shy creatures. However, not too many people seem inclined to put the matter to a test, for a moray's looks belie any suggestion that they are timid. And there are many stories of unprovoked moray eel attacks on divers.

Waterfront reports have it that moray eels may be easily killed with a severe blow on the tail, but this is unsubstantiated. Moray eels are not recognized as a food product in this part of the world, although they are eaten in India, South America, and the Orient. The meat is rich, hard to digest, oily, and in certain species, quite tasty. However, it is said that other species have meat that is tough, tasteless, and poisonous. Apparently, such matters vary from species to species and from one area to another.

Conger eels stand in contrast with the moray eel in this respect, for they are heavily fished commercially in European waters. English records, dating as far back as the turn of the century, show that congers weighing almost 200 pounds were caught.

In the conger eels, a line runs the length of either side; in color they range from grayish to black, becoming lighter toward the underside of the belly. They have an air bladder and rather large pectoral fins that extend completely around the end of the tail. Normally, the conger grows to a length of no more than 8 feet. As is the case with most eels, little or nothing is known about their sex cycle. An interesting peculiarity about conger eels is the fact that the jaws, even when severed from the body, will continue to lock on to anything thrust into the mouth. This dangerous peculiarity is evident even after a considerable period of time. Fishermen use hook and line, baited with squid or fish chunks, to snare conger eels. Once inboard, the heads are chopped off and thrown overboard. "On more than one occasion," said Sandy Churchton, who skippers a bottom trawler out of Grimsby, England, "I've brought up a discarded eel head, only to find the teeth clamped shut on a small fish that had been eating at the remains."

The conger, like numerous other large eels, has a bulldog tenacity to its bite. Once they strike, it is sure to lock on. However, before taking a fisherman's hook, a conger will nibble endlessly at the bait. Finally, when assured that it won't fight back, it will clamp down hard and hold fast.

Mr. Churchton, when asked if he knew of any dangerous peculiarity inherent in large eels, immediately answered that he was impressed by one mannerism. "There're only a few fish that'll willfully go after a fisherman once boated; the moray and conger eels will do just that. Another fish with the same nasty habit is the mako shark.

"Once you got a big eel or a mako aboard, watch out; them devils will come after the nearest man even if he's half across a good-sized fishing boat.

"Most large fish brought inboard suffer from shock and kind of just flip and flop aimlessly. I've had them big eels leap out of the hold on my fish boat and instead of trying for the side, they try for one of my hands. Which ain't real fearsome until this happens in a heavy sea. An angry ten-foot mako or moray eel thrashing around on a slippery deck don't offer me no tea party."

Even when viewed in the murky underwater depths, the green moray appears threatening. Photo, Rex Gary Schmidt, United States Fish and Wildlife Service, Washington, D.C.

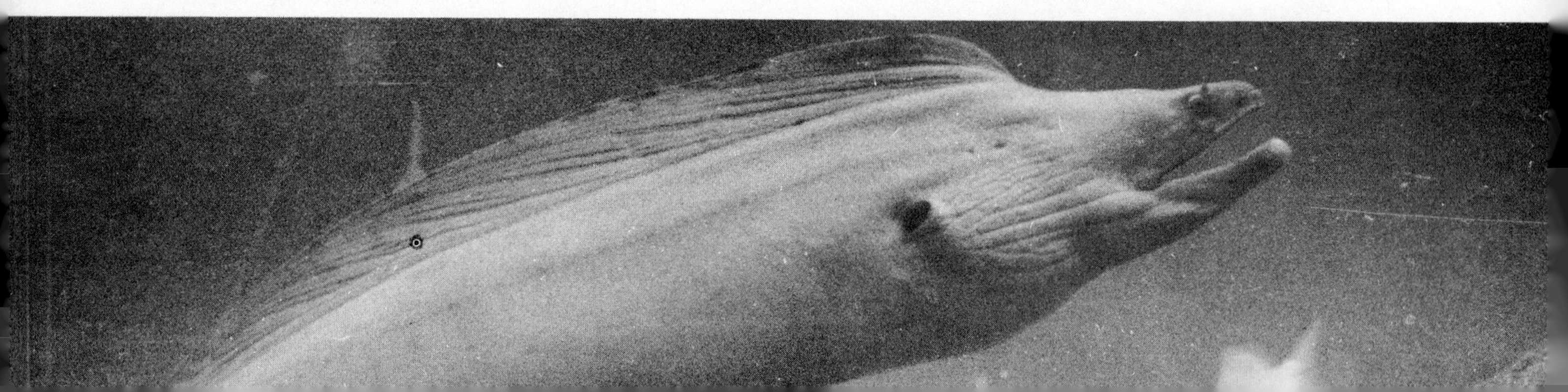

The bite of a moray eel, seen in this rare action picture, can be deadly. Photo, United States Naval Oceanographic Office, Suitland, Maryland.

Since there are so many kinds of large eels in the oceans of the world, each with its own variations, foibles, and peculiarities, it is well that we look at one particular species: the American eel (*Angilla rostrata*), which, although smaller than the moray, is better known to most fishermen.

The American eel is brownish yellow, with a dirty white underbelly. It is found in coastal streams from western Greenland to Central and South America. It has been said to grow over 5 feet in length, with 4 feet being the average. These eels migrate to a region between Bermuda and the Bahamas to mate and die. The eggs hatch into a ribbonlike, transparent form called a leptocephali. Over a period of about eleven months, this leptocephalid is carried to the eastern coast of the United States. Here it changes to a body shape that resembles the parent; but at this time, it is only about 3½ inches long, is still transparent, and has a peculiar row of black spots running the length of its body. The young, called *glass eels*, enter rivers and streams by the millions and gradually work their way upstream. It has been estimated that it is possible for them to take as long as twenty years to reach full maturity. When mature, the American eel leaves the coastal waters and starts the migration to Bermuda waters. During this process, the eel stops eating and changes its color to a black back with silvery sides; the eyes of the male seem to double in size. Following the mating period, about which almost nothing is known for certain, the eel dies.

Like most eels, this one is primarily a night feeder; it spends the daylight hours buried in sand or mud. It will eat almost anything, either alive or dead. It is itself excellent eating and presents some sport if taken on light tackle.

The specific area from which a moray originates gives the eel its distinctive colors. Photo, John G. Shedd Aquarium, Chicago.

This photograph of a large Atlantic octopus shows clearly the animal's muscular arms, pouch-shaped body, and large, highly developed eyes. Photo, Marineland of Florida.

7

Octopus and Giant Squid

Both the octopus and the giant squid are members of the class Cephalopoda, which includes a wide variety of divisions. Cephalopods (the name comes from a Greek word meaning "head foot") come in extremely large sizes (as previously noted in the first section of this book) and minute sizes. "Head foot," first thought to have been used by fishermen in prehistoric Greece, dates back to about 4000 B.C., the Neolithic period. The term is certainly appropriate and quite obvious because the octopus and the squid seem to have feet or arms connected directly to their heads.

Ironically, although one or another form of cephalopod existed 400 to 500 million years ago, and although cephalopods are presently among the most populous fish in the sea, there is very little factual information known about them. They have nevertheless been the subject of more imaginative writing than most other creatures of the deep. Homer wrote of the octopus in wildly descriptive terms. Pliny the Elder wrote about an octopus that repeatedly raided fishponds in Spain. Local fishermen set their dogs on the intruder, thereupon stirring up a great fight in which the cephalopod scourged the animals with the ends of its tentacles.

In 1555, Olaus Magnus authored a vivid descrip-

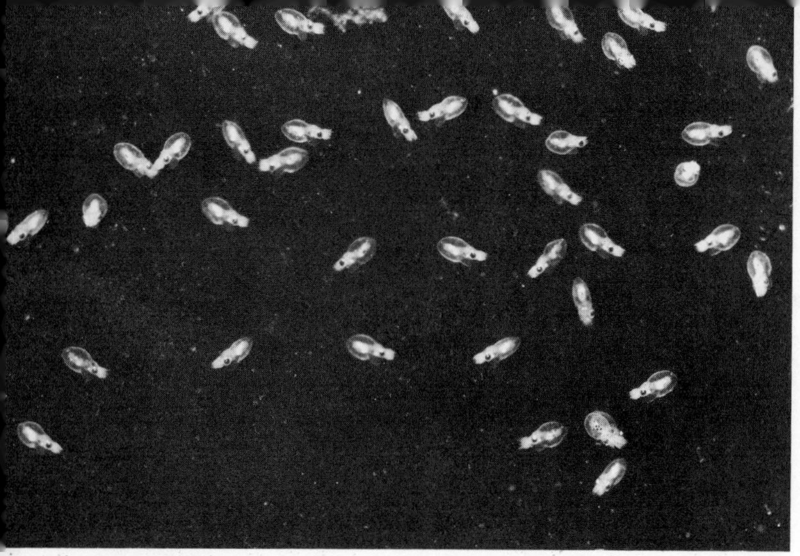

An octopus may lay thousands of eggs. The embryos, shown here, are often devoured by predators. Photo, Marineland of Florida.

tion of something horribly like an octopus or supergiant squid. Herman Melville, in his masterpiece *Moby-Dick*, first published in 1851, told of the whaling ship *Pequod* coming upon what could only have been a massive octopus. And that wildly inventive French naturalist, Pierre Denys de Montfort, who was head of the Museum of Natural History in Paris, told of a mass of squidlike creatures that came to the surface of the sea and sank ten men-of-war in one night. All this can be topped off with the exciting story *Twenty Thousand Leagues Under the Sea,* written by Jules Verne in 1870, which tells of a great battle between a giant cephalopod and the submarine *Nautilus.*

Cephalopods may be divided into two general categories: the octopus and the squid. The common octopus (*Octopus vulgaris*) is an underwater-breathing type, having eight arms that appear to flare directly off the head. Its body resembles an inflated sack in which are set two highly developed eyes that are both large and expressive. Each of the eight arms of an octopus is equipped with suckers that are used to grab its prey; a horny beak (sometimes likened to a parrot's) bites and injects a poisonous secretion. Octopuses move by forcibly expelling water through a funnel or siphon and also by pulling themselves along with their eight arms. Frequently, they will build underwater barricades of large stones in which to live; most, however, manage to find themselves caves

Octopuses belong to the class Cephalopoda, which also includes the squid, nautilus, and cuttlefish. Photo, Bob Noble, Marineland of the Pacific.

A female octopus is extremely dangerous when guarding her eggs. In this photograph, the eggs can be seen hanging from an overhead rock. Photo, Marineland of Florida.

in which to set up housekeeping. Octopus eggs are enclosed in tough capsules and attached to an overhead rock, where the female guards them carefully.

Again, it should be made clear that in the octopus family there are so many subfamilies and so many of any one genus that only an expert can tell them apart. Therefore, we must confine our discussion to generalities. The paramount feature we are interested in here is the danger the octopus presents to man, if indeed it is a hazard to humans. Here are two pieces of evidence to verify the contention that they are a danger.

Some years ago, a story was recorded in the United States Navy Main Library concerning a salvage diver named Lee Jones. Apparently this man was an offshore diver with better than ten years of experience in working on the bottom. On this particular occasion, Jones had only been on the bottom about ten minutes when he felt something tugging on his lifelines. Looking upward, he received a frightening shock. In the gloom of the ocean's floor, he could discern a massive dark glob crawling down his air hose. It took him some time to realize that the creature coming slowly toward him was what he had heard called *octopus.* Never having seen one before, and possessing no prior knowledge of how to deal with such a situation, Jones simply backed into a corner of the shipwreck on which he had been working to sweat out developments. In those days, there was no such thing as an underwater telephone; consequently, a diver was entirely on his own once he went beneath the surface of the water. In this instance, Jones could not signal topside because the octopus had by now fouled his lines with its long tentacles.

To the diver, watching in horror as the massive beast crept closer and closer, it seemed as if the octopus "tried every spot in vain for the purpose of sucking juice out of the thick air hose that it grasped." Then an arm slowly crept around the

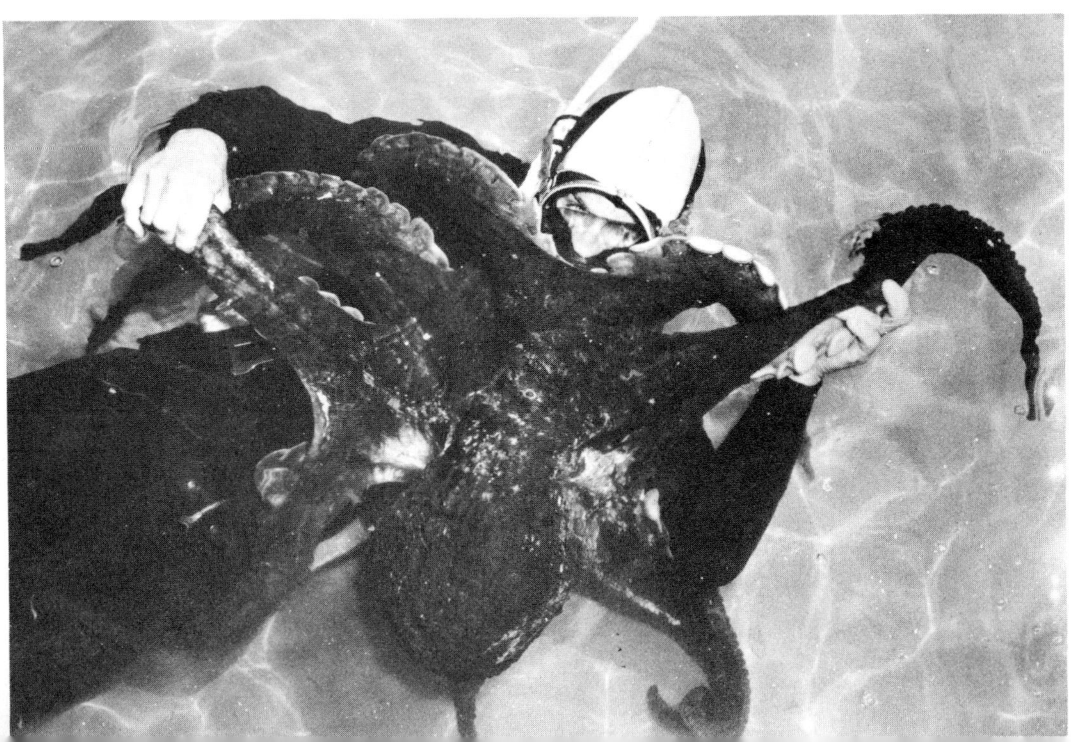

A diver shows how to grasp an octopus. Photo, N. W. Marineland, Inc., Seattle, Washington.

The eye of an octopus is one of the most highly evolved found in lower animals. Physically, it is like that of a man. Photo, Marineland of Florida.

diver's shoulders, and another gave an exploratory tug at his heavily leaded feet. That was too much for Jones. Inside the diver's helmet, below the glass port, on the right side, there was a chin button for emergency use in releasing air. Jones hit this button with his chin and a stream of air bubbled violently upward. The octopus took off, and Jones signaled to be pulled topside.

Once on the deck of the steam launch being used as a diver's tender, Jones examined his air hose and gave this report: "My one-inch vulcanized rubber tube, reinforced by numerous layers of tightly knit cotton, had been slashed at intervals by what looked like knife cuts. The cuts only ran as deep as the outside layer of rubber, but numbered 58 along a distance of over 23 feet."

This story emphasizes the dangerous aspects of any octopus; it is possessed of an overpowering sense of curiosity, and it has the ability to bite deep, which in many cases involves a venom.

This immediately raises the question: How dangerous is the bite of an octopus? The Commonwealth Serum Laboratory in Melbourne, Australia, records an instance where a pretty little blue-ringed octopus, no bigger than a silver dollar, bit a young soldier and within one hour the man was dead. Dr. Robert Bowman, a marine biologist at Brisbane University, said: "An increasing number of such octopuses are being reported from Caloundra (approximately 100 miles north of Brisbane) to the Gold Coast, an area to the south of the state capital."

These two stories indicate that the octopus is indeed a dangerous creature.

Man has used the octopus for food since the earliest recorded history. Aristotle (384–322 B.C.), the Greek philosopher, historian, and writer, tells of the octopus being fished commercially for food. He was the first man to relate the myth that octopus can pry clams apart by placing stones between the valves. Although no such thing is true, modern scientists have discovered a remarkable trait. *Octopus vulgaris* learn easily and retain what they have been taught, but they have only a limited repertory when it comes to manipulating objects. They cannot, for example, do what Aristotle claimed for them: pry open an object, such as a tin can with a lid or a transparent bait box with a hinged top. They can, however, squeeze through any opening that is large enough to accommodate its parrotlike beak, since that is the only bony part of its body. This means that an octopus of considerable dimension can disappear through a baseball-sized hole in a tank. Furthermore, they can climb, run, jump, and fight on dry land. They also possess the ability to squirt a

The eight arms of the octopus were once a single foot. Photo, Marineland of the Pacific.

cloud of ink into the water when attacked. Using this cloud as a shield, they can rapidly change color, becoming translucent, and make an escape while almost totally invisible. Octopuses generally move backward, but they can drive themselves in any direction should the need arise.

There is one current fact about the octopus that is worthy of note. Recently, an infestation of octopuses off the eastern coast of Australia has become a menace to the once-lucrative lobster fisheries that abound in that area. Frantic research has revealed that shark-meshing operations, established along bathing beaches as a means of protecting swimmers, have killed so many predators of the octopus that young cephalopods are now prospering in dangerous numbers. Once again we see how man, in tampering with the delicate balances established by nature, has tipped the scales in the wrong direction.

Now let us look at the squid. Here, as with the octopus, we find a number of species, sizes, and shapes. It is size that makes the giant squid a definite menace to man. The following incident involving a giant squid has been investigated and reported by the Seamen's Church Institute of New York.

During April, 1875, three Irish fishermen were in a small boat off the Emerald Isle. Having shot their lines, they had little to do other than bide their time and observe what was going on around them. One of the men pointed out the oddity of a flock of gulls hovering over a large mass floating on the surface of the ocean. More out of curiosity than anything else, the men rowed over to the area and were astonished to find a giant squid basking on the surface.

After a consultation, the men decided the squid would be a valuable prize. They attacked the great beast by severing one of its long arms. At once, the giant squid roared into action. It thrashed about and sprayed the area with water and black dye. Carefully, the fishermen timed their numerous approaches on the beast, each time chopping off one more of its long tentacles. For several hours, they kept after the beast, until finally the determined fishermen were able to close in for a kill. They cut off a portion of the head, only to have the rest of the body sink like a stone. They

This fully developed octopus is a dwarf octopus (Octopus joubini). Photo, Miami Seaquarium, Miami, Florida.

proudly returned to their native village, where their catch created a considerable stir. The local police sergeant, a man named Thomas O'Connor, carefully preserved the pieces of the giant squid and sent them along to the museum of the Royal Dublin Society, where it was calculated that each tentacle measured 15 or more feet in length and 15 inches in diameter at the base. There was no way to figure the size of the head, so that the overall length could only be estimated at between 50 and 60 feet.

The common cuttlefish (*Sepia officinalis*) and squid are called *decapods*, which means "ten-footed," because they possess two more arms than the octopus. This pair of attachments is quite remarkable in concept. They are kept well concealed until needed to seize a prey, at which time the appendages flash out. The aim of a squid is extremely fast and very precise.

One of the most remarkable features of the common squid is its ability to change colors. It can vary from one hue to another instantly and seems to employ this ability in many ways. Very

Squid are jet-propelled. They take in water to obtain oxygen and drive it out through a siphon for locomotion. Photo, Wometco Miami Seaquarium, Miami, Florida.

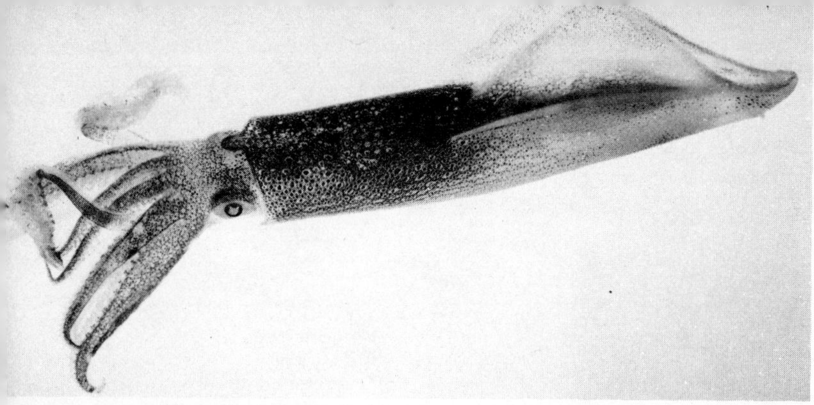

All varieties of squid have certain features in common: the horny plate, the mantle that extends into fins at one end, and the sucker-bearing arms at the other extremity. Photo, Marineland of Florida.

appropriately, red seems to indicate anger; when these decapods are faced with an enemy, they turn red. Annoyance is indicated by a series of rapidly changing colors. Further, the squid, like the octopus, can eject a great squirt of inklike substance with which to cloud the water. Behind this self-made cloud, it will either hide or make its escape by changing color to produce the effect of invisibility.

The squid family, including the giant squid, has a pair of highly developed eyes that have a human look about them. The squid's ability to glare at an unwanted intruder, such as a diver, can be very disconcerting. A deep-sea salvage diver, a member of Siebe, Gorman & Company's staff, had a terrifying experience some years ago while working on the coast of Plataea, Greece. The location of his descent was close to a precipitous cliff with an almost-sheer drop of 100 feet and was very cavernous at the bottom. The sea was wonderfully clear, and never before had he worked underwater in such beautiful surroundings. The bed, for the most part a firm yellow sand, was relieved here and there by rocks 10 to 15 feet high, which stood sentinellike at intervals as far as the eye could see, the more remote ones taking on a ghostly aspect as they gradually disappeared in the watery haze. Submarine weeds as tall as himself and taller were waving all around him, and the floor was carpeted with beautiful varicolored growths. The whole scene was lit by the rays of a powerful sun that, streaming through the water, glorified everything.

Siebe, Gorman & Company reported Mr. Burwood's experience in this fashion. "In and out of the caves were swimming fish of gorgeous colours, and the diver was so interested in these that he followed some of them into one of the caves. In groping about in the dimmer light his hand came momentarily in contact with something which he described as being like an elephant's trunk, but far larger—a rubber-like slimy thing that heaved at his touch and sickened him. This feeling was succeeded by an instinctive warning that he was in the presence of an enemy to be feared. His fears were intensified by the appearance, directed full upon him, of a pair of the most horrible eyes that could be imagined—eyes which seemed to concentrate in their gaze everything malignant and hateful. The lids were alternately raised and lowered, curtain-like, halfway over the eyes, imparting to the latter a still more evil expression. The diver felt that he was becoming, as it were, hypnotized, and only a movement on the part of the dread creature, which was evidently preparing to attack, recalled him to a full sense of his danger. Still facing the monster he retreated cautiously from the cave and, regaining the open water, immediately signalled that he was coming up. Preferring to risk breaking the decompression rules to an unequal encounter with a giant squid —for such it was—he inflated his diving dress and shot straight to the surface without a stop. Fortunately, no harm came to him, owing to the fact that the whole adventure from the time he descended till his return to the surface, lasted only a very few minutes."

The giant squid resides at great depths, and its arch enemy is the sperm whale (*Physeter catodon*), which feeds on the animal. Frequently, captured sperm whales are found to carry scars from encounters with giant squid. The suckers of these squid are toothed-rimmed, as a result of which they leave massive scars. Such scars have been found to measure 3 inches across, which would indicate a squid of perhaps 50 or more feet in length.

The giant squid has a long history of verified

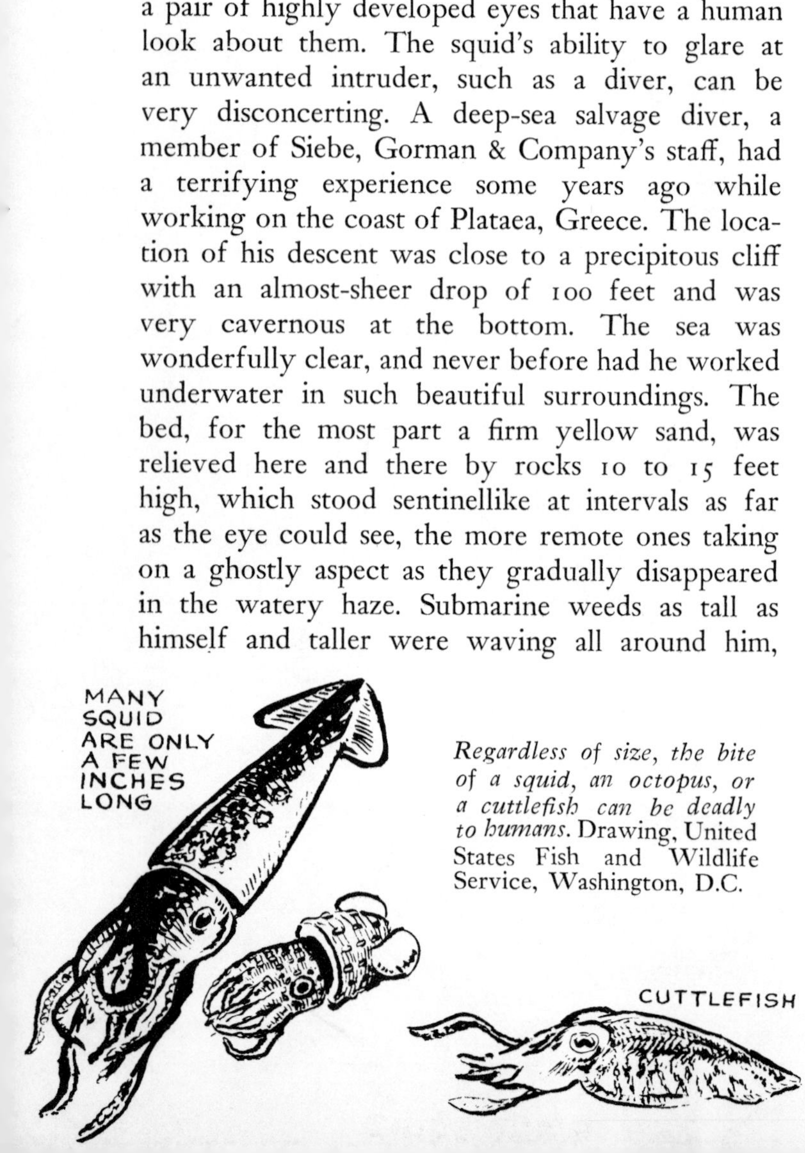

Regardless of size, the bite of a squid, an octopus, or a cuttlefish can be deadly to humans. Drawing, United States Fish and Wildlife Service, Washington, D.C.

Squid are shown here swarming together for mating purposes. Photo, Marineland of the Pacific.

appearances. In 1847, the north countries of Europe experienced the first of a number of giant squid happenings when one was washed ashore in Zeeland. In 1854, another was stranded at the Skaw; and in 1861, the same thing happened on one of the Shetland Islands. It was also in 1861 that the now-famous sighting by the French corvette *Alecton* occurred. On November 30, a giant squid was sighted swimming on the surface of the ocean 120 miles off the coast of the Canary Islands. After cannonading the creature without results, the ship's commander, Lieutenant Frederic-Marie Bouyer, gave the order to capture the giant squid. The crew harpooned it once and got a noose over its tail. A sudden lurch of the ship tore all this loose, and the creature sank, leaving them with only a portion of the tail fin. From what was salvaged, it was estimated that the creature had indeed been a giant squid of enormous size.

A number of giant squid incidents took place along the shores of Newfoundland. Several giant squid were washed ashore; some of them attacked boats. The *Canadian Illustrated News*, on October 27, 1877, carried a complete story on one of these occasions. In part, the story related that a heavy gale had swept across the shores of Saint John's, Newfoundland, and a giant squid became stranded, in an exhausted condition, at Catalina, on the northern shore of Trinity Bay. "The tail had got fast on a rock," the newspaper reported. The article went on to say: "In its desperate efforts to escape, the ten arms darted about in all directions, lashing the water into foam, the thirty-foot tentacles in particular making lively play as it shot them out and endeavored to get a purchase with their powerful suckers, so as to drag itself into deep water. It was only when it became exhausted and the tide receded that the fishermen ventured to approach it. It died soon after the ebb of the tide, which left it high and dry on the beach."

The South American Humboldt Current squid (*Ommastrephes gigas*) frequently weighs 300 pounds or more. Its arms reach to 12 or 15 feet in length. These squid are known to be extremely dangerous when hooked and quite risky to bring inboard if taken from a small boat. Unquestionably, should the opportunity present itself, these mammoth-sized squid will bite, wreck small craft, or make off with a human.

Two squid are shown here joined in sexual union. Photo, Marineland of the Pacific.

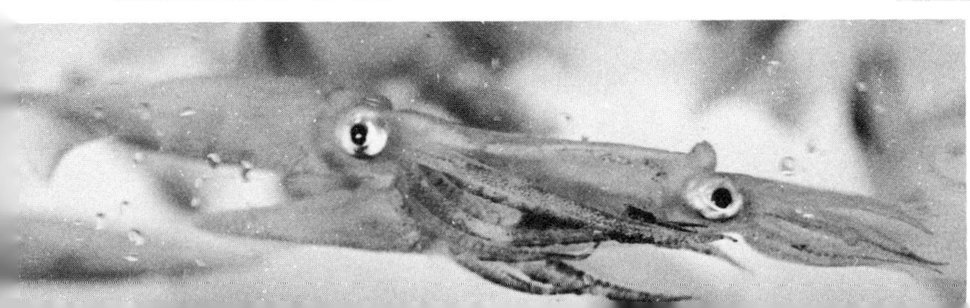

This sketch showing the parrotlike beak of a squid (upper left) *was drawn over a hundred years ago.* Drawing, Bureau of Commercial Fisheries, Washington, D.C.

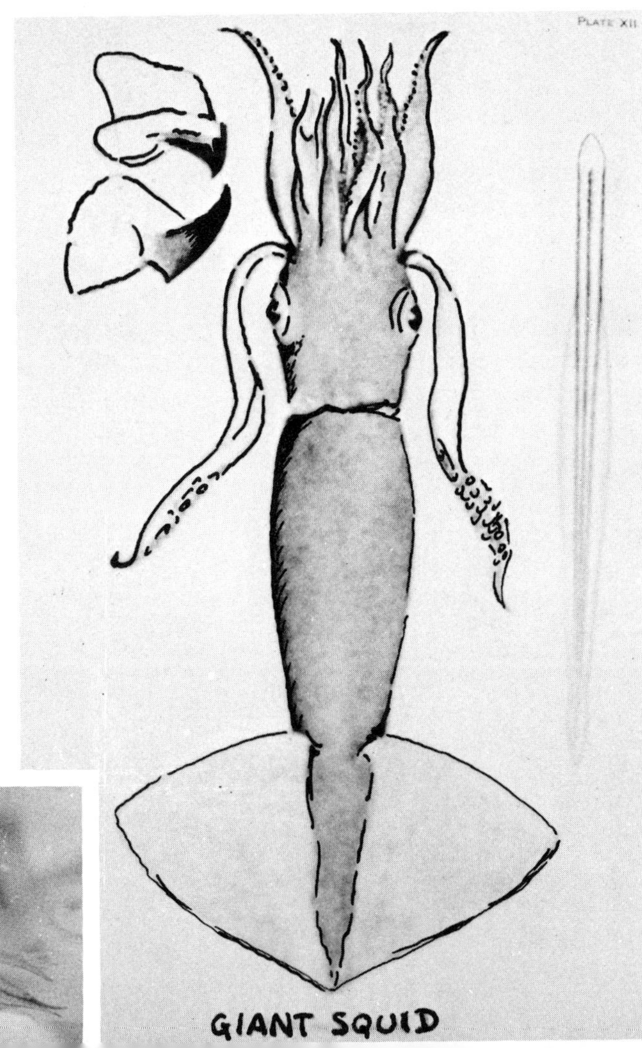

A record-sized Mola mola. Photo, Library of Congress, Washington, D.C.

SECTION TWO

Dangerous Through Size

8

Mola

Mola mola (family Molidae; genus *Mola*), or the ocean sunfish, as it is commonly called, is another massive fish about which very little is known. There are two general varieties of mola: the round-tailed class and the sharp-tailed class. The only difference between these two varieties is the shape of the tail. In general configuration and weight, it would seem that they are the same. When seen in the water, they look alike, although an expert could probably detect a slight difference in the shape and positioning of the fins.

They are flat, disklike fish that swim, not in a horizontal position as rays do, but in a vertical manner. They are dark gray, with a positive silvery sheen, that tapers into a dusky white at the bottom. They may be found floating or swimming gently on the surface of almost any temperate or tropical ocean. It has been established that an 8-foot ocean sunfish will weigh about 1,800 pounds. Maurice Burton, science editor for the *Illustrated London News*, has stated that these ocean sunfish are to be found weighing in the neighborhood of 2,000 pounds and measuring at least 10 feet in length. It is known that they have a spinal cord no longer than 1 inch and that they are very sluggish and completely stupid. They drift with the warm currents. Ocean sunfish are believed to live close to the surface, going into the depths only when driven to do so by hunger. They are capable of feeding at depths of close to 1,000 feet and subsist mainly on eel larvae, crustaceans, small fish, and jellyfish.

They have no known commercial or food value. Upon occasion, commercial divers have turned in reports of annoyances created by these massive fish. Because they have small mouths, they do not present any danger except through weight. When jabbed with a gaff, they are inclined to leave a diver alone. About the only thing startling that can be said concerning the ocean sunfish is that it groans when harpooned and brought alongside a ship.

The flesh is generally white, exceedingly tough, with a great amount of elasticity; it has an unpleasant smell. This ability to stretch or bounce is caused by a high proportion of elastic fiber, interlaced with muscle and fat. During the days of sail ships, mariners told of seeing natives fashion balls out of the elastic flesh with which they would play games that depended on its bouncing characteristics. At night, the *Mola mola* displays an amazing amount of phosphorescence, which has earned it the name *moonfish* and has often caused sailors to claim that they have at last seen a true sea monster.

The *Mola mola*, when viewed on the surface, is frequently noted to be asleep. When swimming, so it has been reported, it will quite often turn round and round like a wheel. It has also been reported by natives to be found drifting with its head and eyes just above the surface.

Native fishermen say that the liver is fat and heavy with oil. The oil is of a low grade and suitable for rough commercial purposes only.

Scientists say that there is probably no other fish alive that is so full of parasites as the *Mola mola*, for it is massively infested with tiny animals, both internally and externally. The flesh and internal organs swarm with entozoa, and the skin is literally covered with thousands of crustacean parasites.

Several native tribes throughout the tropical world have been known to call *Mola mola* "head fish." This, they say, is because the giant fish appears to be nothing more than the forward part of a massive marine animal.

Generally, these massive oddities are found singly, far out in open water; however, there have been recorded instances of schooling, probably during certain seasons, for mating purposes. Such periods create a rare sight for those fortunate enough to happen upon the scene. These great round beasts are inclined to lie almost fully on one side in order to sun themselves more effectively and are almost indifferent to the presence of man. Eight, ten, or a dozen such mounds basking in the sun is an impressive sight. One account, written during the early 1900s, describes such an occasion.

"Some half-dozen of these great fish were sighted lolling about within one square mile, the largest showing as much above water as the side of a husky bull.

"We decided to take one with iron and lowered a boat manned by four oarsmen. A harpoon was driven at the beast a dozen times, but came loose after each try. The animal, while apparently aware that it was being attacked, failed to display sufficient intelligence to sink, which is a feat they do very well upon occasion. After repeated trys, a hook was finally snagged in the gill cover and the sun fish was towed back to the ship. There, with block and tackle, it was raised to the deck for dissection."

The story continues by saying that the flesh of this ocean sunfish was undersheathed by a layer of tough, grisly flesh that was almost a half-foot thick. Obviously, it was this depth of elastic material that made the fish impervious to strikes by a harpooner. The writer also noted that rifle fire had been employed on another of these beasts but had obviously failed to penetrate its tough inner hide.

On opening the animal, its stomach was found to contain parts of silver ling, which is a deepwater fish. Consequently, it was again surmised that this variety of sunfish does feed at depths exceeding 600 feet.

As a danger, the *Mola mola* cannot be discounted, for it is a slow but powerful swimmer. With so much weight and a great store of strength, it can be a deadly threat under certain circumstances. For instance, one was accidentally caught in a deep trawl net and had to be swung aboard ship in order to be freed and cast back into the sea. The powerful creature suddenly began to show amazing liveliness and took to thrashing about the deck. Before it could be pushed overboard, it had torn a sturdy iron capstan out of the deck, flattened 30 feet of bulwark plating, and ripped loose a goodly section of the starboard sheer strake.

Very little is known about the mating habits and the day-by-day existence of the *Mola mola.* Many captured specimens have contained gigantic masses of tiny eggs; more than 5 million were counted in one female. When hatched, each egg would yield only the barest splinter of a fish, which, because of the raw, open element in which it finds itself, has very little chance of survival.

The designation *sunfish* should not be confused with numerous varieties of freshwater game fish of the same name, which are relatively small and in no way related to *Mola mola* in size or growth. The true name of *Mola mola* is Latin and means "millstone," which is a fair description of its great girth.

The largest mola ever weighed, at least with any degree of accuracy, was reported by a member of the Game Fish Committee of the International Oceanographic Foundation. One such animal was stranded on the beach at Comodoro Rivadavia, Argentina. It weighed 2,640 pounds, and the report describes it as a fish of slow movement ordinarily, "but [it] can leap completely out of the water despite its peculiar shape and gigantic size."

The mola have been seen trailing great numbers of jellyfish, upon which they were obviously feeding as the tide took both along at a leisurely pace. Sea lions suddenly appeared upon the scene and began to throw the great fish across the water. The mola, because of their massive weight, were stunned when they crashed against the surface of the ocean. Instantly, sea gulls swooped in to peck out the eyes of the inert fish, which, upon regaining consciousness, showed obvious signs of dying. Smaller fish, crabs, and swarms of shrimplike animals closed in and eventually devoured the tremendous creatures.

An interesting undersea activity has been observed in connection with the mola. In certain sections of the sea, there are "cleaning stations," where large fish go to be cleansed of barnacles and parasites by swarms of small fish and tiny shrimp. This small marine life eats away all foreign bodies around the trunk, within the gill covers, and even inside the mouth of the mola. When this great animal arrives at a cleaning station, it stops swimming, tilts at a peculiar angle, and appears to give certain signals with its fins. This performance on the part of the *Mola mola* seems to indicate that it is ready to be serviced, for almost at once, a number of small fish move in to chew away at the parasites. The process appears to involve a trancelike condition that lasts for a given period of time. At the end of this time, the little fish and shrimp abruptly dart away; the ocean sunfish flutters back into motion and swims leisurely on its way.

Mola mola *are also called sunfish.* Drawing by the author.

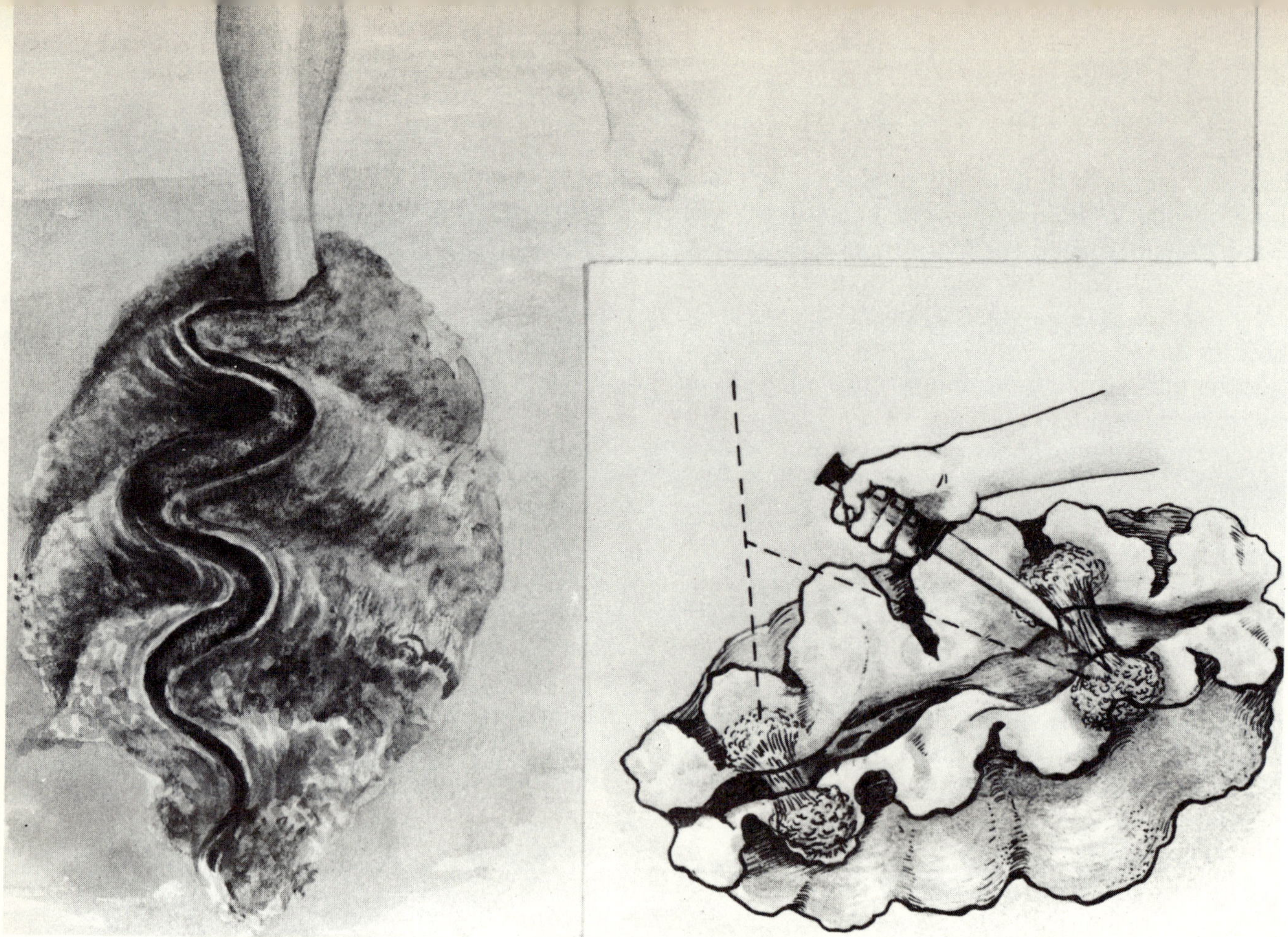

Giant clams (Tridacna gigas) *are said to trap divers who accidentally step into their open shells.* Drawing, United States Naval Oceanographic Office, Suitland, Md.

9

Giant Clam

The giant clam (*Tridacna gigas*) is said to be one of the world's largest living mollusks. A mollusk is an invertebrate animal possessing several distinct characteristics: a soft, unsegmented body, a protective shell that it secretes itself; a muscular foot; gills; and a strip of horny material covered with teeth, called a *radula,* that is used for feeding itself.

The shells of giant clams have been said to measure as much as 6 feet across the lips, with an entire weight of as much as 500 or more pounds. They are found in Australian waters and in certain portions of the Indian Ocean, notably in pearling grounds, where they are a source of danger to divers. In a booklet entitled *Science and The Sea,* the Maritime Safety Division of the Naval Oceanographic Office says that this mollusk, if prodded or stepped on while open, will close with a vicelike grip. Divers and others have been reported trapped by the giant clam. Release can be gained by inserting a knife between the valves and severing the large transverse muscles that hold the two halves together.

A beautiful pair of shells from a giant clam is used as ceremonial fixtures in the church of St. Sulpice in Paris, France. They are said to have been a gift from the Republic of Venice to Francis I in the late 1800s.

One startling thing about these giant animals is that they are also part plant. This combination of animal life and plant life is called *symbiosis*, which means that there is a dependent relationship that is beneficial to both participants. Generally speaking, the plant flourishes within the animal host's skin and is thus able to capitalize on the carbon dioxide given off by the animal; the animal host then receives the benefit of oxygen produced by the plant. In the case of the giant clam, symbiosis is intensified by the fact that the large mollusk has in its skin lenses that itensify sunlight; thereby the growth and spread of plant life are speeded up, which then causes a greater supply of oxygen.

It should be noted that these giant clams are quite common in their home environment. Although they are seldom seen by average people, their history dates back many centuries to Indian Ocean pearl divers who are reported to have been trapped by these monsters. One ancient scroll put it in this fashion: "Whereof he knew not whence the danger cometh, a limb was crushed within the vice of a shelled mischief-maker and thereupon he suffocated with surrounding waters."

Despite exhaustive research in this matter, not one death attributable to a man being trapped by a giant clam has been authenticated. In all fairness, it should be first acknowledged that any pearl diver trapped by a giant clam has little chance of returning alive to tell of the event. Nevertheless, these massive beasts are not able to snap their clam shells closed, as is so often depicted in horror stories. They first must force out a considerable volume of sea water as the two massive shells are brought together. Although this takes only a few seconds, it is sufficient time for a diver to withdraw a foot or hand.

In the National Library in Bombay, there is an artifact of an ancient bit of mythology about the giant clam. A human hand was said to have been found within one great clam shell; it had become completely coated with mother-of-pearl. This relic, although extraordinarily beautiful, was believed to possess terrible powers of evil.

Native fishermen, in order to fill orders for tourists who desire a trophy, capture a selected great clam through a simple ruse. They jab a pole deep into the meaty folds of a giant clam and wait those few moments it takes for the shells to draw together. The pole is then fastened to an overhead block and tackle, hoisting the specimen aboard.

A fur seal in its natural Alaskan habitat. Photo, Library of Congress, Washington, D.C.

10

Walrus, Sea Lions, Seals, and Sea Elephants

The walrus, sea lion, seal, and sea elephant are all members of the carnivorous aquatic suborder Pinnipedia. Pinnipeds have modified flippers. Within this suborder are many families, such as bladdernose seal, the Baikal seal, and the bearded seal. Because their limbs are directed backward, pinnipeds make slow, laborious progress on land. But in water pinnipeds are excellent swimmers, possessing great speed and agility.

Because of their immense size, walrus are hunted by Eskimos for their large hides, their vast amounts of edible flesh, and their copious supply of oil. However, it is to the Alaskan fur seal that Eskimo hunters look for clothing.

Sea elephants, sea lions, and seals constitute a sizable number of pinnipeds; in world population, it has been estimated that there are no less than 25 million individuals. All are known to be meat eaters; some include limited amounts of seaweed in their diets. Their food generally consists of fish, octopus, squid, shellfish, seabirds, plankton, smaller seals, and other marine animals that might provide meat. The most surprising thing about pinnipeds, which are found from pole to pole, is that only two are classified as being extremely dangerous to man: the walrus and the leopard seal.

The walrus is a gregarious aquatic mammal that

is found only in Arctic waters. Males frequently weigh well over a ton and measure up to 15 feet in length. The tough, thick, well-wrinkled hide has little or no hair and has been found to make excellent boat covering, tent shelters, and mukluks (boots). Walrus, both male and female, possess downcurved tusks that grow as long as 36 inches and are used to dig up such food as crustaceans and mollusks. These tusks, which are smaller in the female, are also used as weapons. At one time, they were hunted ruthlessly for their ivory and oil, which activities gave rise to fantastic stories about man's conflict with these wild beasts. Although game laws now protect the walrus, Eskimos are still permitted to hunt them for food under certain restrictions. One story, possibly originating 1,200 years ago, was recited to this author by a famous Eskimo chief named Ti-yuk, tells of a fierce battle between a walrus and the greatest hunter of all times, a native named A-puk-een-a. This brave young Eskimo, in the dead of winter, would go far out to sea in his tiny kayak in order to find a walrus or a whale that he could harpoon and bring back to his village for food. During numerous severe winters, he kept the entire population from starving and was therefore honored as a great hunter. One winter's day, while out hunting, he harpooned a walrus and a mighty battle ensued. As the battle raged hour after hour, the birds and the whales and the sharks all gathered about to watch. At last the young hunter won the contest and fastened the supposedly dead walrus to the back of his kayak. As they neared shore the walrus suddenly regained its strength and killed the hunter by stabbing him with one of its great tusks. A seabird went to the village and told the people what had happened. The Eskimo men quickly got into a sturdy umiak (boat) and caught the walrus, which they then brought back to the village, where the women skinned it alive.

Such an act was a terrible insult to the supreme god of all sea animals, who immediately caused a great flood to engulf the Eskimo village.

"Waters," Ti-yuk described the situation, "go higher and higher; all low place used by village covered by water. No more village. That big punishment never go away and leave only two hills."

Pointing solemnly across the water, Ti-yuk said, "That one now called Big Diomede Island and that one over there Little Diomede Island."

The big, blubberous walrus works for his food at depths as great as 300 feet and requires at least 100 pounds per day in order to live. Conservationists have faced a massive task when dealing with the walrus: its gigantic weight. A 12- to 15-foot bull, may weigh well over 3,000 pounds, and it is no small task, working on the fringe of an ice floe, out of a small boat, without block and tackle, to capture and dissect such an animal.

The beautiful ivory tusks of both male and female walrus are worth about $200 apiece when decorated. The largest might be 3 feet in length, weigh about 26 pounds, and sell for approximately $3 a pound on the black market "raw" (undecorated). Hunters simply slaughter these great wonders of nature, chop out the tusks, and leave the rest of the carcass to spoil. If the female is killed, her pup is left to starve to death.

There are two distinct species of walrus: the Atlantic walrus and the Pacific walrus. The Atlantic walrus abounds in the Arctic, near Russian shores and the Laptev Sea, southward to Greenland and Spitsbergen, and even as far south as the northern reaches of Canada. In the summer, the walrus inhabit the open waters of the Arctic; but as winter begins to close in, they move slowly southward. The early Pilgrims reported seeing walrus as far south as Massachusetts.

The second species is the Pacific walrus, which is found in waters adjacent to Russia and Alaska.

A Pacific walrus. Drawing, United States Naval Oceanographic Office, Suitland, Md.

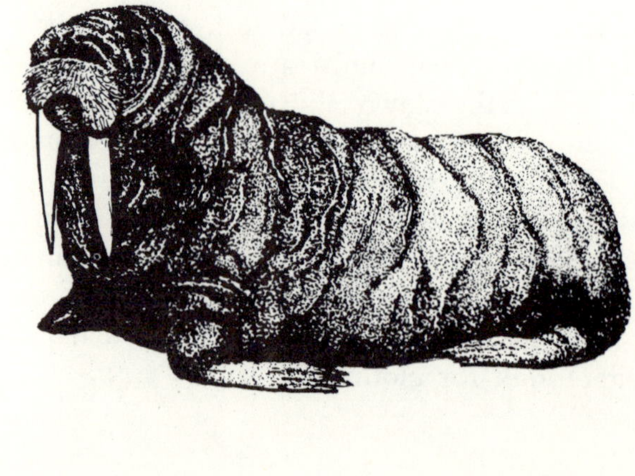

Walrus in their natural habitat. Photo, Field Museum of Natural History, Chicago, Illinois.

The mighty animal lives on ice floes that drift southward into the Bering Sea in the fall of each year. They then come as far south as the Pribilof Islands, the Aleutian Islands, and even around to the Pacific shores of Canada. In spring, the journey is reversed.

Fossilized portions of skull bones and tusks indicate that these animals populated this same area at least 50,000 years ago. Because the walrus is nomadic and tends to live in a harsh, cold environment, his life cycle has been very difficult to study. A great deal of the information presently available has come from Eskimo hunters. These hunters claim that mating occurs in the spring; after a thirteen-month gestation period, the pups are born during April, May, and June. However, it has been observed that some few may be born out of season.

Pups weigh up to 150 pounds, depending on sex, and have the ability to swim immediately. A nursing period of one to two years follows, and the young remain with the mother for several months following weaning. The males become sexually mature at five years of age, with the females maturing a slight bit earlier. Full-grown bulls examined on the Pribilof Islands weighed up to 3,500 pounds. Females are generally estimated to weigh about two-thirds as much as the males.

One of the more fascinating features of the gigantic beast is its large food requirement. Nursing youngsters also eat solids; thus each beast represents a massive consumption of food. One scientific expedition reported that a six-month-old calf consumed 3 or 4 pounds of milk each day, plus about 20 pounds of solids. A young 1,200-pound walrus was estimated as downing at least 60 pounds of food per day, with the average increasing to a minimum of 100 pounds as the animal grew in size. We must assume that the bigger they become, the more food they consume.

As the walrus uses its tusks to plow up the bottom of the sea in search of food, protruding pads on either side of the muzzle channel the food into the mouth. These pads, which serve as lips, are covered with tough bristles that undoubtedly aid in breaking loose bivalve mollusks, worms, and crustaceans from their grip on the floor of the ocean. Walrus also occasionally eat seals, and these bristly mouth pads apparently assist in tearing chunks of meat out of the victim.

Grazing habits have been said to vary widely. According to Eskimo hunters, of seventy-one bull walrus killed, over three-quarters had empty stomachs, and the remainder had packed in about 100 pounds of food. There is no explanation that can be offered for this fasting-gorging cycle.

Because of its great bulk, the walrus has few enemies worthy of concern. Killer whales (*Orcinus orca*), however, are a definite threat. The killer whales catch the walrus on the bottom while they are feeding or when they are swimming out in the open and slaughter them there. Eskimos say, that, from time to time, the killer whales will attack a herd of walrus in well-organized schools. Apparently, the whales first take the fight out of a walrus by tossing it into the air; when it crashes down against the surface of the water, it is stunned. These Eskimo hunters say that the killer whales somehow manage to throw even the largest walrus a considerable distance, which requires the tossing of solid bulk that weighs over 2,000 pounds. On one such bloody occasion, the walrus were so stricken with fear that they all piled up on an ice floe; more than two hundred died by being either smothered or crushed.

The question that baffles science is: How can the walrus manage to plunge so speedily to great depths and still live? Many assumptions have been made, for example a slowing of heartbeats and the curtailment of blood flow to certain organs of the body. These contentions have yet to be proved.

The word *walrus* comes from ancient Norse hunters, who, in searching the Atlantic for the great animal, came to refer to them as *valross*, which means "whale horse."

Leopard seals (*Hydrurga leptonyx*) are the other pinnipeds that are a specific danger to man. Large male specimens grow to a length of over 10 feet and a weight of about 600 pounds. This particular breed is found in areas adjacent to either pole, the southernmost tip of South America, Australia, New Zealand, and South Africa; a few are scattered throughout isolated areas of the South Atlantic, Pacific, and Indian oceans. This powerfully aggressive beast actually shuns mankind but will attack when inadvertently brought face to face with a tormentor.

A northern sea lion. Drawing, United States Naval Oceanographic Office, Suitland, Md.

Although the females are not as testy as the males, they are bigger in both length and weight. A large-sized female will measure 12 feet long and weigh half a ton. Leopard seals have pointed snouts and incurved teeth, which are useful in biting and shaking prey into smaller chunks that can be more easily swallowed.

Leopard seals subsist on penguins, seabirds, fish, and smaller seals. These seals present a danger to commercial fishermen, who occasionally find them enmeshed in their gear. To untangle the angry animal is to come within the reach of dangerous fangs and crushing weight. Also, leopard seals have been known to attack small fishing boats when they think there is danger to their pups.

The leopard seal is a quarrelsome beast that lives within a murderous fraternity. On solid ice or ground it is prey to more agile earthborne beasts such as polar bears or packs of sled dogs. If it seeks to find safety in the ocean, it immediately becomes subject to attack by sharks and killer whales. The sex life of the leopard seal is also a source of constant combat because it is based on the harem system. This means that the male must battle to reach a position of competitive authority. Once this has been accomplished, he must then fight to win himself a group of females, following which he is required to stay on the alert against theft. It is a cruel existence.

California sea lions. Drawing, United States Fish and Wildlife Service, Washington, D.C.

As a result of these combined hostile factors, those few humans, mostly Eskimos, who inadvertently come in contact with leopard seals experience a good deal of difficulty in handling them. Much of what is known about these fascinating animals comes to us from the Eskimos. Other men rarely have any extended opportunity to observe their behavior firsthand.

All pinnipeds pose a threat to mankind. It is not because of their extreme weight (walrus), nor because of their caninelike fangs (leopard seal), but because of the depletion the entire Pinnipedia suborder cause in our fishing resources.

The Canadian seal and the Scottish gray seal are known to be the carriers of a parasitic roundworm called *nematode,* which can be transmitted to many commercially valuable fish. Although the resultant worms do not infect man, they nevertheless destroy fillets obtained from cod and salmon. Fishermen argue that if the seal population were reduced, the number of infectious worms would be lessened, thereby cutting down on spoiled fillets.

In addition, seals unquestionably deplete available stocks of fish. They are known to make heavy inroads on black cod, halibut, and salmon. There is no established rule of thumb on how much Pinnipedia consume in any given period, since they vary in weight from almost 3,000 pounds down to 50 pounds, and each, of course, devours an amount commensurate with its size.

An example of this wide weight variation is the difference between the Steller's sea lion (named for George W. Steller, the man who "discovered" the

Steller's sea lions. Photo, Bureau of Sport Fisheries and Wildlife, Washington, D.C.

The southern elephant seal is the largest of all seals. Its proboscis can be inflated to a length of 2 feet. Photo, Field Museum of Natural History, Chicago, Illinois.

Weddell seals (Leptonychotes weddelli). Photo, Field Museum of Natural History, Chicago, Illinois.

Harbor seals (Phoca vitulina). Photo, Field Museum of Natural History, Chicago, Illinois.

Northern fur seals (Callorhinus ursinus). Photo, Field Museum of Natural History, Chicago, Illinois.

A monk seal with her pups. Photo, Bureau of Sport Fisheries and Wildlife, Washington, D.C.

animal in 1741) and the Canadian seal. The Steller's sea lion may grow to a length of 13 feet and attain a weight over 3,000 pounds. A Canadian, or harbor, seal reaches a maximum length of only 6 feet and weighs no more than 250 pounds. However, seals are traveling animals; consequently, they consume a wide variety of fishes that are sought by numerous methods of commercial fishing. The Steller's sea lion ranges from the Bering Sea, which is bounded by Russia and Alaska, south to the lower tip of California. The Canadian seal covers an area from the Bering Sea as far south as Mexico. Either or both of these pinnipeds consume vast amounts of shellfish, squid, and fish.

Statistics compiled by the University of Washington reveal some alarming facts. In a study on sea lions and their salmon consumption, which was mathematically computed to be one-fifth of an animal's body weight per day, it was determined that these predators ate twenty-two times the weight of all salmon caught by fishermen in central and western Alaska during 1958.

The same source said that Canadian seals probably eat 2.5 percent of the annual salmon catch and

A herd of fur seals. Photo, Bureau of Sport Fisheries and Wildlife, Washington, D.C.

4 percent of the annual herring catch. Nor is such devastation limited to Pacific seals; Scottish fishermen say that in their neighboring waters, at least 100,000 tons of commercially salable fish are devoured every year.

It has been suggested that Pinnipedia be fully commercialized by developing latent markets for their many valuable parts. Orientals are said to use seal whiskers, which average from 12 to 14 inches in length, to clean opium pipes. These same people use the sex organs of seals as an aphrodisiac and value the liver as something to be mixed and consumed with alcohol. Eskimos and Indians say that seal meat tastes good and that the hides unquestionably are useful for their leather. From 1951 to 1958, 36,000 seals were killed in Alaska. The farming of seals might readily produce immense numbers and a correspondingly large market for their meat, hides and other parts.

Elephant seals (*Mirounga angustirostris*) are among the largest mammals in the sea; bulls sometimes weigh over 5,000 pounds, grow to over 16 feet in length, and have a girth (circumference) of about 18 feet. Until very recent years, these ponderous seals were thought to be extinct. Then a small colony was discovered on an isolated Mexican island. Protective laws were enacted and herds were carefully nurtured. Today these tremendous animals are found on the Channel Islands off Santa Barbara and along the shores of Lower California in Mexico. In winter, they can be found on Vancouver Island, in British Columbia; a few have been noted as far north as southern Alaska.

Elephant seals come by their title naturally, for they possess a nose, or more properly a proboscis, that verges on being elephantine. This appendage has fascinated scientists, for it apparently serves a series of important functions in the life of the bull sea elephant. The female, who does not have this tubular attachment, must be lured by the remarkably loud noises produced by her 2-ton swain, for blowing trumpetlike calls through his nose is one of the more important aspects of the mating behavior of a bull sea elephant.

The miniature-sized elephant trunk hangs over the muzzle of the beast like a limp, deflated sack. When the great sea elephant yawns, which he does frequently while sunning himself on some quiet beach, this pouch of a nose almost falls into the great open mouth. The proboscis has a downward curve, so when the sea elephant inflates the appendage, opens its mouth, and then snorts, the resultant impact is like firing a cannon into a steel barrel. Audio tests show that the blast of a sea elephant can be heard for a mile or more. And herein appears to be one of its functions: to frighten away enemies before they dare to close for an attack.

The sea elephant is a harem seal. The bull gathers for himself all the females he can adequately protect. The mating season extends for a period of about six weeks, during which time the male does not eat and stands constant guard against intruders. However, poor eyesight plagues the bull sea elephant; his eyes are better suited for seeing underwater than on dry land. As a result, hunters can easily move in on the herd by crawling on hands and knees. However, when the hunter stands erect, thunder breaks loose; the bull instantly recognizes the presence of an enemy and blasts a challenge through his trunklike nose. If the hunter drops to his knees again, the roaring ceases.

Bull sea elephants stage one of the world's most spectacular competitive bouts when engaged in collecting a harem. They begin by trumpeting challenges at one another through their noses, which, in most instances, appears to settle the matter. Inevitably, this bluff is not always successful, and the two bulls smash together in a massive force of ripping, barking, frothing, wild-eyed blubber. The ground literally trembles at the impact. In a few moments, the battle is over, and the vanquished waddles off with strips of raw hide hanging free. No fatalities have ever been reported in such affairs of honor, although ghastly wounds are obviously inflicted.

From what has been observed, these massive sea elephants have specific offshore burial locations where they go to die quietly when their time has come. What triggers a signal that death is at hand, or at what age this comes about, no one seems to know for sure. However, certain of these sites have been found, and fangs and other bony parts carpet the floor of the ocean. Some have been there for so many thousands of years that they have turned to stone.

11

Manatee, Dugong, and Sea Cow

The manatee, also known as a sea cow, is a heavy-bodied mammal ranging in length from about 9 to 15 feet. There are three species known to man: one lives in waters around Florida, Central America, and the West Indies; the second, in the rivers of South America; and the third, in the coastal lagoons of West Africa.

Manatees have a distinct body makeup. The torso is sausage-shaped, without legs, but with a pair of weak flippers near the head, and a tail that is round and flat. They are herbivorous, aquatic, and toothless. The upper lip of a manatee is cleft, so that it forms two thick pads, each studded with coarse, tough hairs. With this appendage it can root out, crop, and gather vast amounts of seaweed for necessary fodder. Because of its elephantlike hide, excellent oil, and edible flesh, the manatee has been hunted almost to extinction; but it is now protected in Florida by strict laws.

The adults of the species propel themselves underwater by means of powerful strokes of their tails; the young, however, swim with their flippers. In color, all are dark gray to black. Manatees frequently enter gentle streams and are valued for their work in clearing away underwater weeds in canals that might otherwise become choked with growth.

The young manatees are born underwater but

are immediately brought to the surface by the mother. Normally, a pup will nurse underwater for about eighteen months. The American manatee exists in Atlantic coastal regions from North Carolina to the upper portions of South America. In recent years, manatees have found a haven in the Florida Everglades and even within the city limits of Miami.

The manatee of South America, or more specifically of the Amazon River basin, is slightly smaller than its American counterpart and lives in fresh water only.

The African species of manatee is a bit larger than the American variety and is said to be more aggressive; it is therefore more dangerous.

Although all varieties of manatee have been actively hunted around the world by professional fishermen and Indians, little is known about the animal. It lives underwater, amidst seaweed, and shows only a small portion of its head for brief periods. The manatee, when it ceases all motion, sinks slowly to the bottom. Therefore, it is not easy to observe when it is nesting or in heavy sleep. As a consequence, much of its life cycle is lost to those who wish to study it. These same qualities, however, make it easy to stalk if the hunter is patient and equipped with proper weapons.

One interesting aspect is that certain manatees carry numerous scars, yet they have rarely been seen in a fight. These scars are in the form of gouged-out portions of the body, chopped tails, or body lacerations. Officials in the U.S. Bureau of Sport Fisheries and Wildlife say that one of the great threats to manatee populations is the speedboat. The animal is difficult to see when it is near the surface and slow to react to oncoming craft, as a consequence of which they are frequently run over and hit by high-speed propellers. These body scars could be the result of such collisions. On the other hand, the markings seem to be far too numerous to have come only from occasional watercraft. Therefore, some scientists surmise that they are the results of encounters with larger marine life, such as the sawfish or the octopus.

It has been surmised by some observers that manatees seem to gather with the coming of darkness because, in the early hours of morning, they appear on the surface in numbers, closely packed, to luxuriate in the sun.

Little has been accurately observed about the sex life of the manatee, other than the fact they do nuzzle one another affectionately in a form of courtship. They have been noted coming together with a kisslike embrace that brings them muzzle to muzzle.

A manatee.

The manatee, for all its gentleness and harmlessness, can be a danger because of size. A mother, nursing a young pup, is capable of fetching a vicious swipe of its broad tail. Or a sleeping manatee, gently resting on the bottom, if suddenly disturbed by a scuba diver, is quite likely to deliver a deadly butt.

The dugong, which is also known as a sea cow, differs only a little from the manatee. It, too, is a herbivorous aquatic mammal, but it resides along the shores of the Red Sea and the Indian Ocean. Dugongs are of the same order as the manatee (Sirenia) and are thought to be the source of mermaid stories that once abounded in these areas.

The dugong has a brownish or grayish body that ranges up to 10 feet in length, flexible front flippers, and a two-lobed tail.

Dugongs are not amphibious; they belong entirely to the sea. They possess thick, elephantlike hide that is used for making footgear. When cured, the hide generally becomes black, wrinkled, and extremely tough. The meat, when properly cooked, is highly palatable, although it tends to be tough. The oil is of a good, smokeless quality. The fat, which underlies an inch-thick hide, frequently runs to 9 inches in depth. Natives attribute some marvelous qualities to this buttery material. They say it is delicious as a spread, will not easily turn rancid, and possesses an agreeable odor. It has been likened in taste to sweet almonds, but for those not adjusted to it as a food, it acts as a gentle purgative. Because of these many valuable features, man has mercilessly hunted the dugong, usually in a small boat or canoe. However, the dugong, when finally realizing that the herd is being attacked, will occasionally attempt to upset a hunter's boat with smashing tail blows.

A strong bond of affection seems to exist between pairs of dugongs, for if one animal is annoyed, attacked, or killed by man, it will not be abandoned by another. As a result of this characteristic, after the first dugong is captured, the second is easy prey.

It has been noted that the dugong is infested with a tiny insect resembling a louse. This minute bug abounds in the natural crevices of the hide, under the flippers, udders, and anus. When these insects succeed in boring through the hide, wartlike protuberances result. It appears that sea gulls consider these mites a delicacy, for they have been seen perching aboard partially submerged dugongs and busily pecking at the hide.

Native fishermen admit that one way of locating the dugong for killing is by observing the actions of sea gulls. As a result of this assistance and the ease with which the beast can be killed, the dugongs, like the manatees, are on the verge of extinction.

Because sea cows (either manatees or dugongs), do not exist in European waters, their recorded history is relatively recent. They were first noted by Columbus. Later Ponce de León mentioned killing some that he referred to as "sea wolves." Mayan Indians are known to have used the hides of manatees for covering on the canoes they built, and even today some South American Indians value drippings from the eyes of sea cows as an aphrodisiac.

A sea cow. Drawing by the author.

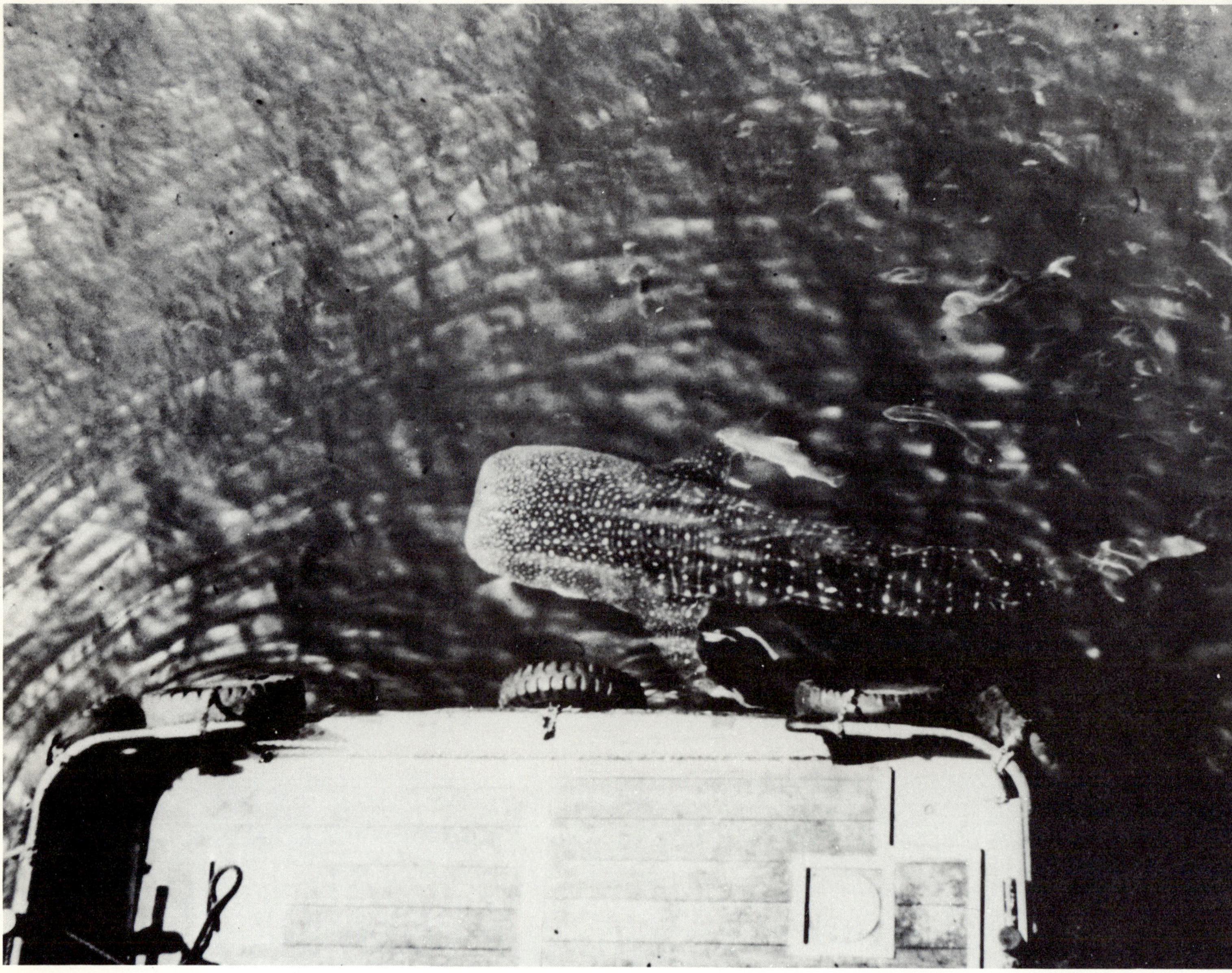

First picture ever to show a whale shark accompanied by smaller sharks. This photograph was taken from the top of an oil rig. Photo, Interspark Magazine, Atlantic Richfield Company.

12

Whale Shark and Basking Shark

The whale shark, genus *Rhincodon,* is the largest shark alive today. It is known to reach a length of 60 feet or more and attains a weight of well over 2 tons. It has a wide variety of names in as many different countries; in India it is the mhor; in the Philippines, chacon; and in Panama, tintoreva. One of the most amazing facts about this gargantua of the seas is that it feeds on tiny planktonic creatures and on small crustaceans. The whale shark strains these relatively minute bits of sea life out of the ocean water by use of gill rakers. However, it has such an enormous mouth that much larger objects, including men, are said to have been ingested. In India, there are numerous stories about parts of humans found in the stomachs of these gigantic beasts. Almost identical stories are prevalent in Japan and the Philippines. Since none of these stories can be substantiated, all must be discounted as myth. On the other hand, numerous unusual objects have been found in the stomachs of whale sharks. An old boot, a tin bucket, part of a wooden oar, and a woman's pocketbook were taken from a whale shark harpooned in the Celebes Sea, off Borneo, by a commercial Japanese fishing boat.

The whale shark itself has been known to fishermen for a great many years. However, because it was mistaken for a whale in most instances, there is really very little known about this massive brute. No information exists on its breeding habits or its life cycle; its longevity is completely unknown, as are its migratory customs. Based on scientific findings, it is known that the whale shark is viviparous, which means that it produces living young, instead of eggs, within the mother's body.

No known economic advantages can be attributed to the whale shark. Those who have sampled portions of its meat say that it is inedible, and its hide has not been found satisfactory for leather. Some mariners have voiced the opinion that the liver might yield a goodly amount of commercially valuable oil. The one big difficulty in this respect seems to be in locating a constant supply of the whale shark itself. They are hard to find.

The only known species of whale shark was first discovered in 1828, when a large one was harpooned by Dr. Andrew Smith, an army surgeon, in Table Bay, South Africa. After extensive examination, the doctor was able to differentiate it from an ordinary whale. About twenty years elapsed before another specimen was taken, and this created so much interest that it was skinned, stuffed, and presented to a museum.

Today, whale sharks are as scarce as ever. Normally, they cruise slowly along at surface level and only seldom appear to go to any considerable depth. They are cylindrical to oval in shape and possess a broad, blunt head and a long, scythelike tail. There seems little doubt that if the tail were to be used as a flail, it would inflict considerable damage. Since the whale shark is so rare, there are no records reflecting that it has ever willfully inflicted damage on man. On the other hand, there are records of steel-hulled ships having sliced across whale sharks as they lay near the surface of the water.

"One moonlight night," reported seaman Johnothan Werner, of the 900-ton coastal freighter *Hekla,* William Hansen fleet, "I was on lookout, standing my post at the bow. Suddenly an enormous black mound surfaced dead ahead. Before I could signal the bridge, we hit dead on.

"On the bridge, somebody turned loose a big arc-light and in the glare it was plain we'd hit a big animal of some sort. At first, I thought for sure it was a whale and then I noticed how it was all peppered with white spots. This was the give-away. Later we looked it up in a book the captain had and it said as how it was a whale shark.

"Whatever happened to that whale shark, I can't say for sure. It appeared to be gashed bad, but it was hard to see, even in the spot-light. Then it just sort of sank and we kept on going."

The upper part of a whale shark's body seems to be lined with ridges that run lengthwise. The eyes are very small for the size of the body, and the large gash of a mouth is at the exact end of the head. Inside the mouth, the jaws are well lined with small, curved teeth that form a very severe rasp. The breathing holes are small and situated behind the eyes. The gill openings are wide; four are located above the pectoral fin. The gill rakers in a whale shark are long and close-set. The second dorsal fin is above the anal fin, which is entirely free of the caudal fin. The pectoral fins are slightly sickle-shaped and quite long.

The whale shark is dark gray to greenish brown, almost to the point of being nondescript in color, with a pattern of white or yellow spots covering the upper portions. As the hide goes around the lower portions of the body, it fades to a dirty yellow white at the bottom.

Although the whale shark is widely distributed in tropical waters of all oceans, few have ever been caught, measured, and weighed. It is only chance that brings a ship upon one of these curious beasts. Consequently, a research vessel has seldom set out to specifically locate and study the life of the whale shark. One report states that a 38-foot whale shark was caught and found to weigh 26,594 pounds.

Fishermen who have harpooned whale sharks report that their staying power is prodigious. On being ironed or wounded, they will dive to a great depth or make a high-speed run, towing whatever vessel is on the end of the line. In 1948, an inshore trawler from Barcelona sank an iron in a whale shark and was given a high-speed tow for several hours before the great beast succumbed. The captain of this 50-foot vessel made an interesting observation: "Number one iron went in easy and the barb held at a deep mark. It was the only one that we could make penetrate.

"I personally believe that this whale shark could somehow constrict its back muscles and bounce off further harpoons.

"We must have tried to sink a dozen additional harpoons and not a one would penetrate."

The whale shark is not the only member of the shark family having horizontal ridges in the trunk of its body. The zebra shark has these ridges, or keels, running lengthwise along the back and sides. However, the zebra shark seldom seems to exceed 6 or 7 feet in length.

The scientific nomenclature of the whale shark, *Rhineodon typus*, has been taken from two Greek words meaning "file" and "tooth." This was inspired by the rasplike tooth bands within the oral cavity.

The basking shark (*Cetorhinus maximus*) is another beast that is extremely large and proportionately harmless to man. Any danger inherent in these great marine animals, which can grow to 40 feet in length, rests entirely in a sweep of the tail. If this powerful appendage should happen to hit a small boat, it could conceivably shatter the hull.

As the name implies, the basking shark likes to lie on top of the water and soak up the sun. Sometimes it rolls itself over and exposes its lighter-colored underside to the direct rays of the sun. Mostly, though, it is found in temperate water with one great fin and an ample portion of back protruding above the surface. In this position, it looks just like a very large and voracious man-eater slowly prowling the seas. This is not the case, for the basking shark feeds on minute shrimp, tiny crustaceans, and many other small marine creatures. It feeds by opening its enormous mouth, swimming leisurely through a school of tiny marine animals, sucking in a massive amount of sea water, and forcing it out the gills. In the process, the food-ladened water is first forced through sievelike appendages, or gill rakes. These strainers actually retain all eatables and allow the clear water to continue over the gills. The basking shark thereby takes both oxygen and food in one gulp.

The snout of the basking shark has a blunt point that overhangs its gigantic mouth. The mouth is lined with numerous tiny teeth set in several rows within each jaw. The gill clefts, or outside slits, run almost entirely around the body of the animal, thereby allowing it to pump an immense amount of sea water through the life-sustaining system.

In color, the basking shark is dark, a deep blue or grayish black that generally lightens on the underside. This colossal shark is found in temperate waters of all the oceans and is especially numerous in the North Atlantic. Thus, the basking shark is the largest shark in temperate waters, and the whale shark is the largest in tropical waters. It is a totally inoffensive brute that is slow and extremely sluggish. Most often it appears to be a loner, but on some infrequent occasion, basking sharks have been observed in small groups of twos, threes, and fours. They are even known to swarm in numbers up to 100 or more, which probably has something to do with the sex cycle. However, very little is known about this monster or its actions. A further peculiarity about their propagation processes is that no extremely young basking sharks have ever been found. It is thought that the young are born within the parent and, upon emerging as tiny pups, immediately go to the floor of the ocean, where they can find haven amidst the rocks.

The basking shark is said to have migratory habits, and it has been established that they do migrate along the Irish coast during spring, with a northern terminus, during August, adjacent to the coast of Norway.

Basking sharks may have inspired some sea monster sightings. When traveling in pairs they usually follow in exact alignment, one close behind the other. This forms a silhouette with two dorsal fins and two tail fins over 80 or 100 feet in length. If the leading basking shark has its enormous mouth open and its head slightly projected above the surface of the ocean, there is then a well-coordinated dark mass that is serpentine in configuration, possesses numerous

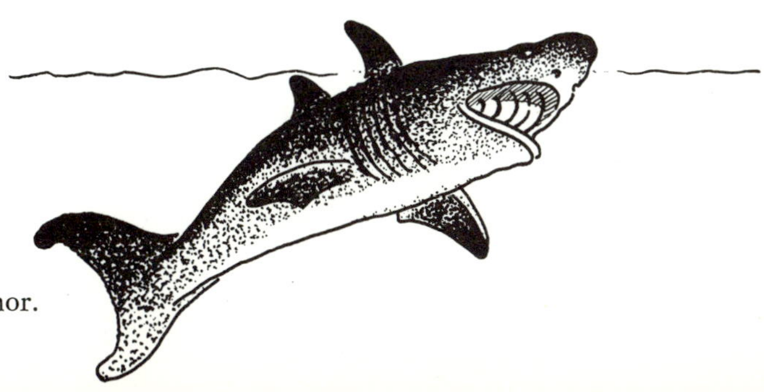

A basking shark. Drawing by the author.

mounds along the back and a large head, and stretches for 100 feet.

Stranded basking sharks (as is also true of whale sharks) sometimes die and decay before they are discovered. A landsman might then happen upon the remains of such a beast and, not being trained in the sciences, report finding a strange beast, covered with hair, that possesses a horse's head. When examined by an expert, however, the sea monster turns out to be a basking shark. The tough muscle fibers break away from the flesh and look like coarse hair. Because the skeleton is cartilaginous, with the massive jaws making up the more prominent part, the last remains are a tough frame with a grotesque head. At this late stage, an observer might well assume that he is looking at the remains of a sea monster.

The basking shark has never been considered of any significant commercial value. Its meat is of poor quality, and its hide is not as sturdy as that of many other sharks. The one aspect worthy of commercial consideration is the amount of oil to be obtained from the liver. Up to 200 gallons of low-grade industrial oil can be secured from the liver of one basking shark; the all-time record was 400 gallons from a single liver. The basking shark was once hunted commercially in Scandinavian waters but now has disappeared from those seas.

"They are easy to approach," Captain Hans Larson, a retired whaler from Oslo, said, "but if you sink an iron, that first one has got to be astern of the dorsal fin, and it better bight deep, else that fis'll take ever and a day to die. He'll lug you from here to the Arctic and back, and if you try to close for a second strike, that critter'll scuttle your boat with one blow of its tail."

This old man of the sea should know. By his own count, he has captured more than fifty basking sharks, twelve whale sharks, and "so many whales a man can't keep log on them all."

A leaping whale. Photo, United States Naval Oceanographic Office, Washington, D.C.

13
Whales

Before discussing the individual characteristics of specific whales, let us make some general observations about these colossuses. Whales are mammals of the order Cetacea and are to be found in all seas throughout the world. They are ideally insulated, with up to 12 inches of blubber; they have hairless skin, forelimbs that have become appendages for swimming, and a tail that has developed into horizontal flukes suitable for propulsion only. Whales have one or two blowholes used for breathing; certain whales are able to hold their breath for as much as one hour. Spouting, which can sometimes be observed for many miles, is the result of warm spent air hitting the colder atmosphere on the outside. The young are generally born one at a time, although twins have been observed, and suckle the mother for milk. There are two principal categories of whales: Those that have baleen in their mouths are called Mysticeti whales, and those that have teeth are called Odontoceti whales. The latter whales have up to 100 teeth, depending on the species. Baleen (whalebone) is a flexible, rodlike material suspended from the upper jaws of the whale in plates or well-organized rows. These baleen whales, when feeding, swim with their mouths open, thereby permitting vast quantities of sea water to flow through their jaws. This sea water they then strain with their baleen, which retains copious quantities of small marine life that they consume.

Whalebone and whale oil were of great commercial value at one time but have now been replaced, at least to a great extent, by plastics and petroleum oil. Whales were first hunted from land as early as the tenth century by people of Spain and France. They were being hunted at sea by the fourteenth century in Newfoundland waters. However, it was not until the sixteenth century that large numbers of whales were hunted. Spitsbergen, the first commercial whaling site, was established by the Dutch in the beginning of the seventeenth century. Shortly thereafter, an English company, called Muscovy, was organized to give the Dutch competition in the whaling business. But this company collapsed in 1625, and the Dutch gained ascendancy.

Comparison sizes of some species of whales. Drawing, United States Fish and Wildlife Service, Washington, D.C.

America began whaling in the middle of the seventeenth century, with the industry centering on Long Island and Cape Cod. As the industry grew, the center shifted to Nantucket and then to New Bedford. Most of the intimate knowledge about whales that we possess today stems from this thriving enterprise and the knowledgeable captains who skippered the old-time whaling ships.

Let us look at one kind of whale that still lives in seas around the world, the right whale.

One interesting feature about the right whale (genus *Balaena*) is its name, which goes back to the days when little or nothing was known of whales except that some were good for oil, and some were not. This particular whale, having certain distinct features, was the "right" whale; it was good for a considerable amount of oil.

Right whales have a long lower lip that projects upward along the sides; they have noticeably arched upper jaws that set within the lower lip when closed. The interior of the mouth is well hung with very valuable baleen. This family (Balaenidae) includes the Greenland right whale (*Balaena mysticetus*), black right whale (*Eubalaena glacialis*), and pygmy right whale (*Neobalaena marginata*).

The Greenland right whale can be found in areas adjacent to Greenland. This whale does very little migrating, confining any change of location to areas within the Arctic regions. Whaling records indicate that some members of this species have reached a length of 100 feet; however, the average large Greenland right whale measures between 60 and 70 feet.

The Greenland right whale, being confined to a certain area, was relatively easy to locate, hunt, and kill. Literally tens of thousands of them were destroyed by American sailing fleets that prowled the Bering Sea, Baffin Bay, Davis Strait, and the shores of Greenland.

Being a baleen whale, the Greenland right whale feeds on small marine animals and is ideally constructed for this purpose. It possesses a large, cavernous head that has a lower jaw geared to flap down, thereby acting as a scoop. Tons of water are driven into the mouth as the animal swims through schools of tiny crustaceans or plankton. The water is then drained through the baleen blades, and the resultant deposit of food is swallowed. Because of the large head, the mouth cavity is gigantic in proportion to the rest of the body.

The period of gestation is about the same as it is for humans, about nine months, sometimes a little less, sometimes a little longer. Actual mating,

Alaskan whaling station about the turn of the century. Photo, United States Archives, Washington, D.C.

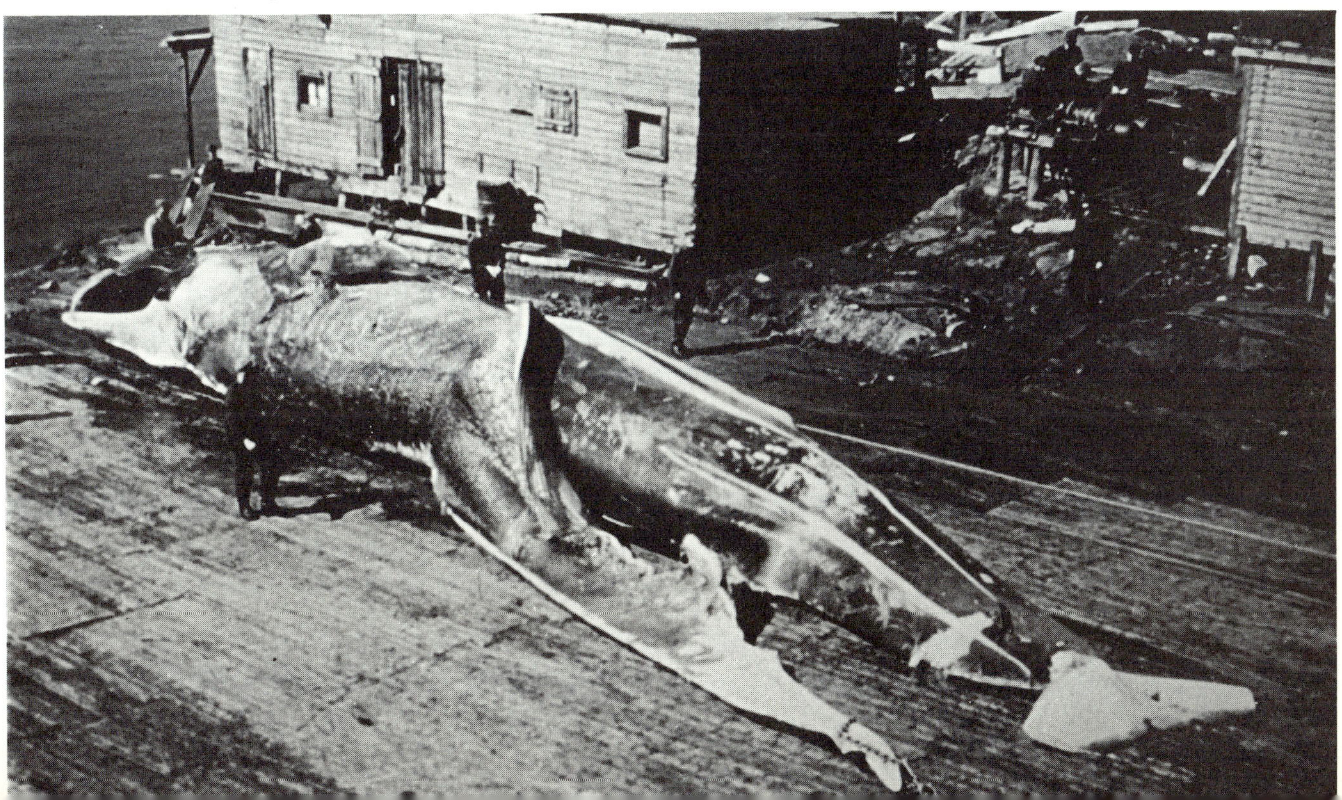

Cutting a right whale. This photograph was taken in 1902. Photo, Library of Congress, Washington, D.C.

or pairing, takes place toward the end of the summer. Calving begins during the spring and is generally confined to single births, following which calves suckle for milk.

During the old days, when rowboats were launched from square-riggers, the Greenland right was an ideal whale to hunt. It was relatively slow, did not show any apprehension at the presence of rowboats, and once dead due to exhaustion, remained afloat for long periods. The latter point was a factor of great importance because a good deal of time was required to return such a dead monster to the mother ship. If it sank before this could be accomplished, all was wasted. (Today, this problem does not exist. Whales are hunted from steel ships. The whalers fire a projectile-driven harpoon, and once the animal dies, compressed air is forced into the body to make it buoyant.)

During the early days of whaling, market prices for whale oil and baleen were fantastically high. The oil was used in household lighting, street lamps, soap manufacturing, textile production, paint processing, rope manufacturing, leather tanning, and in innumerable other industrial enterprises. Baleen was used in women's corsets, parasols, furniture, fishing rods, hunting accouterments, and many other goods that required a lightweight rod of great strength and flexibility. One individual whale produced about 1½ tons of baleen, which on the European market brought as much as $10,000 for a single ton. Because of its many uses, baleen was in constant demand and therefore was never cheap. There obviously had to be a great number of whales slaughtered each year in order to meet the demands of the market.

The history of whale hunting for Greenland right whales is a vivid example of man's plundering of the sea. When they were first hunted, around 1611, these giants inhabited all available. bays and inlets. Because they were not familiar with the treachery of man, the great whales were easy to approach and as easy to kill. The bodies were brought ashore and processed in hastily rigged shoreside plants. By 1630, so many right whales had been killed that they were no longer to be found along the coasts. It only took until 1720 for man to drive the right whales into the far reaches of the Arctic. Between the years 1846 and 1851, a fleet of about five hundred ships hunted right whales for baleen and oil, inflicting a terrible slaughter on the herds. By 1913, the Greenland right whale had become a rarity. The slaughter of right whales has been so extensive that during the 1934–1935 whaling season, out of almost 40,000 whales taken, only 4 were right whales. It is sad to note that in the early days of whaling, this beautiful animal of the sea was once so plentiful that it was referred to as a "common whale."

We have used the Greenland right whale to typify all the right whale species, but there are variations in right whales that inhabit different oceans of the world. The Pacific right differs from an Australian right or a South Atlantic right. These differences are minor. The name of identical right whales varies according to locale. The Dutch call a right whale a *Nordkaper;* in certain areas of the north, they are called *Biscayan* whales; old-time French whalers call them *Sarde;* and the Icelandic people know them as *Sletbag.*

One right whale does deserve special notice.

As recently as 1954, whales like the one shown here were killed simply because they consumed too much fish from the open sea. Photo, United States Air Force.

This is the pygmy right whale, which, as the name implies, is a much smaller version of its large namesake. The pygmy whale attains a maximum growth of 20 feet. Its baleen is shorter, yellowish, and not nearly so copious. In color, the pygmy right whale is much lighter than the Greenland right whale. Its back fin is small, curved, and distinctive to its own kind.

Pygmy right whales are said to be rather plentiful and would very likely be worth hunting commercially if more were known about them. As of today, practically nothing is known about their feeding or sex cycles nor about their migratory tendencies, except that the pygmy right whale inhabits southern seas and is well known around New Zealand, Australia, and South America. In discussing the pygmy right whale as a "small" mammal, it should be kept in mind that a 20-foot beast, when encountered in the water, can be a highly dangerous antagonist if needlessly provoked by man.

Another prominent group of whales is the fin whale, or rorqual (which means having pleats). Fin whales are most notable for the immense size that some of their species attain. They may be divided into four species: the blue whale, the finback whale, the sei whale, and the little piked whale. These classifications are based on broad generalities and are subject to many variations in certain suborders.

Let us take a closer look at each of these categories of rorquals, or fin whales. The blue whale is the largest animal in the world today. It is quite common to find a blue whale that measures up to 100 feet in length. As may be expected from its name, this whale is slate blue, with white patches on the ends of the flippers. However, the depth of the blue color varies considerably among individuals. The dorsal fin, which is low and relatively small, is placed far back on the body. The ventral grooves that run the length of the throat and chest vary from eighty to about a hundred in number. The flippers are long and tapering, possessing a convex lower border. The head of the blue whale is small compared with that of the right whale, being only about one-fifth of the entire body. The girth of the blue whale is less than that of the right whale, which gives it a more streamlined appearance. The snout of the blue whale, which has a beaked look, fits between the lower jaws when the mouth is closed. The baleen of the blue whale is jet black.

Pairing generally takes place during June, July, and August. The gestation period is about one year, and the young are approximately 24 feet in length at birth. Following birth, the calf suckles for a period of up to seven months, growing rapidly as it does so. Within a two-year period, the blue whale reaches physical and sexual maturity and measures as long as 75 feet.

Over the years, whalers regarded the blue whale as one of their principal catches. They developed a rule of thumb: 1 ton of weight per foot of length. In other words, a 98-foot blue whale would weigh approximately 98 tons.

Because of the continuous slaughter inflicted on all whales, the blue has withdrawn to nature's more favorable habitats: the Arctic and the Antarctic. In these cold environments, food is plentiful, and the big blues require a lot of sustenance. Their diet consists almost entirely of a minute shrimplike animal called *krill,* which is found in immense shoals, especially in the Antarctic. They eat in the same way right whales eat. They migrate into warmer waters for breeding.

One affliction suffered by blue whales is a mysterious sore or pitting of the skin that is noticeable on the flanks and tail of all fin whales. These hollowed-out marks are generally about 3 inches in diameter and 1 inch deep. They are sunk in the blubber and appear on captured whales as fresh sores or as healed scars. Within recent years, this wound has been likened to the mosquito bite in humans. It is theorized that one of the shark family is the culprit.

The second of the rorquals mentioned is the finback whale, or common rorqual. It is also known as the finner whale. Although this whale is smaller in size than the blue whale, it is nevertheless a faster swimmer. Its speed has been said to be incredible. Color is the easily established feature that separates the finback whale from all others. It is light gray on top and turns to white on the bottom. Oddly enough, the left forward part of finner whales is colored in gray, and the lower jaw on the right side is white. This difference in coloration extends to the inside of the mouth and jaw. The lower sides of the flukes and flippers are gray, and this color on the underside of the body usually extends from chin to tail flukes. Baleen in the finback whale is a pale yellow.

The finback whale inhabits those areas common to all whales. It is thought to reach a maximum length of about 82 feet, but the Antarctic species is larger in size than the Arctic finback. This difference is unquestionably the result of more plentiful food; krill literally superabounds in the Antarctic. However, the finback whale has a varied diet, eating some smaller fish.

The sex cycle and other reproductive facts are about the same as those of the blue whale. The size of the young at birth, obviously, is somewhat smaller, being about 20 feet for the average calf. Unlike the blue whale, the finback whale generally moves in schools, called *stims,* that may contain as few as one or two, up to over a hundred. This gregarious behavior necessitates a difference in whale-hunting techniques. A school of finback whales moves forward in a well-organized direction, and any hunting must be done at greater speeds.

The third of the rorquals mentioned is the sei whale. Again, size is the main difference between this whale and either the finback whale or the blue whale. Sei whales never exceed a maximum length of 60 feet. The color of the upper body is blue black, with a white underside. This white area frequently runs from chin to tail flukes, but it varies as it nears the tail, sometimes extending all the way to the flukes, at other times not going much beyond the afterend of the belly.

The sei whale, being so much smaller than the right, blue, and finback whales, is less often hunted. As a consequence, fewer on-the-spot observations have been compiled by interested whalemen. In Japanese, the name of this whale means "sardine whale." This name refers to the stomach contents of captured specimens. This variation in diet is more pronounced than in other whales, as is the sei's method of diving. When sei whales sound, they do not hump up like blue whales (which they so closely resemble), but simply go down headfirst. They show very little of their bodies as they plunge. Normally, in feeding, they move slowly, rolling back and forth, but not turning on the side. When struck with a harpoon, a sei whale shoots forward with a short burst of intense speed, which has been clocked at up to 40 miles an hour. Once the edge has been taken off the initial rush, it settles into a slow, constant pull.

A baby narwhal at New York Zoological Park. Photo, New York Zoological Park, Bronx, New York.

Our fourth rorqual is the little piked whale. This is the smallest of the rorquals. It generally reaches a maximum length of about 30 feet. It is blue gray on top and gray on bottom, including the underside of the flukes. Its baleen is only about 1 foot in length and is yellowish white in color. Calves of the piked whale measure about 8 to 10 feet in length and are born during the peak of the winter season.

Mariners frequently mistake the little piked whale for the killer whale (*Orcinus orca*), which also attains a length of about 30 feet. Very little is known about the migratory habits of the piked whale, but it has been established that in all areas of the world, there are seasons when it becomes numerous.

Now let us estimate the longevity of whales in general. Scientists have only recently been able to calculate this through an analysis of two features: growth patterns of the baleen (which is not a very accurate method) and the growth of "ear plugs." For example, blue whales have a canal running from the opening in the ear to the eardrum. When they sound, this is blocked by a wax ear plug that shows definite growth lines. By using both these methods, a fairly accurate age can be given any one whale. It has been calculated that blue whales live to be about fifty years of age.

There are two other whales that merit special attention: the humpback whale and the California gray whale. The humpback whale is most conspicuous for its unusually long flippers, which extend about one-third the total body length. It is also conspicuous because of nodules or bumps that pepper the head and flippers. In general configuration, the humpback whale is both shorter and more thickset than the finback whale.

The pleating along the throat runs from the chin to the navel, with a greater distance between each fold than in other whales. A shorter, broader beak is also noticeable in the humpback whale. The male normally grows to about 50 feet in length. The young humpback is black all over, but as it matures, it turns white on the underside. The flippers and the flukes are white underneath. The oil yield of a humpback whale is less than half that of a blue whale.

The humpback, like other whales, is found all over the world. It likes to prowl harbors, bays, inlets, and relatively shallow offshore areas. Because of this inclination, there are fewer strandings of humpback whales to be noted. In all likelihood, the humpback whale is more adept at shallow-water maneuvers and, therefore, is less likely to become fouled with the bottom.

For food, the humpback whale generally relies on krill but seems to supplement this diet with small fish. A few rare instances have been reported of humpback whales ingesting seabirds and larger fish such as codfish and herring. Sexual pairing takes place during the winter months, with gestation being slightly less than one year. The calf, when born, has most often been measured at about 16 feet. Young whales suckle, which does not seem to interfere with the parent again pairing. It has been estimated on good authority that the humpback whale can produce one youngster each year; this differs from other whales, most of which give birth every other year.

The humpback whale is notorious for its expressive lovemaking. The male goes through a whirlwind of leaping, plunging, splashing, and fondling the female with its flippers. Upon occasion it will roll onto its back and literally hug the female to its chest, while rolling back and forth in sheer pleasure. Normally the humpback whale confines itself to shallow dives, but when it sounds, it may remain under for thirty minutes or more.

The humpback whale, because of its inclination

to fish inshore areas, is perhaps the easiest whale to take with a harpoon. It is neither a fierce beast nor one that is frightened off by the presence of man. Since the humpback whale is a relatively slow swimmer under normal conditions, it is easily overtaken by a small boat.

The male California gray whale grows to about 45 feet in length; the female is slightly longer. It possesses a rather small head. The dorsal surface is noticeably convex and runs to the beginning of the tail region; there is, however, no dorsal fin. A series of low humps or bumps (as many as twelve) start at the stock of the tail and run as far as the flukes.

Although California gray whales resemble both the right whale and the rorquals, there is a slight difference in the thickness of the flukes and the tail.

The California gray likes to feed close inshore. Because it enjoys water that is relatively shallow and therefore needs only a modest supply of air, the pleats in the throat are not conspicuous. Such folds are the result of a bellowslike effect; when inflated in order to hold a reserve of air, they are smooth; when on the surface in a deflated position, they pleat. The California gray whale has four pleats (at most) and these are rarely longer than 5 feet.

The baleen in the California gray whale is rather distinctive in that it possesses thick individual blades and has a yellowish tint. The blades extend to about 16 inches in length, and there are approximately 150 to 175 blades to a side. The eye is near the end of the lips, and both it and the external opening of the ear are almost in a direct line with the inboard end of the fluke.

The baleen of a whalebone whale is admirably suited for the straining purpose for which it is intended. Photo, Library of Congress, Washington, D.C.

One distinctive feature about the California gray whale, is that on the outside, it is heavily encrusted with parasites. Added to these are numerous varieties of barnacles, which embed themselves quite deeply into the surface flesh.

The shoreline areas that the California gray whale inhabits run from the Arctic south to the shores of California. On the far side of the Pacific,

Eskimos flensing a whale. Notice the thick fattiness of the outer skin. Photo, Library of Congress, Washington, D.C.

it is found as far down as the southern tip of Japan, in which area it is known as the "devil fish." In the Orient, both in Korea and Japan, this species of whale was hunted over 150 years ago. Being a shallow-water species, it was hunted from small boats and towed ashore for butchering. One established method prescribed that the calf be harpooned first and drawn ashore. The distraught parent would then come close inshore to assist the youngster, making its capture and butchering an easier task.

The California gray whale's performance in thick Arctic ice is worth noting. If an opening in the ice is not of sufficient size to allow the animal the luxury of surfacing in its entirety, it will poke its snout into the air and breath in an upright position. Occasionally, under such circumstances, it will rest its chin on the edge of an ice floe before submerging again. However, its most noticeable personality trait is its love of inshore surf. The California gray whale has frequently been noted playing in the waves in shallow waters. They seem to enjoy the activity, for they roll over and over in the waves and luxuriate in the gentle swells just outside the breaker line. If they become stranded with a low tide, they patiently await the return of high water.

Because of its custom of feeding inshore, this species of whale was avidly hunted. Its proximity to the shoreline made its butchering an easy matter, and for many years it afforded Californians a lucrative business. In a relatively few years, it became hard to find and was thought to be extinct. Later, it was discovered that some survived on the Japanese side of the Pacific. Slowly, these beautiful and delightful-to-watch animals have been allowed to propagate, until they now may be seen in their old haunts along the coast of California.

Also worthy of note are the sperm whale and beaked whale. They belong to the family Ziphiidae, the second largest family of whales. They have teeth instead of baleen. This classification of toothed mammals includes quite a few other cetaceans, such as the pilot whale, plus all the porpoise family, including the killer whale. They cannot be categorized en masse by the number of teeth, for this varies from the narwhal's single tooth, which projects through the upper lip into a long horn, to the porpoise's many solid teeth in both the upper and lower jaws. The toothed whales also differ from the baleen whales in that they possess a smaller mouth. Baleen whales need a large mouth to take in the vast quantities of water out of which they strain food, but toothed whales need no such large aperture to grasp their prey. Their lower jaws are straight. Also, their skulls have a peculiarly rounded shape not found in the baleen whales.

There are two types of sperm whales: the sperm whale and the pygmy sperm whale. The principal difference is size. Each has a pool of transparent oil that rests in the forepart of the head. The single blowhole, common to both, is situated to the left forward part of the head. However, in the pygmy sperm whale, the content of the oil in the head is so meager that it is of no commercial value.

The sperm whale, sometimes referred to as a cachalot whale, is one of the most common and most widely distributed of all whales. The name "sperm" was applied by early whalers, who thought the oil, within the head of a cachalot,

Cutting in on a sperm whale. This picture was taken from old-time square-rigger in 1903. Photo, Library of Congress, Washington, D.C.

was in actual fact spermaceti used for spawning. The word *cachalot* means "big tooth." Such a name must have been suggested by the big teeth that line the lower jaw.

It is the sperm whale that made the American whaler the historic figure he is today. At one time, during the height of the whaling industry, in 1837, almost five hundred ships sailed the seas in search of sperm whales. During the previous decade, these whalers produced an estimated 50 million barrels of oil. Obviously, there was a tremendous slaughter of the whale population of the world. The sperm whale was pursued so relentlessly that it became hard to find. Fortunately, at just about this time oil began to be extracted from the earth, and this meant that whale oil was no longer a necessity.

The sperm whale was the aristocrat of all cetaceans to the whalers of this era. The log of the famous whaler *Charles W. Morgan* noted on her first voyage in 1841: "Saw more right whales, an excellent harvest for those who wish to harvest them, but we are after what is considered higher game." The skipper meant the sperm whale, and records kept on this grand old ship indicate that during this voyage alone, she took fifty-nine sperm whales. When it is considered that in the active life of the *Charles W. Morgan,* thirty-six recorded voyages were made, that suggests a lot of sperm whales were killed.

Sperm whales are famous for producing ambergris. The word is a combination of two French words, *ambre* and *gris,* meaning "gray amber." It is a waxlike substance that comes from the intestine of the sperm whale. Generally, it contains the horny beaks of cuttlefish or octopuses, principal foods of the sperm whale. It is thought that these nondigestible animal parts form the basis for ambergris. Lighter than water, it is frequently found washed ashore in gray, black, or variegated masses. These masses are most often small; however, pieces weighing several hundred pounds have been found. In ancient times, ambergris was thought to have miraculous powers as a medicine. Now, however, it has a high value in the manufacture of perfumes. Since it is soluble in alcohol and has its own sweet, lingering odor, ambergris serves well as a vehicle for other scents.

While feeding, the sperm whale moves along at about 3 or 4 miles an hour. However, when disturbed by whalemen, it can swim as rapidly as 15 miles an hour. The sperm whale grows to a length of about 60 feet, with the female being noticeably smaller. In one or two cases, however, sperm whales have been measured up to 75 feet in length. The gestation period runs to about one year, and the newborn measure approximately 12 or 14 feet in length. The sperm whale does not have a dorsal fin, but shows a series of lumps along the afterend of the back. The very generous-sized tail flukes have been measured at widths of up to 15 feet from tip to tip. Generally, except in very old whales, the skin is smooth. Unlike rorquals, it does not have furrows in the throat region, but has numerous wrinkles instead. The skin of older sperm whales tends to wrinkle all over.

Because the diet of the sperm whale is so dependent on octopus and squid (although a 12-foot shark was once taken from the gullet of a large sperm whale), the lower jaw is of great importance. It is rather long, thin, and almost delicate in relation to the rest of the head. The teeth, which average about thirty in number, are large and rugged. They fit into individual casings inside the upper jaw, thus forming a completely sealed-in area for any animal caught within the mouth.

There are many stories about great battles between a giant squid or massive octopus and a sperm whale on the surface of the ocean. In truth, these battles are quite common, for the scars on large sperm whales indicate many such struggles. The sperm whale has to go deep to locate one such tasty morsel but has to return to the surface for air as it fights. On the other hand, the giant squid or octopus, being able to obtain oxygen out of water, struggles to stay submerged, in hopes that the sperm whale might be drowned. (Under normal browsing conditions, a sperm whale can stay submerged for about one hour.)

As has been mentioned, a large reservoir of oil is located in the head of the sperm whale. It has been reported on good authority that as many as 15 barrels of oil have been taken from a single sperm whale. The blubber of this whale is more than 1 foot thick and also contains additional oil.

A narwhal. Drawing, United States Naval Oceanographic Office, Suitland, Maryland.

This rare photograph shows a porpoise (a close relative of the whale) at the moment of delivery. Photo, Library of Congress, Washington, D.C.

The pygmy sperm whale differs from its namesake in that it has a dorsal fin, a mouth of distinct configuration, and about fifteen teeth in the lower jaw, each of which is rather slender. The pygmy sperm whale has been known to grow to about 15 feet in length.

The bottle-nosed whale is a relatively small animal. It generally grows to a length of about 30 feet. This species has come by its name because of a bottle-shaped snout. There are two deep grooves running longitudinally on the underside of the throat. The dorsal fin is small and positioned to the rear of the body.

Although this whale is far less dangerous in size, nevertheless it has teeth and because of this is hazardous to man. Furthermore, in view of the scarcity of large oil-bearing whales, these bottle-nosed whales are hunted by man because, in spite of their size, they do contain a worthy amount of oil.

Bottle-nosed whales possess one fatal fault: They will never leave the presence of another bottle-nosed whale that has been wounded. It is not until their companion whale dies that they will move off. This characteristic makes them easy prey for hunters. If a whaleman is careful, he can wound bottle-nosed whales one after the other, going back to finish them off at his leisure. Since bottle-nosed whales move in groups of four to twelve, catching an entire school can add up to a considerable quantity of oil.

Beaked whales are smaller than bottle-nosed whales; they grow to a maximum length of only about 26 feet. There are numerous categories of beaked whales, but none are of importance as a danger to man. They do possess teeth, but because of their relatively placid nature toward man, they represent no known threat.

There is one point that should be mentioned in conclusion. Small whales are frequently mistaken for porpoises. This is especially true of the bottle-nosed whale and the beaked whale, for they do, in many respects, resemble porpoises. Small whales of any variety are of a different family from that of porpoises, and in approaching them or attempting to handle them, this fact should be kept in mind.

White whales. Photo, Bureau of Sport Fisheries and Wildlife, Washington, D.C.

There are approximately thirty species of stingrays. Some grow to as much as 6 feet across the disk. Photo, Marine Studios, Marineland, Florida.

SECTION THREE

Stingers and Shockers of the Sea

14

Stingrays

Stingrays belong to the genus *Raja*, of the family Trygonidae, which includes a wide variety of members, of which we can only give a cross section. Members of this class generally have flat, disklike bodies that taper downward from the head to the long, thin tail. It is the tail that gives this kind of fish its name because the tail contains one or more barbs capable of inflicting a piercing and deadly wound. Frequently, this barb is connected with a poison-secreting gland, and a wound can be fatal. The tail of a stingray apparently has no other function than to inflict a wound, for it does not aid in swimming, maneuvering, or in fishing for food. Near the end of the tail, there are generally two small dorsal fins; the pectoral fins extend along the entire side and as far forward as the snout. This snout is supported by cartilage, which forms a definite hump or ridge. On either side of the ridge is an eye, and if this fish is struck between the eyes with force sufficient to snap the ridge of cartilage, death will occur instantly. Native fishermen, pulling in their nets from along the floor of the ocean, quite often catch stingrays and use a hammer or heel of a soda bottle to crack the cartilage.

Stingrays grow up to 10 feet in length, with

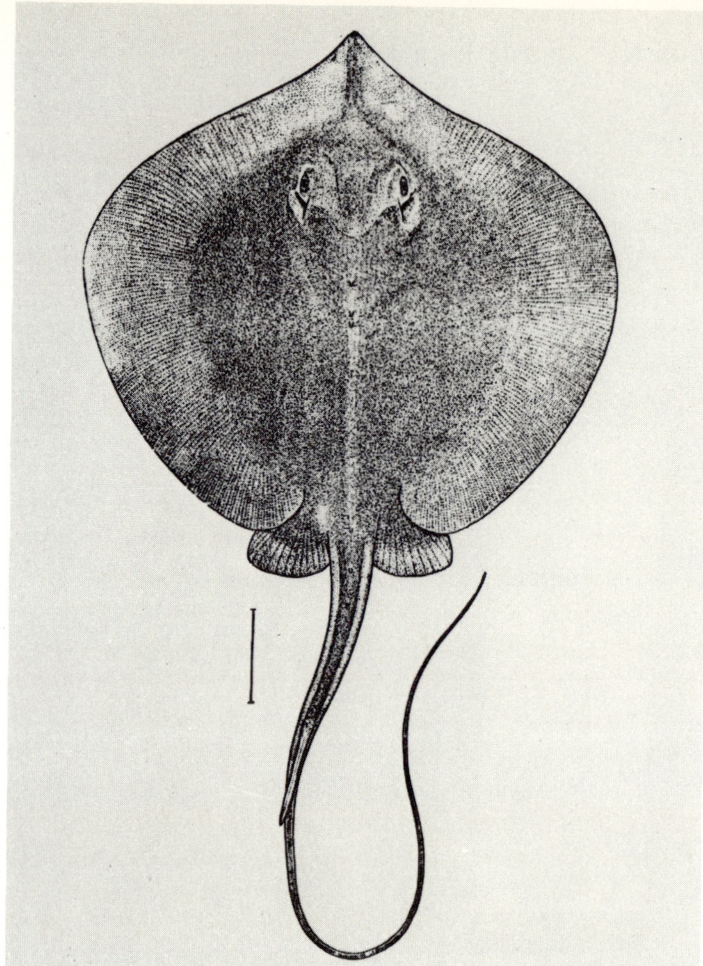

The southern stingray is found from the Carolinas to Brazil. Drawing, Bureau of Commercial Fisheries, Washington, D.C.

many being smaller. There are so many variations that fishermen quite frequently lump them under the heading "skates."

Since stingrays are bottom-living fish, most wounds experienced by men are in the feet and come about through inadvertently stepping on one of the creatures. Consequently, people are warned to shuffle their feet when walking in ocean water because rays are timid fish, more inclined to flee than fight.

Because they are admirably suited for living on the floor of the sea, stingrays are hard to detect. Both eyes are on top of the body, which is dark in color; the mouth is on the underside, which is entirely white. When resting quietly on the floor of the ocean, the stingray uses its "wings," or outer edges of the pectoral fins, to throw sand across its back. In this manner, it becomes almost entirely hidden from the eye. When an attractive fish comes by, the stingray flips through the water and flops down on its victim; in this position, the stingray actually smothers its victim before consuming it. It is amazing to see the precise manner in which a clumsy-appearing stingray can maneuver when in the process of feeding, although they are not at all fast swimmers. Locomotion is gained through a symmetrical movement of the wings, which gives them the appearance of flying rather than swimming. They apparently will eat any living thing that exists on the floor of the ocean, sometimes even another of their own species.

Differences in sex are very pronounced. In fact,

the difference is so marked that in some varieties, the male and female would appear to be entirely different kinds of fish.

The eyes of stingrays (and of other rays, as well) have been the subject of considerable study. They are different from the eyes of most fish in that they have a veillike arrangement to control the amount of light falling directly on the pupil. This is necessary because stingrays occasionally surface in order to loll about in the sunshine. These optical curtains are designed to protect the sensitive iris against direct rays of the sun.

Female stingrays lay eggs that are contained in tough little envelopes known as "mermaids' purses" when washed ashore. Normally, it takes up to fifteen months for eggs to hatch, during which time the developing fish lives on the contents of the egg. For many years, in fact until just recently, rays were considered an annoyance to fishermen; today, certain species are viewed as tasty and a large market exists for their meat. In Europe, for many centuries, certain species were sought with rod and line. Never considered a game fish, rays generally plummet to the bottom when hooked and stay there. It is said that they can adhere to the bottom very strongly, at least until the head is forced to come up off the ground, at which time they sail toward the surface like a kite. In 1927, a stingray weighing 250 pounds was reportedly hooked and reeled in off the coast of Spain.

Most stingrays are saltwater fish, but some do live along the estuaries of rivers in southern climes. All are thought to be warmwater fish.

Whip-tailed stingray possess a long, thin tail that resembles a buggy whip. They are different from other rays in that they may use their tail as just that, a whip. Upon numerous occasions, they have been observed lashing fish to sections before eating them. However, there are at least thirty types of whiprays, and all are known to possess poisonous barbs, which they use upon very little provocation. The flesh of the whipray has been found to be unpalatable. The tail, in ancient days at least, was carefully oiled, made flexible, and used as a bullwhip.

Eagle rays and true eagle rays are two different species, despite the similarity of their names, and grow to a much larger size than most of the stingray varieties. They have been known to come close to 1,000 pounds in weight and measure up to 15 feet in length. They are generally less dangerous than smaller stingrays because there is little or no chance of ever stepping on one. Nevertheless, they are of considerable danger to fishermen, for when caught in a net, they become quite vicious and react with tremendous fury. Such a heavy animal, lashing a long tail in furious fashion, can cause a great deal of damage to fishnets, boats, and men. One interesting thing about the larger rays is the gigantic force of their teeth. It has been shown that some sizable clam shells, requiring a force of at least 1,000 pounds to crack, apparently split easily in the mouth of a great eagle ray.

The so-called electric eel (Electrophorus electricus) *is not a true eel, but a member of a different order.* Photo, Marineland of Florida.

15

Electric Rays, Electric Eels, and Electric Catfish

Electric rays belong to a genus commonly called *Torpedo*. Although they are mentioned in the most ancient of histories, there is surprisingly little factual information known about them. Because they are of no commercial value, they have been mostly ignored. Few scientists have studied them.

Oppian of Corycus, the Greek poet of the second century A.D., in five books on fishing called *Halieutica*, spoke of the torpedo ray in a rather lighthearted vein, saying that "on a sudden, it would eject poisoned charm." Since electricity, as such, was an unknown quantity, it is not surprising that ancient writers are rather vague about the cause of that which was emitted by a torpedo ray. Aristotle mentioned the electric ray, observing, at the time, that it caught its prey by means of a stupefying apparatus of the mouth or at the back of the head and ejecting "a sort of poison, or elixir, yet being of neither." Fossil remains of the torpedo ray date back 150 million years. Ancient Roman art depicts torpedo ray along with other edible varieties of marine life, indicating that they were considered eatable during the great days of the Roman Empire. Today, the consensus seems to be that their meat is tasteless. The torpedo ray (*Torpedo nobiliana*) is purple brown to dark brown on the topside, some people claiming the color to be a true black. Like most other rays, it is white on the underside. They are flat, round, and short-tailed. They grow up to 6 feet in length and can weigh several hundred pounds. They are known to be common in the western Atlantic, from Cuba to Nova Scotia. They are also quite plentiful in the Mediterranean and both sides of the Atlantic and Pacific. Torpedo rays live on the bottom, partially buried in silt or sand, and consume passing fish. This type of ray is thought to be most plentiful at a depth of about 200 feet or more, but some few have been taken in knee-deep waters.

Torpedo rays are a danger to man because of the shock they can send through the water. The electric organs are situated on the back and consist of numerous tubes insulated from one another by fibrous tissue. Each tube is divided into a number of tiny compartments and contains a clear, jelly-like substance. As many as five hundred of these tubes go into making up a power pack for a single organ. In its turn, each organ is connected with minute tendrils leading to a main controlling nerve, which in turn goes to a control center in the brain. The outlet system has both a negative and a positive side, with the current running from the upper end to the lower end. The entire system is based on a concentration of electric muscle power. In other words, electric organs have developed in the torpedo ray at an expense to muscle movement. Instead of using the power to move muscles, it is gathered into storage bins for use as an expendable outside force.

The question, about any electric fish or other electric marine animal is: How much power can it release? In actual truth, the first flash or burst of electricity is always the greatest, with ever-decreasing charges going out in extremely rapid order, until all power packs become depleted. The animal must then rest and recuperate for quite some time in order to regain strength.

Professional fishermen say that the shock can be felt through a net that is thoroughly wet, or when water is poured on an electric fish, eel, or ray out of a metal bucket. The shock in either case is not severe, obviously because the contact is not good, but the telltale sensation is present. However, on numerous occasions a human has stepped on a rather large torpedo ray and been knocked down and stunned. At the University of Tokyo, scientists are beginning to conduct extensive research on electric marine animals. One eminent researcher, Professor Koizumi, likened the maximum power of a full-grown 6-foot torpedo ray to a sudden charge of up to about 250 volts.

Electric shock serves the torpedo ray in two ways: as a defensive weapon and as a source of food. In many instances, freshly swallowed fish and eels showing no scars, cuts, bruises, have been found within the stomach of an electric ray. This would imply that the electric fish had first shocked them into submission and then swallowed them.

The young of the torpedo ray are born alive and go through an extensive metamorphosis before their electric organs are fully developed.

Electric eels, Professor Koizumi assures us, are not true eels as previously classified. They are now recognized as being members of their own distinct family. "And a highly dangerous one," the professor hastens to add. "A South American species, reaching a maximum of 10 feet in length, can easily kill a man, for it can discharge a flash of 600

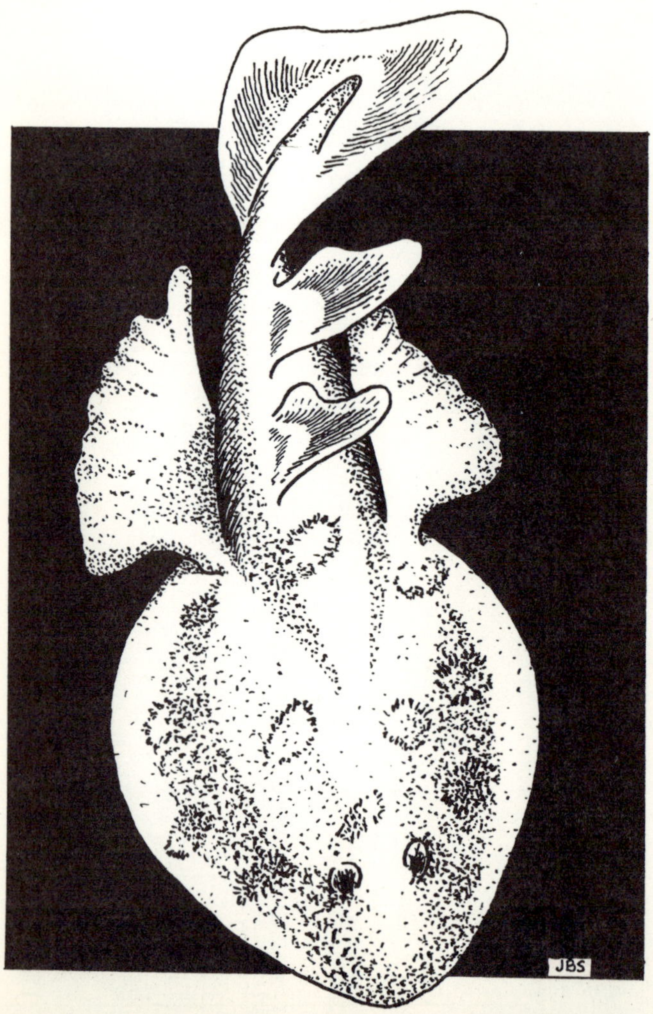

The electric ray (Narcine brasiliensis) *has electric organs between the pectoral fins and the head.* Drawing by the author.

volts. This is about equal to the sudden kick that might be received from fifty automobile batteries."

The electric eel (*Electrophorus electricus*) is not only a dangerous fish but also one about which very little is known. Since this eel is blind when mature, it unquestionably uses electric power much as man uses radar: to locate food, navigate, and detect enemies while underwater. It does not have well-developed gills and therefore must surface for air. The young, regardless of size, produce the same strength of electricity per cell as do larger eels; in other words, electric eels are born with peak power. However, as the fish matures, the cells grow in size while remaining constant in number. Because about 75 percent of the fish is given over to electric power, the discharge put out by a 10-foot electric eel can be enormous and severe. Although the eel cannot control the force of its electricity within each cell, it can regulate the number of cells it invokes. Hence, it will employ only a few cells to stun a small fish but will employ all to kill a man.

The generating plant of the electric eel is much the same as that of the electric ray, except that the nerves supplying the discharge come from the backbone instead of the brain. The large 10-foot specimens, which present man with his greatest danger, inhabit certain marshy, fresh waters of South America. This eel's electric organs are made up of two longitudinal bodies between skin and muscles of the caudal (tail) area; these bodies consist of about 250 cells, supplied by 200 nerves. The electric eel can discharge enough electricity to knock a cow unconscious. It is said that this is not an uncommon event; a cow frequently puts two or more of its feet in the water while drinking, thereby establishing an excellent ground for itself. On good authority, it is reported that natives drive cattle through the water where small electric eels are known to abound. Such action triggers the eels and they are drained of their electricity; they then can be handled with relatively little danger. It is also said that sometimes an unusually large eel is hidden in the water and knocks one of the animals unconscious, and the animal drowns.

Like a man-made electric motor, an electric eel can be short-circuited. This short circuit is created by contacting either end of the eel with an electric conductor. This causes a constant and uncontrollable discharge, which will eventually kill the electric eel.

Little needs to be said in this book about electric catfish and other smaller fishes, such as the stargazer, for their electric output does not pose a serious threat to mankind. They are hardly more than an annoyance.

Historically, the electric catfish is the first electric fish to be mentioned by ancient man. It is found in some of the first Egyptian art and hieroglyphics that date from about 5,000 B.C. The Nile is the home of an electric catfish that grows to 3 or 4 feet in length and can deliver a sizable sting. This discharge has never been known to kill a man or, for that matter, to injure a person seriously.

Electric eels, upon reaching maturity, go blind. Some grow to a length of 10 feet.
Photo, Wometco Miami Seaquarium, Miami, Florida.

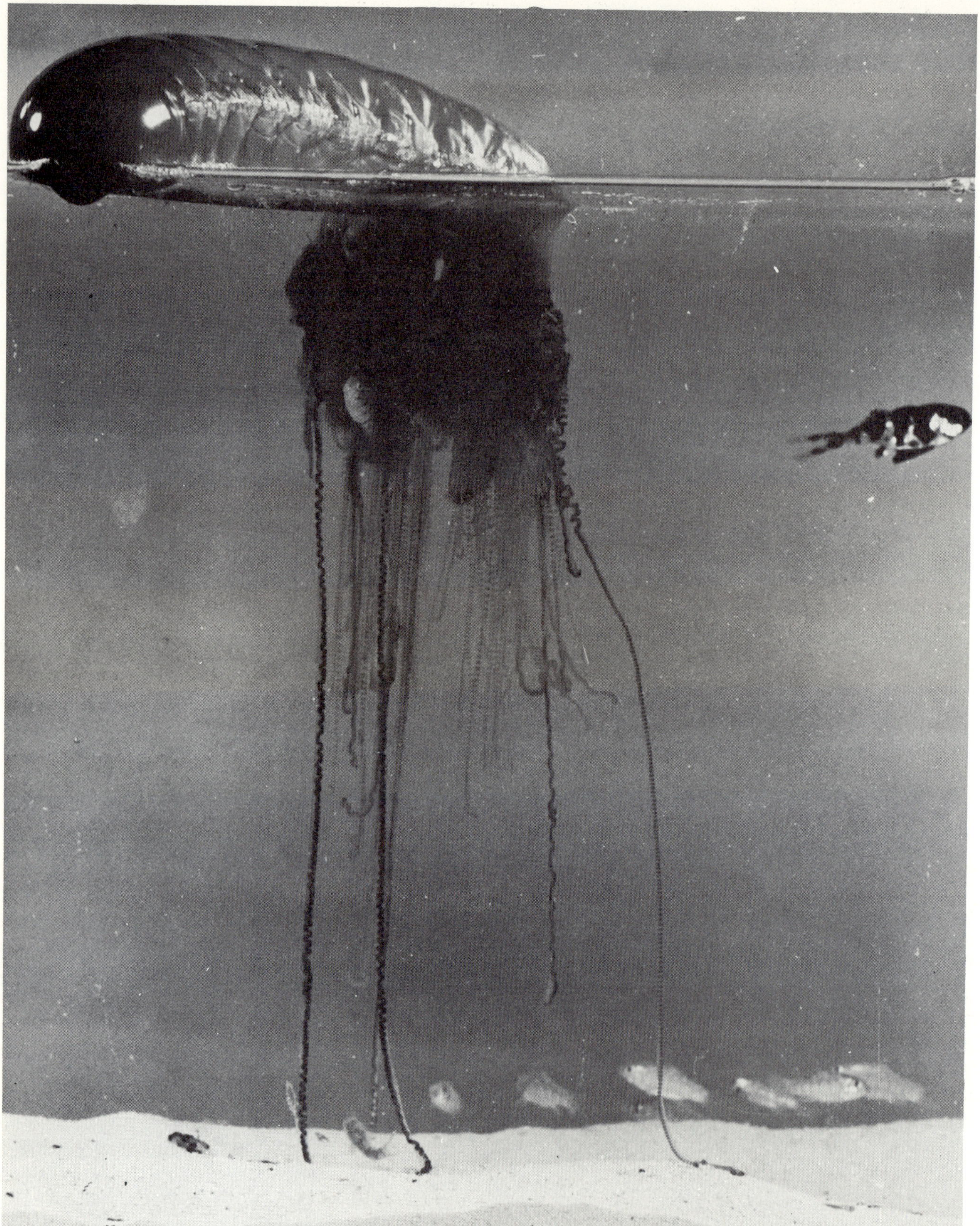

The Portuguese man-of-war is actually a colony of polyps. Both digestive and reproductive polyps hang down from the gas-inflated bag. Photo, Miami Seaquarium, Miami, Florida.

16

Jellyfish

Jellyfish is the common name for free-floating medusae of the class Hydrozoa of coelenterates. The main portion of the jellyfish, generally referred to as the body, is gelatinous, bell-shaped, and almost completely transparent. Ninety percent of this mass is water, yet the animal is able to swim by exercising a series of pulsating movements, to fish for its food, to conduct normal bodily functions, and to reproduce. Like all coelenterates, jellyfish are equipped with stinging cells (nematocysts), with which they overwhelm live food and convey it into their digestive canal.

Jellyfish are quite common throughout the world; the most common are the moon jelly of North America (up to 9 inches in diameter) and the aurelia, or common North Atlantic jellyfish (up to 24 inches in diameter). One of the best known jellyfish is the warmwater colonial member of the genus *Physalia*, the Portuguese man-of-war (*Physalia physalis*).

The United States Naval Oceanographic Office in Washington, D.C., one of the largest oceanographic organizations in the world, publishes a booklet called *Science and the Sea*, which gives a very comprehensive list of all coelenterates. The list of jellyfish includes the following:

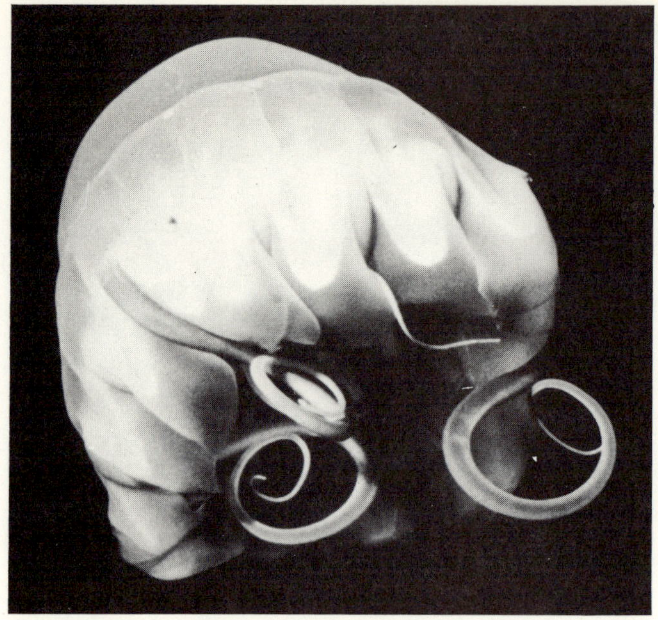

The common jellyfish (Aurelia aurita) *is found all over the world. It can grow to 18 inches in diameter.* Photo, Naval Photographic Center, Washington, D.C.

(a) Sea Wasp, *Carybdea alata,* which inhabits the tropical areas of all oceans.
(b) Sea Wasp, or Box Jelly, *Chiropsalmus quadrigatus,* an especially dangerous inhabitant of Australian and Philippine areas and the Indian Ocean. A less dangerous, but related form, is found in the Atlantic Ocean.
(c) Sea Nettle, *Dactylometra quinquecirrha,* is a widely distributed form which may be found as far north as New England coastal waters, as well as in all tropical sea areas.
(d) Sea Blubber, *Cyanea capillata,* inhabits the North Atlantic and Pacific to the Arctic Ocean. Other stinging species of this genus exist southward into tropical areas.

According to *Science and the Sea,* the tentacles of coelenterates may be sticky for holding or long for entangling, and may be equipped with stinging cells. The stinging cells are situated in the outer layers of tissue on the tentacles and may be likened to tiny hypodermic syringes and needles arranged to inject a paralyzing drug into an unwary victim. These elaborate weapons are cocked and ready to fire on contact. Swimmers who brush against a jellyfish may be stung by literally thousands of these tiny poison weapons.

Symptoms produced by coelenterate stings vary according to kind and locality. Reactions vary from person to person. The wounds are primarily local skin irritations. A few are painful, and ulceration may follow. The sea wasp of the genus *Chiropsalmus* is a most venomous organism, and its sting can cause human death in three to eight minutes.

Symptoms range from a mild prickly or stinging sensation to a throbbing pain; the victim may lose consciousness. The pain may be localized or may radiate to the armpit, groin, or abdomen. Local redness may be followed by inflammatory swelling, blistering, or minute skin hemorrhaging. The victim may experience shock, muscular cramps, numbness, nausea, vomiting, severe backache, frothing at the mouth, constriction of the throat, loss of speech, difficulty in breathing, paralysis, delirium, convulsions, and finally, death.

The usual treatment is directed toward relieving the pain, alleviating the effects of the poison, and controlling shock. Morphine may be used for the pain. Histaminics given by mouth, or in the form of a cream applied locally, may prove valuable in treatment of the rash. Soothing lotions such as olive oil, ethyl alcohol, sodium bicarbonate, and diluted ammonium hydroxide have been used. Artificial

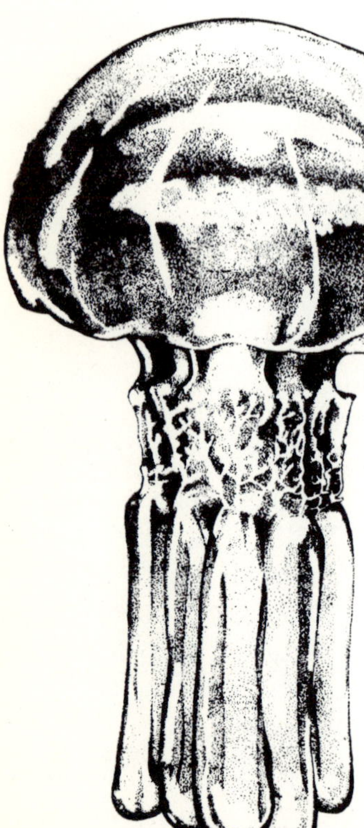

Common jellyfish are called Medusa. Each is a free-swimming polyp that is umbrella-like in shape. The one shown here is called a cabbage bleb. Drawing, United States Naval Oceanographic Office, Suitland, Md.

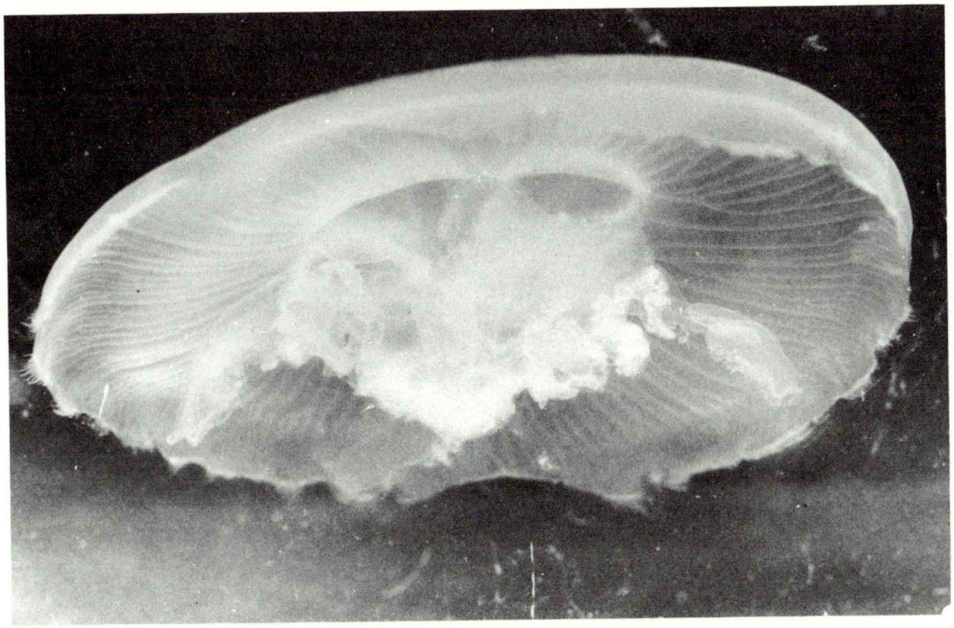

The mouth and tentacles of a true jellyfish are on the underside of its umbrella. Photo, U. Erich Friese, Taronga Zoo, Sydney, Australia.

respiration, stimulants, and other measures may be necessary. Special antidotes are not known.

It is important for swimmers and divers to avoid the original contact with the tentacles. The tentacles may trail as long as 50 feet, so jellyfish must be given a wide berth. Rubber suits and tight-fitting woolen underwear have proved to be useful protection. Dead jellyfish found on the beach are also hazardous because the nematocysts remain potent. Waters that have been storm-lashed may contain many shreds of tentacles and each can inflict a wound. A person who has been stung should leave the water and seek treatment. Washing the poisoned skin area with diluted ammonia or alcohol as soon as possible will be helpful. Mineral oil or baby oil may help to alleviate the stinging sensation.

In certain areas, jellyfish are called sea nettles, and although science finds them useful for a wide variety of venom employed in the production of drugs, bathers are menaced by their summer invasion of beaches. In fact, they are so annoying that the Congress of the United States passed the Federal Jellyfish Act of 1966, aimed at finding a method of eliminating them from coastal areas. Since the passage of this act, almost $2,000,000 has been spent in an effort to learn more about jellyfish and attempt, to some degree at least, to control their appearance each summer. Although much has been learned, little has been accomplished because it is believed that this troublesome species of sea life might directly contribute to the population growth and well-being of oysters, crabs, clams, and fish.

Sea nettles, or jellyfish, are not newcomers to American shores. Indians warned some of the earliest settlers about them. Records show that in 1750, jellyfish killed a swimmer in the Rappahannock River in Virginia. Both the Chesapeake Biological Laboratory in Maryland and the Virginia Institute of Marine Science have done a great deal of research without arriving at satisfactory control methods.

Not all coelenterates live on the ocean surface. Walter Jahn, a scientist with the Naval Oceanographic Office, took a photograph of a very primitive genus called *Umbellula* at 15,900 feet down on the floor of the Atlantic Ocean, almost 350 miles west of the African coast.

Most people believe that the giant jellyfish are to be found in warm waters. This is not so, for in the far north and far south live the real titans. Jon Malakia, an Eskimo hunter from western Greenland, took an oath that he saw a pink (*Cyanea capillata*) jellyfish measuring at least 8 feet across its umbrella, with a 10- or 15-foot sharklike fish ensnared in its tentacles. Some Eskimo hunters have said that such large stinging jellyfish are not uncommon in Arctic waters, and their tentacles have been estimated at over 275 feet in length. It should be noted here that many efforts have been made to weigh massive jellyfish and to secure accurate measurements of their bulk. All known attempts to remove these large jellyfish from the water have proven futile, for they deflate rapidly, break apart easily, and generally disintegrate before they can be handled ashore.

Earlier in this chapter, we mentioned the Portuguese man-of-war jellyfish. Commonly called a floating fisherman, man-of-war (*Physalis physalis*) is a beautifully complex piece of community living that only nature could ever devise. It is a free-floating apparatus belonging to the order Siphonophora, which differentiates it from the true jellyfish in that what appears to be a single entity is, in truth, a complex society.

In most marine animal life, different organs serve different purposes. The mouth, geared with teeth, bites and chews; the throat swallows; the stomach

This large sea nettle is estimated to be 10 feet in length. Photo, Virginia Institute of Marine Science.

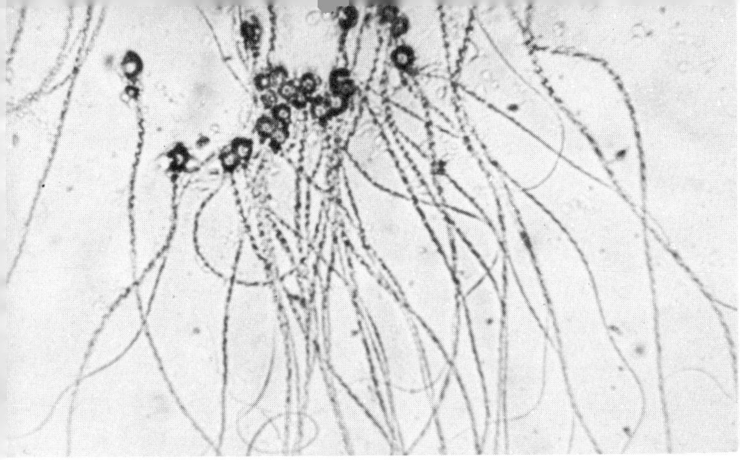

Enlarged view of threadlike tentacles of a Portuguese man-of-war. These delicate but dangerous appendages have been known to trail for 100 or more feet.

breaks the food into workable material for the intestines; and so forth. All, however, belong to one animal. In the Portuguese man-of-war, different animals, all of the same colony, perform separate tasks. There are those that catch food, those that paralyze the catch, those that eat the food, those in charge of swimimng, and so forth. Some scientists have likened the man-of-war to a beehive that has become completely cohesive. There are workers, soldiers, breeders, and so forth.

The Portuguese man-of-war is unquestionably best known because of the beautiful, yet fearsome colored-gas float, or sail, that is seen above water. This very light and buoyant means of propulsion actually subjects the direction of the man-of-war to the whims of the wind. From this colorful float are suspended threadlike tentacles that have been known to extend for 100 or more feet. Stinging cells cover each of these threads, and each, although very minute in itself, is a beautiful and dangerous bit of mechanism. Each cell is equipped with a number of barbs, spines, and hooks that end in needlelike tips supplied with tiny portions of deadly poison. On contact with a foreign body, the whole operation starts with the needle, which pierces the skin and allows the poison to be injected. The bigger the fish or foreign object, the more of these tiny cells are called upon to do their work in order that the enmeshed victim may be quickly paralyzed.

After the potential meal has been made senseless, the tentacles begin to contract. It has been tested and proved in the laboratory that one of these

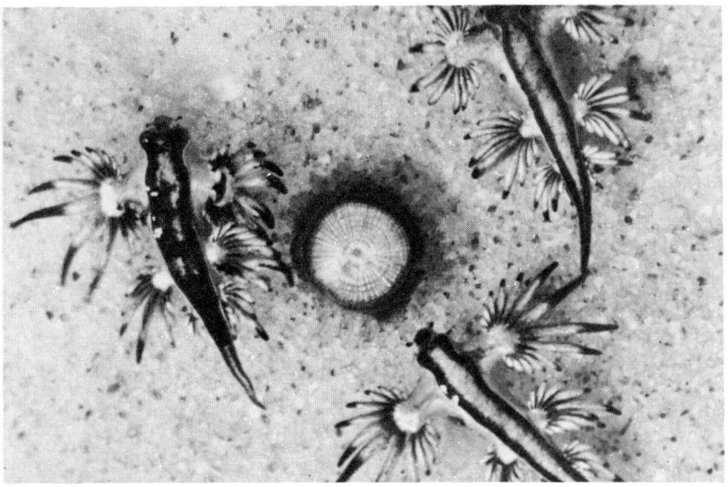

Glaucus marinas, *a form of sea slugs, only a few inches in length, eat the stinging apparatus of jellyfish. This food lodges in special sacs, which in themselves become stinging organs.*

tentacles can reduce itself from a length of 100 feet to a few inches in a matter of seconds. At this point in the contraction, the victim has been drawn to within range of the feeding polyps. These polyps clamp their minute teeth onto the food and begin the process of digestion. The nourishment from this food is then distributed throughout the entire system.

These strange and fearsome objects are not without enemies, for large turtles and massive fish (such as *Mola Mola*) have been observed biting into the gas-filled buoy of the Portuguese man-of-war and actually gobbling the tentacles in great gulps. However, the man-of-war does have friends, for a number of small fish live amidst the tentacles of this jellyfishlike marine animal. In any event, the man-of-war is dangerous to humans and should be avoided at all costs.

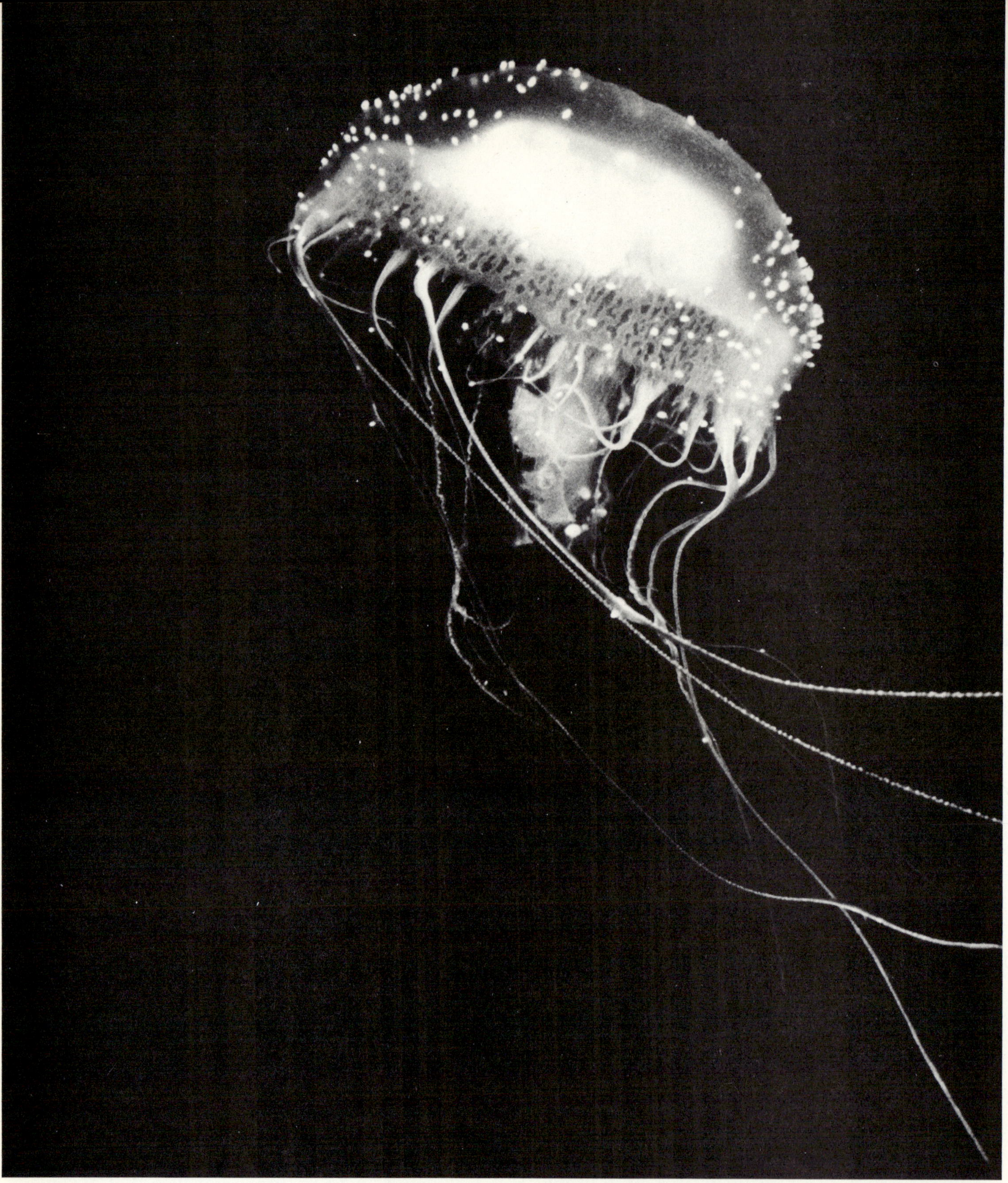

This jellyfish was photographed in McMurdo Sound, Antarctica. Photo, Naval Photographic Center, Washington, D.C.

Some starfish are dangerous to man. This crown of thorns sea star has tiny barbs, each of which is capable of poisoning man. Photo, MacVicar, Navoceano.

In fact, all jellyfish should be avoided by man. Their poison has been likened to that of a cobra; it attacks the nervous system and paralyzes the victim. One Australian species, the box-jelly, has for some years been considered the most deadly marine animal. Almost completely transparent in ocean water, the slightest touch can bring death to a human. Only just recently, doctors at the Babinda District Hospital, North Queensland, managed to save a twelve-year-old boy stung by a box jelly. Upon contact, the victim ceased to breathe, but mouth-to-mouth respiration and external cardiac massage were employed en route to the hospital. Six hours after emergency measures, the boy was out of danger, and two weeks later he was discharged from the hospital, the first person known to have recovered from the sting of a box jelly.

The natural enemy of the poisonous crown of thorns sea star is the painted shrimp. Here one *is pictured in the act of killing a crown of thorns.* Photo, NBC Television, New York City.

Sea anemones have a columnlike body with a flat base and a flattened top, from which bushy tentacles grow.

17

Sea Anemones and Corals That Sting

Sea anemones, like the jellyfish, are coelenterates; however, unlike free-floating jellyfish, they are generally fixed by a base or foot to a rock or other firm body. To be more specific, a sea anemone is a marine polyp of the class Anthozoa, or "flower animal," named for the petallike tentacles that are most often brilliantly colored. Another difference between the anemone and the jellyfish is that the anemone has a partitioned digestive cavity and a body that is more muscular. Although it fixes itself by its basal disk to a rock, shell, or in the sand in shallow water, like the jellyfish, it paralyzes its prey with stinging cells at the ends of tentacles.

There are literally thousands of variations in these living flowers of the sea that are so beautiful to look upon and so deadly to touch: cave dwellers, sand dwellers, gold-spangled, purple, green, large, tiny, brilliantly colored, subdued in hue and so forth. Some, like the color-changing trumpet and the pimplet, can materially alter their appearance.

Generally speaking, sea anemones are saclike animals that possess what could best be described as a cylindrical body. They fasten themselves to something firm with their bottoms. At the upper end, they possess a disk, in which there is a slit of a mouth surrounded by tentacles, generally present in multiples of six. From the upper open-

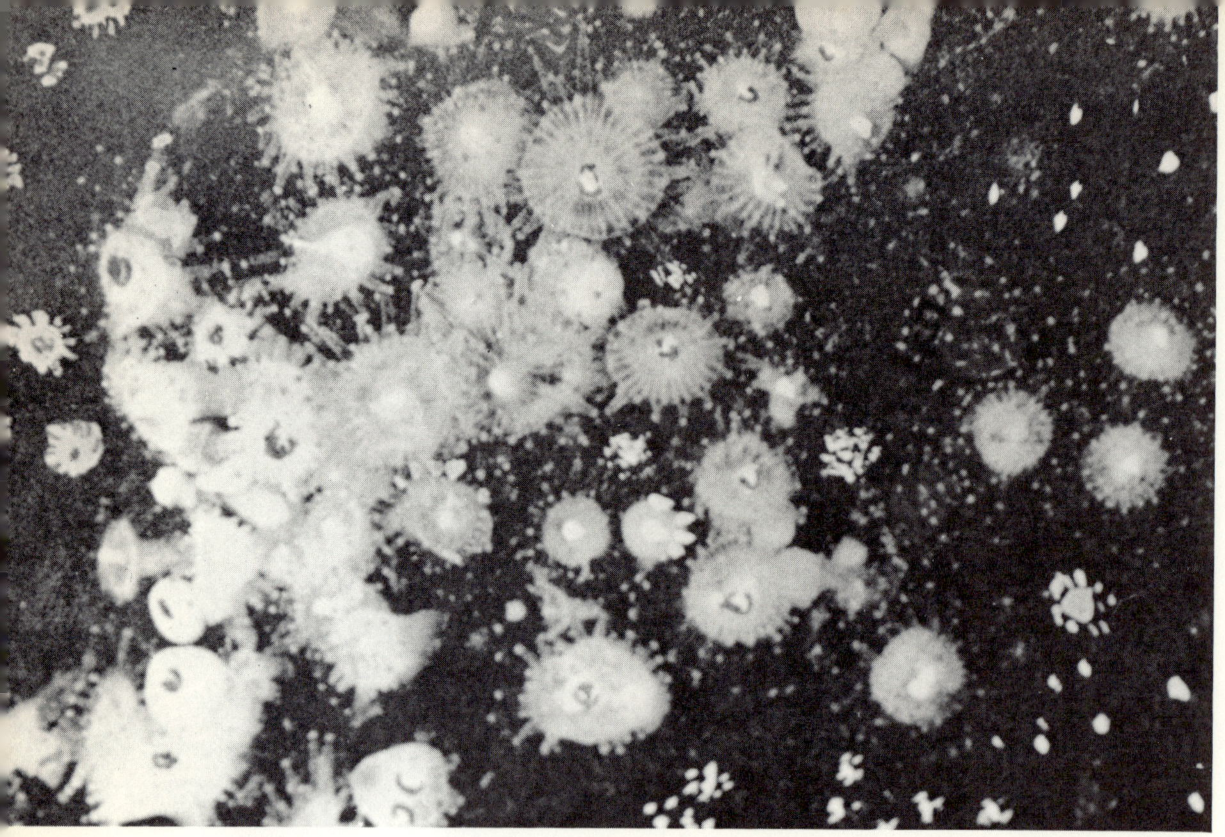

Sea anemones are one of the lowest forms of marine animal life. They have no backbone.

ing, or mouth, a short tubular esophagus runs to the body cavity, this body is divided into a number of sacs. Along the sides of the esophagus run tubes that are always open and always functioning, for they are water tubes that bring in oxygen and flush out waste from the body.

Sea anemones are carnivores whose food is captured through the paralyzing effect of the stinging cells or wrapped by the pliable tentacles and passed into the esophagus and thence along to the stomach for digestion. Those portions of the food that are not digestible are rejected by a reversal of this system. Sea anemones are capable of devouring enormous quantities of food and will expand to great size and multiply at a very rapid rate. If food is scarce, they can reduce themselves to minute forms that in no way lose their general characteristics.

Sea anemones should not be underestimated as a danger to mankind. The United States Oceanographic Office does not take them lightly and on the back of one of their pilot charts, issued regularly to merchant mariners around the world, they present this message.

Coral polyps and sea anemones of the coelenterate group are often mistaken for sponges. In reality they are a more advanced form of life.

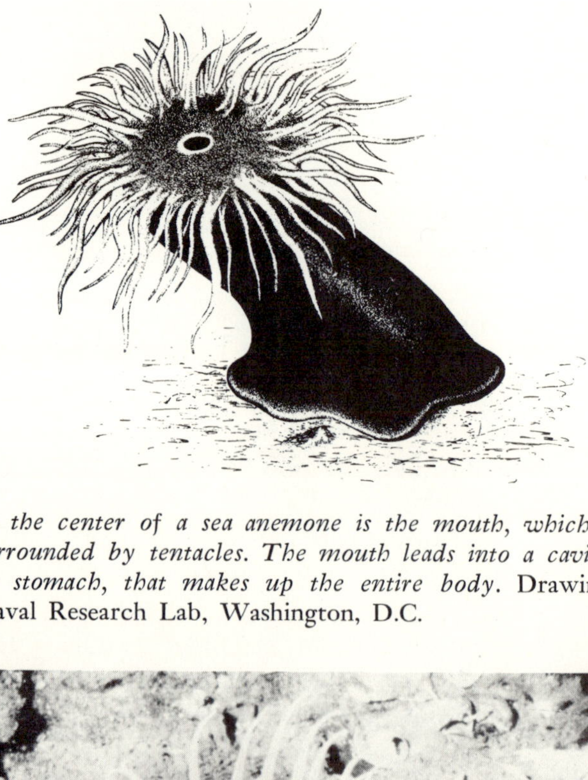

In the center of a sea anemone is the mouth, which is surrounded by tentacles. The mouth leads into a cavity, or stomach, that makes up the entire body. Drawing, Naval Research Lab, Washington, D.C.

A flowerlike sea anemone gently waves its deadly tentacles, each of which is studded with numerous poison darts. Photo, Naval Photographic Center, Washington, D.C.

The coelenterates may be likened to an uncorked flexible bottle, with a circle of tentacles at the mouth designed for capturing food. The points of interest, however, are the remarkable stinging cells which are used for the capture of food or as a defense against enemies. This apparatus consists of a trigger hair which, when touched, actuates a spine followed by a hollow thread through which a paralyzing drug is injected into the victim. The stinging cells, or nematocysts, are the source of many illnesses encountered by skin divers, sponge fishermen, and other marine workers.

According to the Maritime Safety Division, Naval Oceanographic Office, the venomous sea anemones and corals include the following: the elk horn coral (*Acropora palmata*), which inhabits the Florida Keys, Bahamas, and West Indies; the sea anemone (*Actinia equina*), which inhabits the Atlantic from the Gulf of Guinea to the Arctic Ocean and Mediterranean and Black seas, including the Sea of Azov; and the rosy anemone (*Sagartia elegans*), which inhabits eastern Atlantic Ocean waters from Iceland to the African coast and the Mediterranean Sea.

It is only a short step from dealing with anemones to discussing corals, specifically stinging corals. The tiny coral is a flowerlike animal that lives in colonies. The difference between the two is that anemones have only a cellophanelike outside covering, whereas coral supplies itself with a rock-hard protective coating. When the true anemones die, they disintegrate; whereas corals, after death, remain as a hard platform for new generations of corals. In this tiny stepladder fashion, corals have attained construction miracles that man will never be able to equal. Corals have built modest little reefs, which they then have structured into sizable islands, which were then amassed into mighty hills, which they eventually made to form an entire range of mountains.

Nature performs this never-ending task in a stunningly beautiful fashion. There are red, blue, green, brown, yellow, pink, and yes, even black corals. Some are woven like delicate lace; others take the shape of fans; some build themselves into hornlike affairs; one species takes the shape of a man's brains. Some corals are hard, some soft. Some glow in the dark. Regardless of shape, size, or color, most can sting to one degree or another.

Coral polyps extract minerals from the sea and

Corals are actually single-celled animals living in colonies. Photo, United States Naval Oceanographic Office, Suitland, Maryland.

Steps showing a sea anemone in process of catching a fish. Drawing by the author.

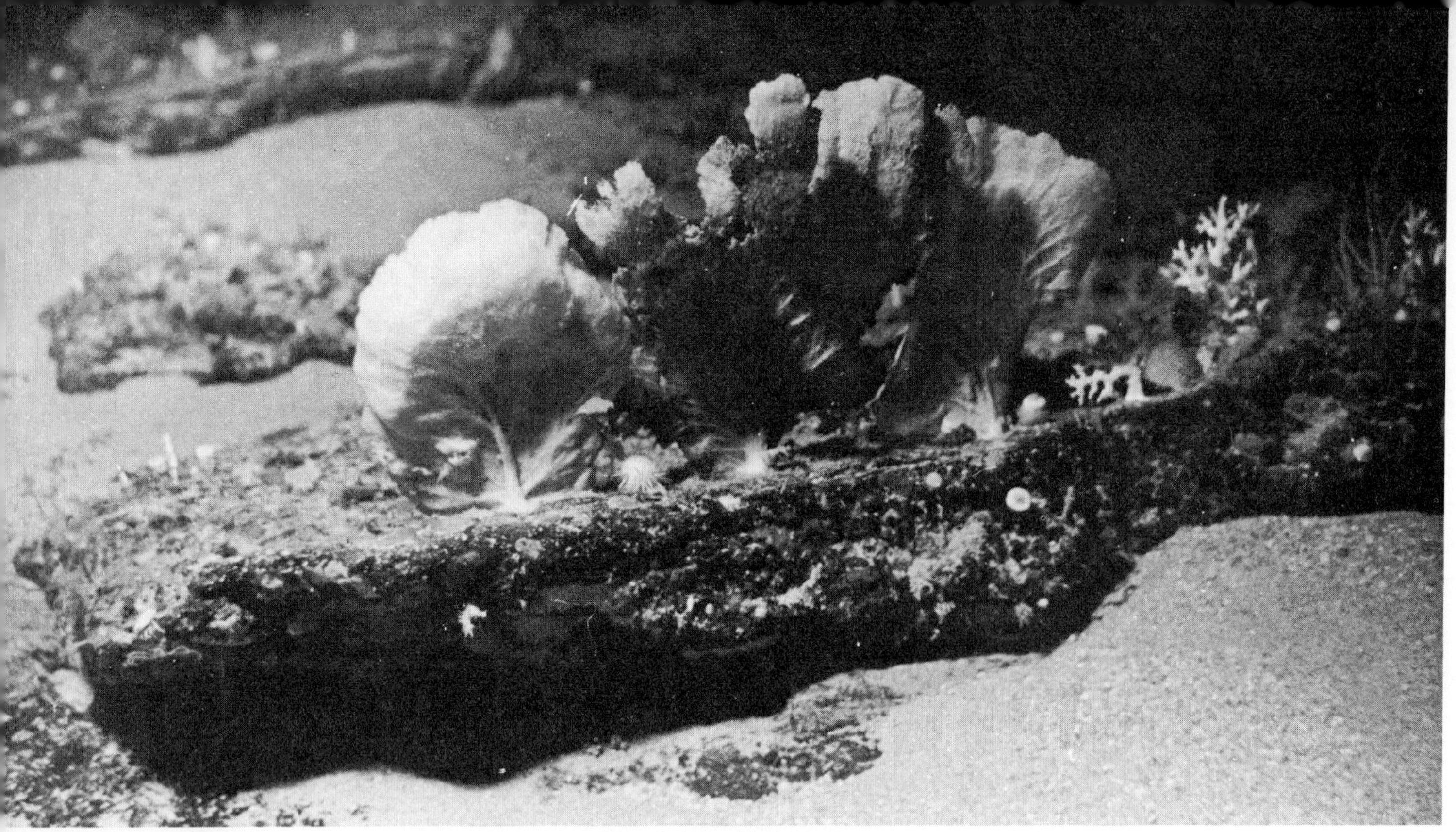

Large stony corals in a reef start as larvae, which change into polyps that become the parents of an entire colony. Photo, Bureau of Commercial Fisheries, Washington, D.C.

Some corals are shaped like flat plates; some, like lacelike fans; and some are round. Photo, Naval Photographic Center, Washington, D.C.

Brain coral (lower right) *is only one of the many unusual species of coral.* Photo, Naval Photographic Center, Washington, D.C.

Stinging corals come in numerous shades and deep colors. Photo, Naval Photographic Center, Washington, D.C.

supplement this with material taken from the billions of lime-covered microscopic animals that are their daily fare. They transform this digestive mixture into a housing as hard as rock and, in so doing, tend to create a glorious undersea habitat for a wonderland of fish life.

Stinging corals have poison cells similar to jellyfish. However, a further danger to man is the fact that the most delicate-appearing corals have razor-sharp edges. Both the poison and the cut are capable of producing red welts or wounds. Both the sting and the cut are slow-healing and painful. Thus, when swimming underwater where it is likely that anemones or corals will be touched or handled, it is best to wear heavy leather gloves.

Sea anemones and corals are not without deadly enemies and trustworthy friends. A tiny fish called a clown fish is friendly. Its bright, clownish colors suggest the appropriateness of the name; on the other hand, there are fishermen who say that it comes by the name *clown* through foolhardy actions. There is one sea anemone that possesses an open face almost three times as large as the largest sunflower. When a small crustacean or fish touches upon its sensitive areas, it will snap closed and devour its catch. Should uneatable parts be involved, like the shell of a crab, it will eventually spit out the hard pieces, sometimes turning itself inside out in the process. The clown fish swims into the open face of the anemone and has even been seen incased in the closed mouth. Yet somehow, it always swims free. Scientists say that the clown fish is an easily recognizable friend and enters the mouth to eat residue food clinging to the anemone's stomach.

An enemy is the parrot fish, so called because his teeth have become fused into a hard, parrot-like beak. This beak the parrot fish uses to bite off chunks of living coral, stinging or otherwise. Apparently the stinging apparatus of the coral is lost on a parrot fish, for it shows nothing but relish with each mouthful.

Recently much has been written about the crown-of-thorns starfish (*Acanthaster planci*). It is a coral-eating animal that looks just like its name. It is mildly poisonous to the touch and crawls slowly across the face of living coral, devouring the tiny inhabitants as it goes. The result is that the coral dies and crumbles into a sandlike substance. Because of a massive infestation of crown of thorns sea stars, hundreds upon hundreds of acres of coral have been ruined. Reefs have begun to crumble; islands have been found to be disintegrating, and nature's entire cycle of land-building coral has been placed in jeopardy.

According to two scientists from the Naval Oceanographic Office, Bernard J. Zahuranec and Willard Vary, who went to Guam to study the impact of the crown of thorns on the coral reef there: "This starfish is destroying coral in the vicinity of Guam at the rate of half-a-mile a month."

A colorful "mock eye" fish feeds amidst a deadly world of stinging coral, sea urchins, and other marine life. Photo, Naval Photographic Center, Washington, D.C.

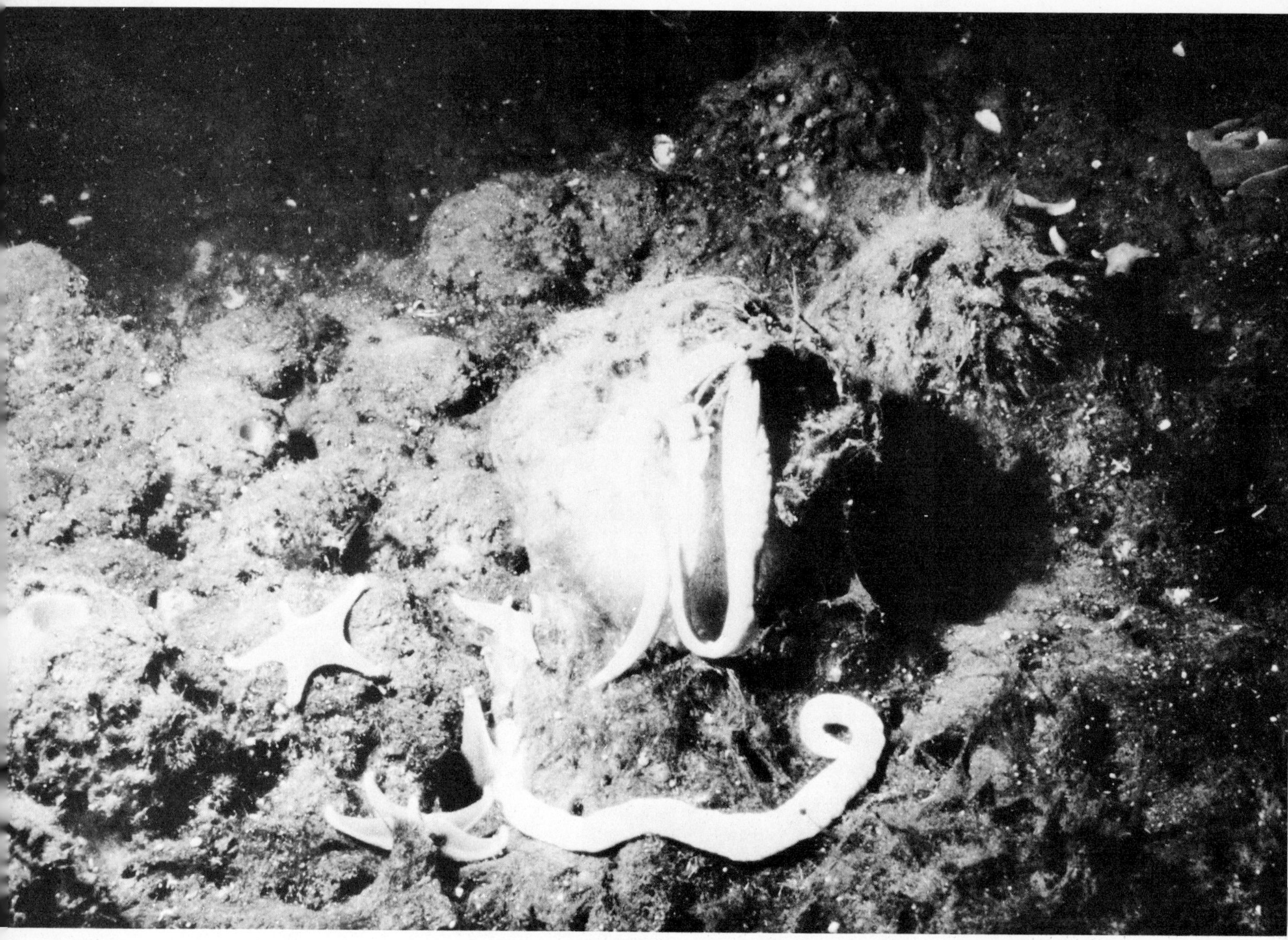

Some sea worms are known to grow up to 90 feet in length. Here a small worm lives amidst deadly corals, poisonous fishes, sponges, and starfish. Photo, Naval Photographic Center, Washington, D.C.

18

Cone Shells, Segmented Worms, and Sea Urchins

Dangerous cone shells are a topic of special interest to anyone fascinated by beautiful sea shells because some of the glistening, brightly colored, and entrancing cone shells are quite capable of inflicting a poisonous bite. The family Conidae, to which they belong, is known to include over five hundred beautiful species.

Because many of these beautiful cone shells were used as money in the early history of various civilizations and therefore have become museum items of public display, a large group of hobbyists is constantly seeking them out. Herein rests the danger, for these cone shells are generally well hidden in sand, beneath a piece of coral or two, with only a small portion of the animal showing. When this exposed portion is viewed through several inches of sea water, it is almost impossible to judge if it is a safe or dangerous part of the animal. Consequently, many people have been stabbed by their poison barbs before they actually touched them.

Live cones belong to the large classification of mollusks, phylum Mollusca. It is estimated that there are 45,000 living species of mollusks. This classification includes invertebrates having a soft body and usually secreting a calcareous shell. A muscular foot is present, which may be modified to serve various functions. Covering at least a portion of the body is a soft skin called the mantle, the outer surface of which secretes the shell. Respiration is accomplished by means of gills or a modified pulmonary sac. Some species have jaws. In four of the five species, food is obtained by use of a rasplike device called a radula. In the cone shells, the radula ribbon is lost and the teeth are modified into hollow, harpoonlike structures.

Mollusks consist of three different classes: Amphineura, Scaphopoda, and Gastropoda. There are about 33,000 members of the class Gastropoda, or gastropods, including cone shells. The most important part of a cone shell is the poison barb. These injectors emerge from the pointed end of the shell and can extend to a length equaling about one-third the length of the shell. However, in some species, the probe can reach far beyond the end of the shell. Although the animal itself is a slow-moving and somewhat lethargic creature, the proboscis is capable of lightning-quick strikes with amazing capabilities to pierce. To make the cone safe for handling, take a stick, stone, dead sea shell, or other inert object and roll the cone shell back and forth; this will cause the animal to withdraw into its shell.

Slor Jones, a missionary-schooled Australian aborigine, said that "people are poisoned not by the finding of our beautiful cone shells, but during

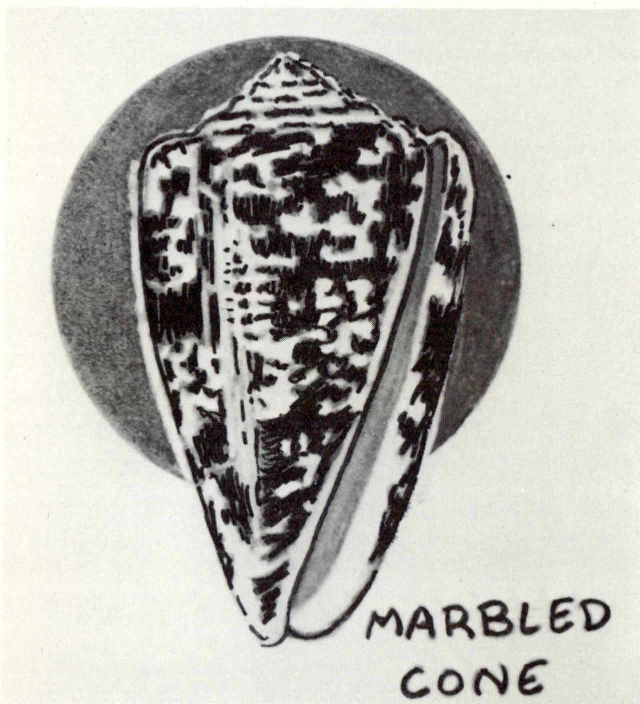

A poisonous marbled cone shell.

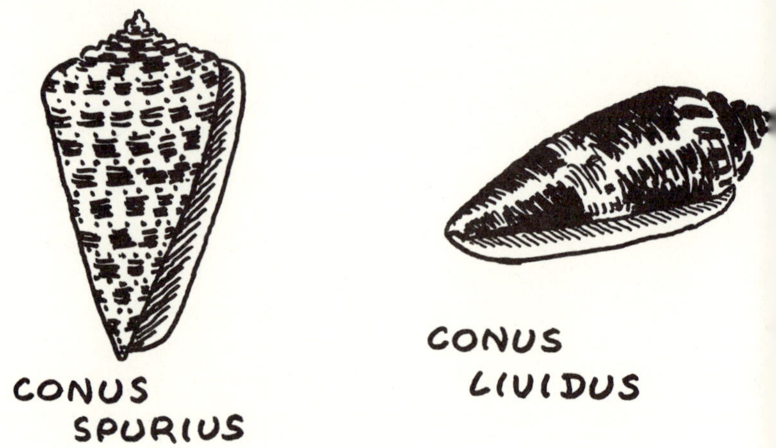

Conus spurius *and* Conus lividus *are known to be toxic.* Drawing by the author.

A drawing of the various parts of a poisonous cone shell. The poison is released through the proboscis. Drawing by the author.

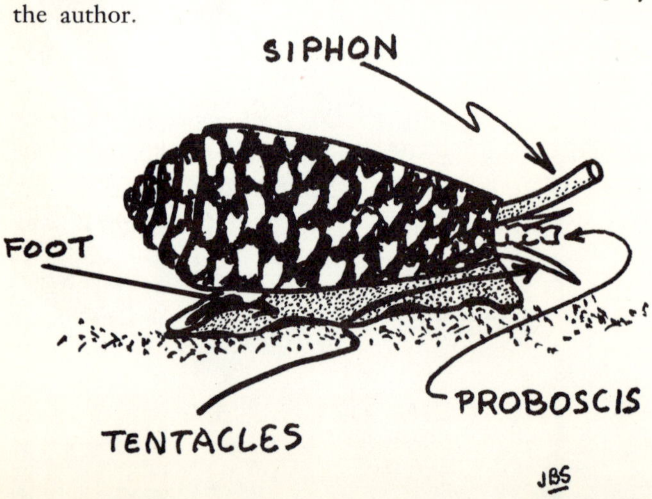

A Juno's volute.

the later handling. For they find shell, careful about placing same in canvas sack and return to hotel. There they thoughtless reach hand into sack and *zing* go poison."

Slor Jones also stressed that amateur shell collectors will hold a newly discovered cone shell in the flat of their hand while studying it in sunlight or under a lamp; this allows the cone shell to dart its probe into the hand, thus causing a painful and serious wound.

According to the Maritime Safety Division of the Naval Oceanographic Office, the following are some of the more dangerous tropical species of cone shells:

(a) Court cone (*Conus aulicus*), which is found from Polynesia to the Indian Ocean.
(b) Geographer cone (*Conus geographus*), which inhabits the Indian Ocean and the Pacific Ocean from Polynesia to eastern Africa.
(c) Marbled cone (*Conus marmoreus*), which ranges from Polynesia westward to the Indian Ocean.
(d) Striated cone (*Conus striatus*), which inhabits the area from Australia to eastern Africa.
(e) Textile cone (*Conus textile*), which ranges from Polynesia to the Red Sea.
(f) Tulip cone (*Conus tulipa*), which ranges from Polynesia to the Red Sea.

The sting of a cone shell, according to the Maritime Safety Division, usually produces a numbness or burning sensation. Numbness and tingling may spread rapidly and become particularly pronounced about the lips and mouth. Respiratory distress is usually absent. Paralysis and coma may follow. Death may be the result of heart failure.

A specific treatment for cone shell stings has not been discovered. Efforts should be devoted to alleviating the pain, combating the effects of the poison, and preventing secondary infection. The wound should be properly irrigated and washed clean with sterile saline, if available. Suction may be applied to remove the poison. A small incision may be required A thirty-minute soaking in hot water or with hot compresses, depending on the location of the sting, has been recommended. Under conditions of unfavorable delay in treatment, the administration of antibiotics may be desirable. The patients should be kept warm, and stimulants may be required. Early hospitalization and treatment is recommended (as in the case of venomous fish sting).

The question then is: What is the best way to handle cone shells? The answer is surprisingly simple. They may be boiled without doing damage to the shells. This is accomplished by dropping them into a cauldron of hot water, preferably when they are partially extended, and letting them boil for about twenty minutes. After this, the meat may easily be removed by winding it from the shell with an ice pick or some other pointed object.

It should be noted that cone shells will fade if exposed to sunlight for a long period of time. This might prove important because in comparing a live shell with an old one, the considerable difference in color and texture may create the false impression that the more recent one is of a different species and therefore not poisonous. A dangerous miscalculation.

Should a cone shell bite occur, it must be kept in mind that all species are endemic to one or more areas. Hence, if a bite happens in a given tract, the particular spot must be noted. This might afford a local doctor some clue as to necessary treatment.

Cone shells, at least of the type with which we have concerned ourselves, are carnivorous. They use their powerful sting to capture fish, worms, and other mollusks. These it will paralyze, envelope, and take back into its shell to digest. Breeding and egg laying are known to take place during summer months. The cone shells move into shallower waters to lay eggs, which presents a further danger to waders in certain areas.

Segmented sea worms have tufted, silky, chitinous bristles in a row along each side. Upon contact

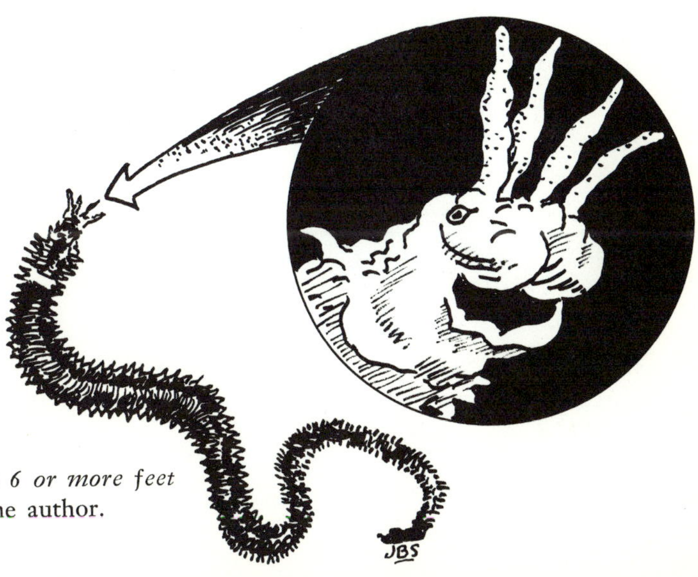

This sea worm (Onuphis teres) *grows to 6 or more feet and has well defined jaws.* Drawing by the author.

The sea urchin comes in spherical, heart, and discoid shapes. Its spines are poisonous to some degree.

A sea urchin is a relative of the common starfish. Photo, Naval Photographic Center, Washington, D.C.

or stimulation of any kind, the bristles rise on end, like the hair on a dog's back. The worm then contracts and presents a defensive ring of armor. The fine bristles penetrate the skin in much the same way cactus spines do. They are difficult to remove and cause a burning sensation. Later, the area becomes inflamed and may swell or become numb. Other worm species have strong jaws that inflict a painful bite. These worms, according to the Maritime Safety Division, Naval Oceanographic Office, up to 12 inches long, may be encountered under rocks or coral.

Representative species of both types are:

1. Bristleworm (*Eurythoë complanata*), which is found in the Gulf of Mexico and throughout the tropical Pacific area.
2. Bloodworm (*Glycera dibranchiata*), which is found on the Carolina coast and northward into Canadian waters.

The bite of the bloodworm may or may not penetrate the skin. The bite is round, with a red dot in the center. It is usually surrounded by a pale area, later becoming hot and swollen, and then turning numb and itchy. A bristle worm contact, on the other hand, produces inflammation, swelling, or numbness that may persist for several days.

Since it is not definitely known whether the bite is poisonous or not, all sea worm bites or stings should be treated in the same manner as fish stings. Bristles are best removed with forceps, since scraping may break the bristles off and complicate matters. An effective method of removal is by the application of adhesive tape over the bristles. After removal, the area may be treated with ammonia or alcohol to alleviate discomfort. Certain sea worms inflict a painful bite and should not be handled, except with protective gloves.

Sea urchins are characterized by a soft round body covered with rather long spines. They are found in large numbers and varieties in shallow coastal waters. Certain sea urchins are of prime danger to the diver or swimmer. They may inflict deep wounds and may be difficult to identify until after a very painful encounter.

All sea urchins have spines, which vary greatly from species to species. Most spines are solid, with blunt or rounded tips, and are not venomous. Others, however, are long, slender, sharp, and hollow, permitting easy and deep entry into flesh. Because of extreme brittleness these spikes may be difficult or impossible to withdraw in one piece. Some of these may be a foot in length and may secrete a deadly venom.

Distributed among the spine tips are small, delicate, globe-shaped, seizing organs called pedicellariae. This sometimes serves as the venom organ

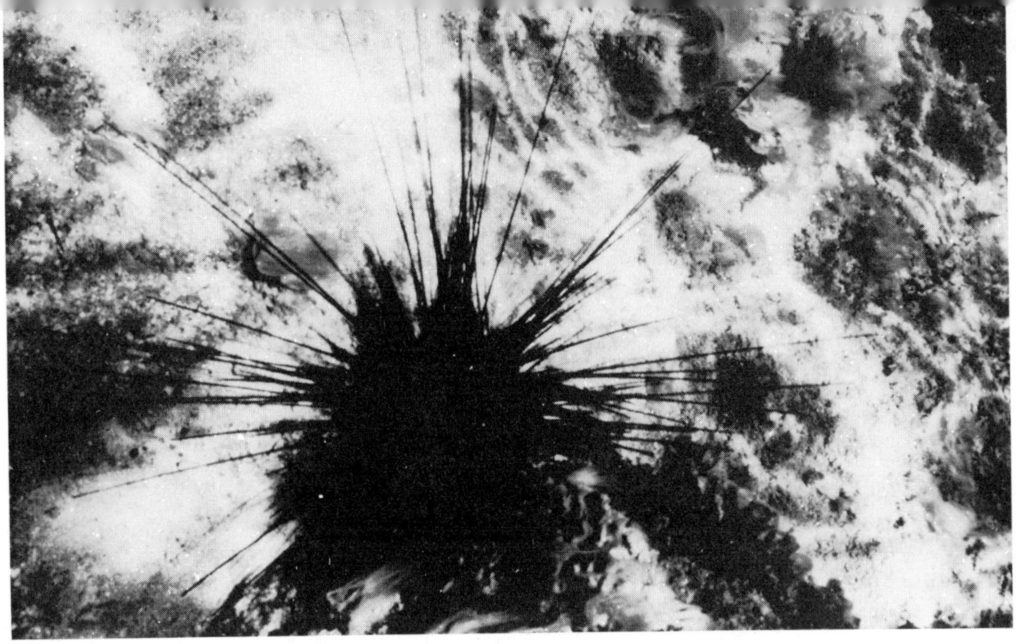

A sea urchin moves slowly along the bottom of the sea on small feet located on its underside. Photo, Naval Photographic Center, Washington, D.C.

and is armed with a set of pincerlike jaws for holding. A sense bristle, or contact, causes a small muscle to contract, releasing venom. A function of the pedicellariae is the sea urchin's self-defense; if any object touches the extended organ, it is immediately seized and poisoned. If the object is large and strong, it will tear away from the urchin, but the pedicellariae remain fast on the object and continue to poison for several hours after being parted from the sea urchin.

There are many species of sea urchins according to the Maritime Safety Division, Naval Oceanographic Office. Here are a few representative ones: the long-spined, or black, sea urchin (*Diadema setosum*), which is distributed throughout the Indian and West Pacific Ocean areas northward to Japan and eastward to Hawaii; the sea urchin (*Toxopneustes pileolus*), which inhabits the Indo-Pacific area from eastern Africa to Melanesia and Japan; the sea urchin (*Toxopneustes elegans*), which inhabits Japanese waters; and the Japanese sea urchin (*Asthenosoma ijimai*), which inhabits the Japanese area southward to Molucca Sea.

When the spines of a sea urchin penetrate the skin, the victim usually experiences an immediate and intense burning sensation. This is followed in a short time by redness, swelling, and a generalized aching sensation. Muscular paralysis has been reported, and secondary infections may ensue.

The sting from sea urchin pedicellariae may produce immediate distress and, in some severe cases, death. The sting produces intense radiating pain, faintness, muscular paralysis, loss of speech, and respiratory distress. However, it may be of short duration, diminishing in fifteen minutes and disappearing in an hour.

Sea urchin stings should be handled in the same way as other venomous stings. Prompt attention must be given to the removal of the pedicellariae because they will continue to introduce venom if not extracted. Although the ordinary spines of some species may be absorbed in the course of a day or two, most will require removal. A purple discoloration in the area of the wound, which is caused by a secretion of the animal, is not a cause of concern.

Sea urchins with long needlelike spines should not be handled. Special protection must be arranged because ordinary leather or canvas gloves and shoes do not afford protection. Marine workers and others must exercise extreme care in working at night without adequate light.

The radial symmetry of a sea urchin is masked by its numerous deadly spines. Photo, Naval Photographic Center, Washington, D.C.

Lion fish, or zebra fish, are brilliantly colored and highly toxic. Photo, Rex Gary Schmidt, Bureau of Sport Fisheries and Wildlife, Washington, D.C.

19

Toadfish, Scorpion Fish, and Sea Snakes

The family Batrachoididae, or toadfish, is characterized by their broad, depressed heads, large mouths, and repulsive appearance. Most species live in oceans, but some are found exclusively in rivers. They live on the bottom, hiding in crevices, burrows, under rocks, debris, among seaweed, or lying buried under a few inches of sand or mud, with only their eyes protruding. They are experts at camouflage and blend in with the background. They are generally sluggish in their movements, but when going after food, they can dart out with great rapidity. They are omnivorous in their eating habits but seem to prefer crustaceans, mollusks, worms, and small fish. Toadfish are generally not valued as food fish.

In studying these dangerous fish, it should be noted that toadfish are also known under such names as oyster crackers, leopard toadfish, bagre, and sapo, among others. Toadfish can be green, brown, yellow, or a mixture of any of these colors. The belly is generally light, and the main body is marked with dark patches and bars.

Toadfish, although small (ranging a little over 1 foot in full length), are doubly hazardous to fishermen. They have a bone-shearing bite that can sever fingers, and they have thorns that are

Toadfish are small bottom-dwelling fish that inhabit warmer waters. Drawing, Naval Research Laboratory, Washington, D.C.

deadly poisonous. Since toadfish are common on sandy or muddy bottoms, they lay their eggs under almost any sunken object, such as old tires, rocks, seaweed, tin cans, or logs. They spawn in late spring and summer, with the male guarding the eggs for approximately three weeks, until they hatch.

Toadfish stay hidden most of the time, but they will dart into the open after passing fish. They are voracious eaters and, all too frequently, become fouled in bottom trawls being worked by commercial fishermen. Since they have no commercial value, these fish must be hand-culled out of the catch and tossed overboard, which is decidedly dangerous. Toadfish, when caught, erect their poisoned fins and generally have to be cut free of the hook. If handled, they will make grunting noises, the significance of which is not known. The flesh of the toadfish is poisonous and should not be eaten.

The way in which toadfish deliver their poison is worthy of study. The animal has four deadly spines: one on either gill cover and two along the upper, or dorsal, edge. These spines are long, thin, and hollow; at the base of each is a tiny bladder-like contrivance that holds the venom. Native fishermen in Mexico, who seem to be somewhat immunized to toadfish stings, say that the pain radiates and is severe; they liken it to the sting of a scorpion.

Toadfish of one variety or another inhabit most warm, torpid waters. The Red Sea abounds with *Barchatus cirrhosus*. The oyster toadfish (*Opsanus tau*) is plentiful along the eastern seaboard from Massachusetts to the West Indies. Others abound in the Mediterranean, Ceylon, Burma, Malaya, and India.

A far more deadly fish, the stonefish (*Synanceja horrida*) is often mistaken for the toadfish. A stonefish is a much rarer fish, although it is found in many of the same waters. It has an array of dorsal fins capable of injecting poison. The Council for Scientific and Industrial Research, Commonwealth of Australia, says that the venom organ in the deadly stonefish varies considerably from one species to the next. They also cite examples of young people, who have stepped upon a stonefish, turned raving mad within minutes, and died within hours. It as a bottom dweller, is most common in shallow waters around portions of Australia, and possesses remarkable abilities to completely camouflage itself and blend into the background.

Scorpion fish belong to a large family (Scorpaenidae) of fish that vary greatly in size and danger.

Certain species of venomous scorpion fish attain large size. In some parts of the world they are valued as food. Drawing, U.S. Department of the Interior, Washington, D.C.

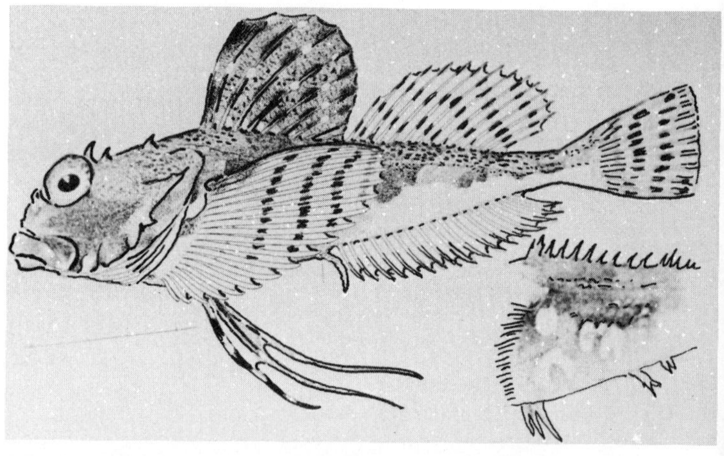

A scorpion fish is seen here resting on an underwater laboratory. These barbed fish seem to gather around any such sea lab on the bottom of the ocean. Photo, United States Naval Photographic Center, Washington, D.C.

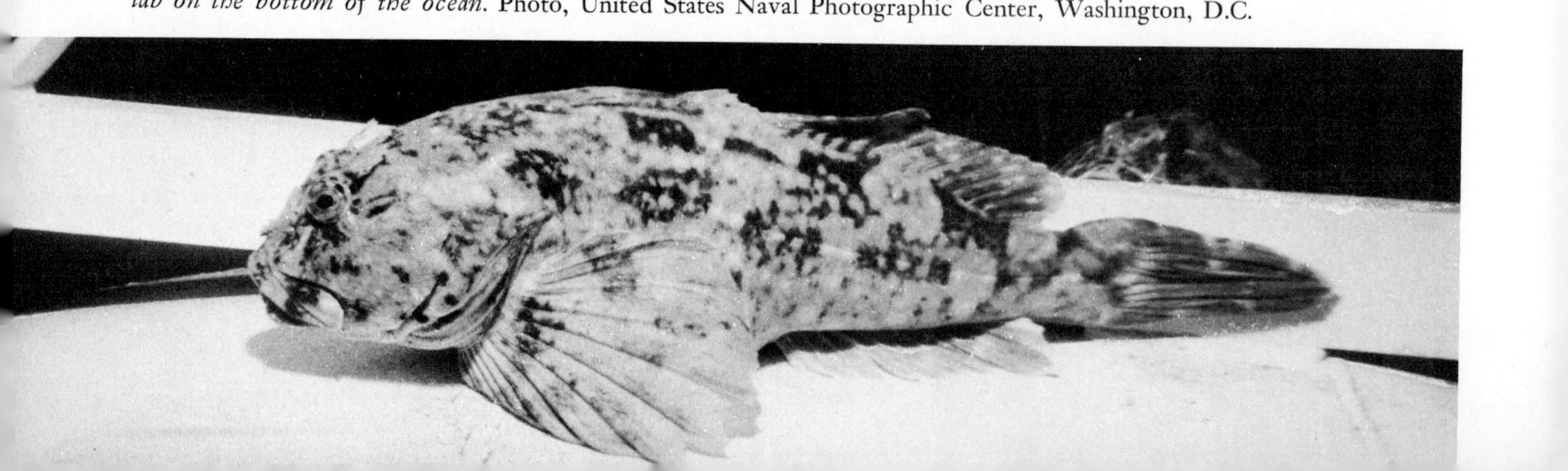

A stonefish is said to be the world's ugliest fish. It is also one of the ocean's most poisonous. Drawing, Naval Research Laboratory, Washington, D.C.

They are found in many tropical and subtropical seas, within the Indo-Pacific complex, the Red Sea, the Indian Ocean, the China Sea, the Sea of Japan, the South Pacific, and especially in the rich ocean waters around Australia. (The stonefish is also a member of the family Scorpaenidae and not of the family Batrachoididae, like the toadfish it so closely resembles.)

The deadly *zebra fish* (*Pterois volitans*) is often said to be the most named fish of all the scorpion fishes and also one of the most beautiful fish in the sea. In the United States, it is also known as a lion-fish or a turkey fish; in the Marshall Islands, it is a ho; in the Philippines, it is a tandan or a lolong; in Australia, it is a red firefish or a butterfly cod; in South Africa, it is devilfish; in England, it is rock perch, butterfly cod, or tiger fish; and in many other parts of the world, it is known by numerous other titles. However, no person will ever deny that the zebra fish, which grows to about 1 foot in length, is a brilliantly colored species of marine life.

Zebra fish parade about coral reefs, spreading their fanlike fins in the same manner as a peacock sauntering around a garden displaying its feathers. Generally, they seem to move in pairs and can be easily provoked into a state of aggressiveness, during which they will attempt to back their spines into their annoyer.

Zebra fish are shallow-water inhabitants and can also be found in crevices among coral, debris, under rocks, or in seaweed that affords them good camouflage protection. When they are discovered, these fish will erect their spiny fins and spread out their gill covers in a manner that makes their anger perfectly clear. They have eighteen venomous spines amid their red, white, blue, and black plumage. There is one poison spine in each ventral fin, three in the anal fin, and thirteen in the dorsal fin. All have their own poisonous tissue, which, upon pressure, will also.break off in the wound along with the spine.

The family Scorpaenidae contain some of the most virulent of all venomous fishes. The sting of any of these fishes will produce serious results, and a few of the stonefishes, may be ranked with the cobra in the deadliness of the poison secreted. Symptoms produced by the sting of a scorpion fish vary to one degree or another, but the pain in all cases seems to be immediate, sharp, and radiating. Pain may cause a victim to thrash in a wild manner, scream, and eventually lose consciousness. The immediate area of the wound may be pale, surrouded by a red zone, swollen, and hot to the touch. Paralysis of an entire arm or leg may result, and intense swelling of a member may impair movements. Many other symptoms may also be present. The victim may die or may recover after months of treatment but with a general impairment to health. Treatment, as with other venomous fish stings, must be administered promptly.

Lion fish, also called zebra fish or turkeyfish, inhabit the Indian Ocean, Red Sea, China Sea, Sea of Japan, and numerous other oceans. Photo, John G. Shedd Aquarium, Chicago, Illinois.

Based on past experience, it has been found that protection against the sting can best be attained by the wearing of heavy leather gloves and avoiding, if possible, any physical contact with the fish. The impulses of the uninitiated to stroke the delicately feathered zebra fish or to poke fingers in mysterious underwater holes are dangerous and should be warned against.

Treatment of venomous fish stings should be directed toward these objectives, according to *Science and the Sea:* alleviate pain, combat the effects of the venom, and prevent follow-up infection. According to this publication, the pain is generally a direct result of the damage caused by the spine or barb. Additional symptoms are produced by the venom and the introduction of slime and foreign matter. Certain stings, such as those of rays and catfish, may produce greater damage if spines are curved or barbed. Such wounds should be carefully washed out or irrigated with cold salt water or with sterile saline, if available.

It is advisable in most cases to make a small incision across the sting wound. Immediate suction for removal of poison should then be applied. But since fish do not inject venom the way snakes do, the results of this suction are not entirely satisfactory.

Use of a ligature, or tight bandage, in the treatment of the fish stings may help. The ligature, if used, should be placed immediately between the sting and the body and as near the wound as possible. Adequate circulation must be allowed every few minutes. Hot soaking of the wound every thirty minutes to an hour should be continued, with the water as hot as the patient can bear without injury. Supportive measures, as described in *The Ship's Medicine Chest and First Aid at Sea*, should be followed as quickly as possible. The assistance of a physician should be obtained, or medical advice requested by radio.

In Part I of this book, many references were made to true marine reptiles, or *sea snakes*. It is only proper that we eliminate any possible confusion of sea serpents of monstrous size and sea snakes as we now know them. The Maritime Safety Division of the Naval Oceanographic Office says that notwithstanding all that has been written on the subject of sea snakes, and the universal fear that most people have of them, there has always been a great deal of misunderstanding concerning their actual existence. Sea snakes are closely allied to the cobra and form a specialized group, adapted by structure and habit to a marine existence. All are poisonous, and many are deadly; however, generally they will not attack without provocation and have often been described as docile in behavior. Fishermen who unintentionally contact sea snakes in their work suffer most of the fatal bites. Fatalities are quite common in the Gulf of Siam and the Philippine area.

Few sea snakes exceed 4 feet in length, but 10-foot speimens have been caught. They are dark above and light below, with crossbands of black, purple, brown, gray, green, or yellow. All the marine species may be distinguished from the land species by a paddle-shaped tail. Unlike eels (with which they are sometimes confused), sea snakes are true snakes and their bodies are covered with scales. The eyes are covered with transparent scales and have no lids. The tongue is slender and forked. Hollow venom fangs are present.

Marine snakes inhabit sheltered coastal waters, particularly the areas near river mouths. They penetrate upstream to the limits of brackish water, and a few species are found in fresh water. They tend to collect close to shore and among coral reefs in the breeding season, when they are more active and sometimes quite aggressive. During this season, fishermen may net as many as a hundred at one time. This presents an especially hazardous condition, since some sea snakes have been reported to possess venom more potent than that of the king cobra.

Swimming is accomplished by a lateral undulation and use of the oarlike tail. A peculiar ability to move backward as well as forward has been reported. Although sea snakes may float on the surface for an extended period of time, they are also able to remain submerged for many hours. Most of their food is obtained underwater and consists of fish, swallowed headfirst. Many of the items of food, such as eels and fish, are quickly killed by the first bite of the snake. Like many other marine animals, sea snakes often feed at night, and many are netted after dark, when they are attracted by artificial lights exhibited by fishing

A sea snake from the tropical Pacific. The detail at the upper right shows the fangs and venom sac (the white area just below the eye). Drawing, United States Naval Oceanographic Office, Suitland, Md.

boats. Individual snakes are frequently taken by line because they will readily accept a baited hook.

The yellow-bellied sea snake (*Pelamis platurus*) ventures beyond coastal waters and is the most widely distributed of all sea snakes. It is found in East Africa and Asian waters northward to Siberia, Japan's eastern coast, the Philippines, and eastward to tropical America. The sea snake *Enhydrina schistosa* is a more aggressive species that ranges from the Persian Gulf eastward to the northern coast of Australia. Other species, nearly fifty in number, inhabit the tropical Pacific and Indian oceans from the Persian Gulf to Samoa. They also follow the warm currents northward into the western Pacific to the Ryukyu Islands and Japan. They are not known to inhabit the Atlantic.

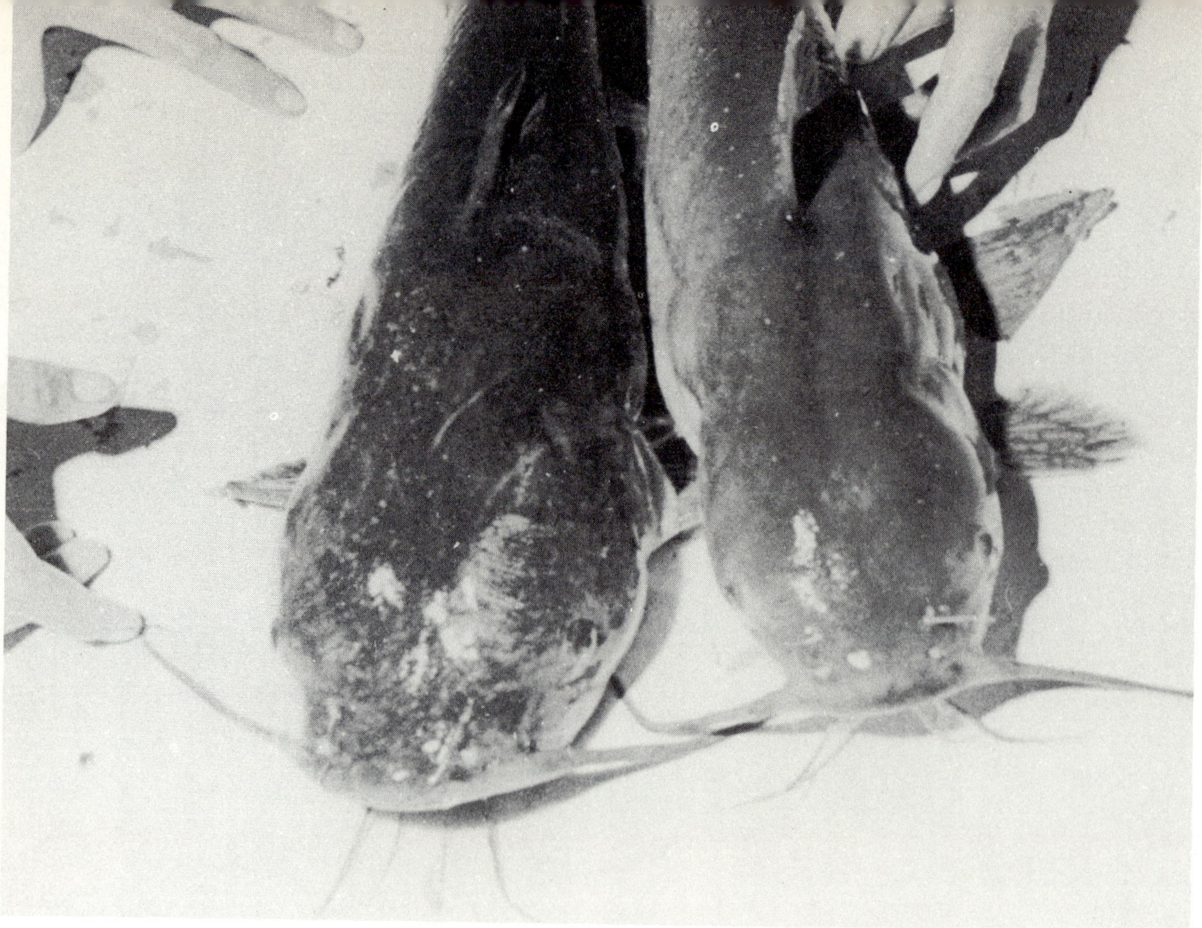

Sea catfish are dangerous to handle because of their sharp rigid spines, which can easily pierce the flesh of an unsuspecting person. Photo, Marineland of Florida.

20

Catfish and Weever Fish

Electric catfish are not the only species capable of injuring a human. There are about a thousand different species of catfish, of numerous sizes and shapes, and with a variety of weapons for defense.

Venomous catfish are provided with a stiff spine in the front part of the dorsal and pectoral fins. Venom glands are located in the outer skin, or sheath, of each. The venomous spine of the catfish is additionally treacherous to the unwary because of a locking device that keeps the spine erect when desired. Certain species have recurved teeth on each spine; this makes venom absorption by the victim more certain.

Although many freshwater catfish are used for food, the saltwater species are not often eaten. A representative marine species of the catfish (*Galeichthys felis*) is found from Cape Cod to the Gulf of Mexico. Another species, the labyrinthic catfish (*Clarias batrachus*), inhabits the waters of India, the Netherlands Indies, and the Philippines. Catfish are abundant in rivers, estuaries, and the open reef areas; they are also particularly numerous in certain large sandy bays.

The wound inflicted by a catfish spine is generally accompanied by an almost-instant stinging, throbbing, or scalding sensation. The pain may radiate or may localize, numbing an arm or other parts. Certain tropical species, such as the oriental catfish (*Plotosus lineatus*), are capable of inflicting an extremely painful wound that may endure for forty-eight hours. A paleness around the wound area is followed by redness and swelling. Gangrene may set in, and a secondary bacterial infection is possible. Death may result from a catfish sting.

As in the case of other fish stings, there are no known antidotes. Care should be taken in handling marine catfish, especially those having sharp, rigid fin species.

Weever fish, all members of the family Trachinidae, are a very venomous fish that live in the temperate zones. They are small marine fishes, less than 18 inches in length, and inhabit mostly sandy or muddy bays. Because of an aggressive temperament, combined with a well-developed poison apparatus, the weever presents a continual source of danger to divers and others. Weevers habitually bury themselves with only part of the head exposed. Upon little or no provocation, they dart out with fins erect and gill covers expanded and strike with unerring accuracy at any target.

The great weever (*Trachinus draco*) may be found along the coast of western Africa, in the Mediterranean Sea, and in the vicinity of the British Isles and Norway. The lesser weever (*Trachinus vipera*) inhabits the North Sea, southward along the European coast, and the Mediterranean Sea.

The armament of the weever consists of venomous dorsal and covered spines. The dorsal spines, five to seven in number, are needle-sharp and are equipped with a venom groove and thin-walled sheath. The bladelike opercular spine has an attached venom gland. The venom is similar to some snake venoms and acts both as a neurotoxin and a hemotoxin.

Venomous sea catfish have a single spine just in front of the dorsal and pectoral fins. Photo, Sport Fisheries and Department of the Interior, Washington, D.C.

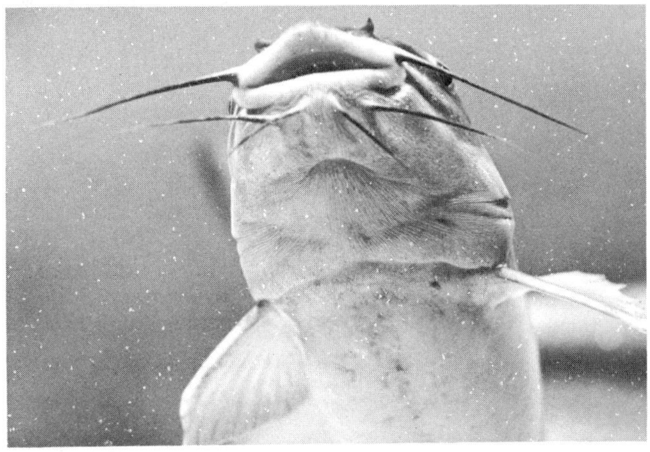

Sea catfish have spines that are unique and doubly dangerous because the fish can lock them into an extended position. Photo, Wometco Miami Seaquarium, Miami, Florida.

A weever wound normally produces instant burning or stabbing pain that gradually spreads in the affected part. Within thirty minutes, pain becomes severe to the extent that the victim loses consciousness. Because of the severity of the pain, the victim may scream and thrash about wildly. Other symptoms follow, such as headache, fever, chills, delirium, nausea, vomiting, sweating, loss of speech, palpitations, and convulsions. Morphine often fails to give any relief. Pain subsides in a few hours and redness and swelling then develop. Secondary infections are common in cases not properly treated. Recovery time extends to several months, depending upon the condition of the patient and the amount of venom received. There are no known antidotes. Immediate first aid and treatment by a physician may save the patient's life.

Since swimmers ordinarily encounter weevers while wading, care must be taken to avoid contact. Adequate footgear should be worn where weevers are known to exist. The fish should neither be antagonized into an attack nor handled in a careless manner, even if it is dead.

Weever fish (family Trachinidae) are small, generally not exceeding 20 inches in length. They dwell on the bottom of flat, muddy, or sandy bays and have daggerlike spines in their fins and gill covers. Drawing by the author.

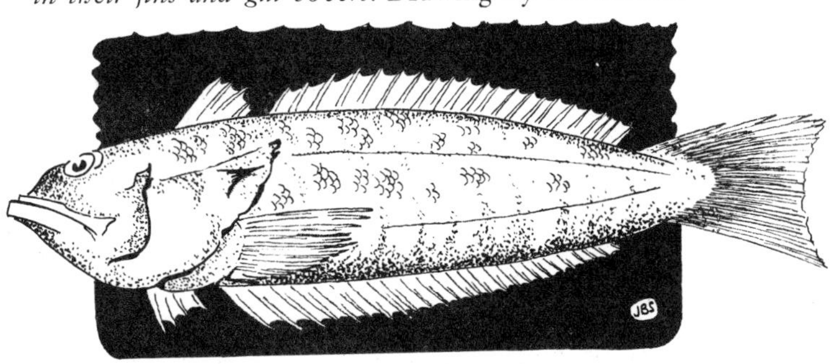

Red tide caused this mass death of fish and horseshoe crabs on the beach at Pass-a-Grille, Florida. Photo, Saint Petersburg Times, Saint Petersburg, Florida.

SECTION FOUR

Poisonous to Eat

21

Pollution

The president of the United States recently addressed the Congress on the matter of waste disposal. He said that "about 48 million tons of dredging, sludge and other materials are annually dumped off the coastlands of the United States. In the New York area alone, the amount of annual dumping would cover all of Manhattan Island to a depth of one foot in two years. Disposal problems of municipalities are becoming worse with increased population, higher per capita wastes, and limited disposal sites.

"We are only beginning to find out the ecological effects of ocean dumping and current disposal technology is not adequate to handle wastes of the volume now being produced. Comprehensive new approaches are necessary if we are to manage this problem expeditiously and wisely."

Every newspaper in the country has recently been carrying banner headlines about the severe pollution of our rivers. Obviously, this runoff reaches the sea, and there it settles into a massive cesspool that must of necessity breed sickness and death amidst valued marine life. When this deadly action is multiplied by all the sewage from the ships at sea and from the flushing action of the Yangtze, Rio Grande, Mississippi, Hudson, and Amazon

This sea bass, or giant grouper, is dragged ashore for burial on a Florida beach. It died from pollution. Photo, Saint Petersburg Times, Saint Petersburg, Florida.

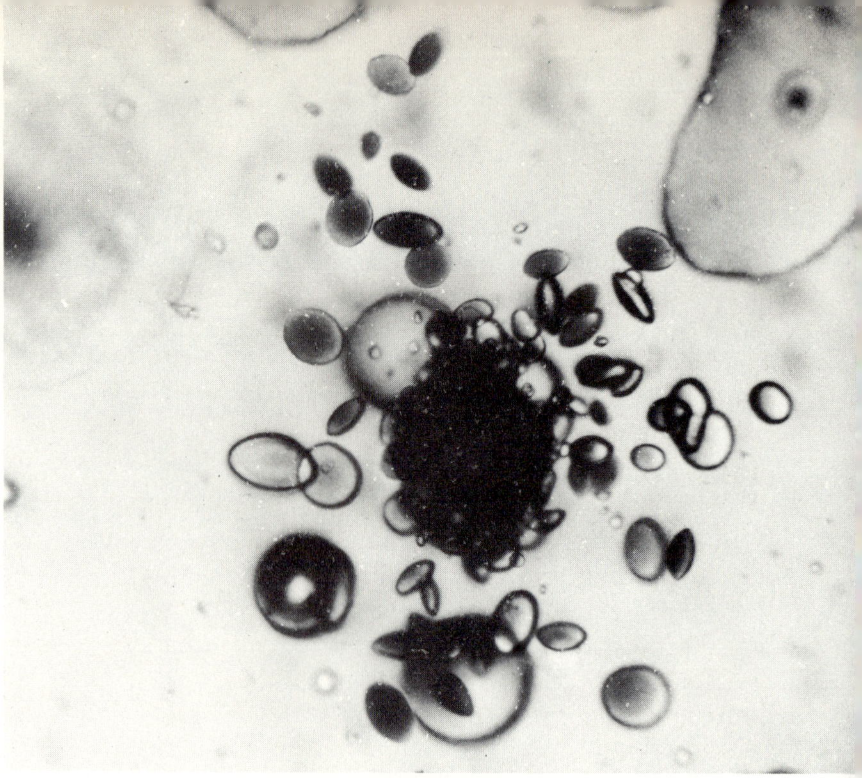

A single drop of oil, magnified 625 times, is shown being devoured by millions of bacteria. Each oval is a colony of more than a million bacteria. Photo, United States Naval Oceanographic Office, Washington, D.C.

rivers, plus untold thousands upon thousands of other waterways, then the massive volume of killing sludge that is hourly pumped into the sea is almost beyond belief.

Light, temperature, salinity, density, and pressure are very delicate factors in the sea. To modify even one of these multiples is to obliterate an entire trend of life. This has been studied very effectively in the vicinity of coral reefs, where fish life abounds. Here it has been noted that the flesh of a specific fish will be edible if the animal lived on one side of a coral reef, yet deadly to man if the animal lived on the opposite side of the reef. The question "why" would appear to be an easy matter to answer; it is not. The passage of poisonous materials created by man's pollution of the seas is a very complex matter. What creates a toxin and how it passes through the food web of the sea are problems man is just beginning to come to grips with.

Pollution creates billions and billions of parasites, and parasites are consumed in one fashion or another by every species of marine life. Furthermore, it has long been established that all marine life, from whales to microscopic one-celled animals, play host to one or more varieties of parasite. Some fish house more than others, and certain of them have parasites that live within the animal's individual cells. Sockeye salmon, for example, are known to house more than fifty different parasites, many of which may be unleashed to become pandemic by the addition of pollutants to their water.

As far as reef fish are concerned, certain parasites attach themselves to the outside of fish and can easily kill the host. Nature, however, has devised parasite-cleaning creatures that remove these minute killers from the gills, fins, skin, and even the inside of the mouth of the affected fish. It is a startling matter to see a comparatively small fish willfully swim into the open jaws of a giant barracuda and, after several moments of tediously picking at the long canine teeth, casually leave. This is a vivid illustration in the balance of nature that man can so easily topple. In Bermuda waters, it was noted that certain man-made pollutants pumped into the vicinity of a specific reef area had drastically damaged the value of once-edible fish. Upon scientific study, it was found that the pollutants had killed off most of the parasite pickers. As a consequence, large fish had abandoned the coral reef, and those who did remain were covered with sores.

Parasites are not the only ones affected by contamination; sharks and even whales are affected. Size is no criterion, nor is it a safeguard. But sharks, fish, and shellfish are different from humans, some people will argue. What is injurious to one might not affect the other. Let us take a hard look at the facts.

In 1953, the medical world became aware of a frightening new human ailment called Minamata disease. The details were sketchy, for it was only vaguely established that the debilitating ailment had been caused by marine organisms. The case fatality rate in the first series of fifty-two persons studied was 33 percent. Age was no factor. Several young boys went spastic, became irrational, and died in fits of apoplexy. Elderly people lost the use of their limbs and minds became paralyzed from muscle spasticity, and died in convulsions.

The affliction is named for the city of Minamata, located on the southwestern coast of Japan, adjacent to the Shiranuhi Sea. This quaint place had a population of approximately 40,000 people, most of whom were dependent upon local industries for their employment. A large chemical and fertilizer plant, which had been producing various chemical products in the region for several decades, began the production of vinyl chloride in 1949. The sewage and chemical waste from this factory were channeled into nearby Shiranuhi Sea until 1950, when a new channel was opened directly into nearby Minamata Bay. The bay is small, and although it was not generally used for commercial fishing, it was utilized by many of the local families as a source of seafood. Following the discharge of the factory wastes into the bay, a series of outbreaks occurred involving a total of eighty-three persons during the years from 1953 to 1960. In each instance, the victims had eaten crabs, shellfish, or fish taken from Minamata Bay. Furthermore, it was observed that cats and fish-eating birds were also affected. Then the investigating physicians made an interesting discovery. They found that the disease could be reproduced in laboratory animals that were fed fish taken from Minamata Bay. The causative agent was then believed to be an organic mercury compound. There appeared to be a correlation between the quantity of mercury in the fish tested and the amount of fish required to produce the disorder in experimental animals. Unusually large amounts of mercury were detected in the mud from Minamata Bay. Shellfish taken from the bay contained significant quantities of mercury. The organs of humans and cats dying simultaneously of Minamata disease and of cats fed shellfish experimentally contained large quantities of mercury. Further field investigations revealed that the mercury probably came from the outflow of the local chemical factory, since large quantities of mercury chloride were utilized as a catalytic agent in the production of vinyl chloride. Since the mercury compound in the seafood could not be extracted with organic acid or basic solvents, it was thought that a simple alkyd-mercury compound may have combined to form a protein complex that was released during the process of digestion.

An extensive United States government report on Minamata disease noted one especially important factor: When the contaminants were placed in water, they were readily accepted by the body of the host fish or shellfish without any visible signs of harm. However, these same contents, when passed on to humans, immediately became lethal.

Recently, the oceanographic community was startled to learn that an "atomic" whale had been discovered. Intense investigation revealed that the massive beast had been feeding on plankton from a nearby nuclear waste-disposal site and had thus become radioactive. It does not take a great deal of imagination to realize that the same fate could befall salmon, shad, tuna, and other choice fishes, thus passing the radioactivity on to man.

Extremely peculiar symptoms occurred in people eating a specific variety of Greenland shark. They developed odd mannerisms, reported hallucinations, and showed signs of neurotic behavior. Once again it was found that the species in question was feeding in areas subject to heavy man-made pollution.

We cannot begin to guess at all the medical consequences of the pollution of our oceans. Veterinary medicine has been studied and practiced for centuries, but the causes of ailments in marine life have gone unnoticed for the most part. True, much is known about worms, scale lice, leeches, bloodsuckers, and thousands of other parasitic animals that infect much of our marine life. Skinning, gutting, chopping off the tail, cutting the head free, and washing will rid marine foods of most of these dangers to man. Also, numerous cysts, such as those caused by worms or protozoans, which develop in the flesh of food fish because of pollution, may be deemed harmless in themselves.

Unfortunately, although we know a great deal about the life cycles of many species of seafood, we understand practically nothing about the development of marine parasites. Many possess an extremely complex development pattern that might begin in snails, move on to lobsters, and mature in fish. Their poisonous aspects could be developed before, in, or after entering one of these hosts.

Most people are somewhat familiar with the expression "red tide." They know it as a blight that turns the sea red and washes great quantities of dead fish onto formerly attractive beaches. But few people understand its true cause. One of the earliest references to it dates from about 1491 B.C. The incident, according to the government publication *Poisonous and Venomous Marine Animals*, is recorded in the Bible (Exodus 7:20–21): "And all the waters that were in the river were turned to blood. And the fish that was in the river died; and the river stank, and the Egyptians could not drink of the water; and there was blood throughout all the land of Egypt."

The expression "red sea" was applied by ancient Greek authors to all the coasts of Arabia, apparently as a result of the numerous red-water blooms that occurred in this region. Red tides are mentioned in the *Iliad*, in the works of Tacitus, and in the logs of a number of sixteenth-century navigators.

Two questions then arise: *What* is a red tide? *Why* does it appear so suddenly?

The answer is that the water becomes discolored by massive numbers of microscopic single-celled organisms, of which there are thought to be about thirty thousand different species. These protozoans may cause ocean waters to turn yellow, brown, green, black, blue, or red. Red is the most common discoloration. The organisms are known to belong to the order Dinoflagellata and cause mass destruction of fish.

In its simplest of forms, this one-celled animal is caused by humans. To be a bit more specific, the normal method for determining the possible public health hazard in fish or shellfish is by the count of a relatively harmless bacteria called coliform. If the count exceeds Food and Drug Administration (FDA) standards, shellfish beds are closed to harvesting.

Runoff from land-based sewage outfalls are the major source of coliform bacteria and dinoflagellates, but ocean dumping of sewage sludge is also significant. The FDA found that ocean-bottom sediments up to six miles from the New York sludge-dump area contained coliform and dinoflagellate counts that exceeded permissible levels. On May 1, 1970, this specific area, 12 miles in diameter, and a similar area off Delaware Bay were closed to shellfishery. Clams harvested for sale in New York City contained coliform bacteria fifty to eighty times higher than the standard set by the FDA.

Hepatitis virus are carried by shellfish. A 1961 outbreak of infectious hepatitis was traced to raw shellfish taken from Raritan Bay in New Jersey. Shellfish have been found to contain polio virus in concentrations at least sixty times that of surrounding waters. White perch have become actively infected with human virus by exposure to human wastes, and they may transmit these infections. Exposure is sufficient for them to develop antibodies to such human diseases as tuberculosis, paratyphoid fever, bacillary dysentery, and a variety of other chronic infections.

Significant economic losses result from ocean pollution, especially in the commercially valuable fish or other seafood that are killed or rendered inedible. This represents serious social and financial losses to the many fishermen who live near the poverty level.

In 1939, the total catch of crabs, lobsters, shrimp, oysters, and scallops was 729 million pounds. Since one-fifth of the nation's 10 million acres of shellfish beds are now closed because of contamination, it can be estimated that the total catch would have been 181 million pounds higher. This estimate is probably low, since the closed areas, in lush estuarine systems close to large cities where they would have been harvested quickly, were particularly productive.

Dinoflagellates are single-celled microscopic organisms. Dense concentrations create red discoloration (red tide) and cause mass death of marine life. Here two species are magnified six hundred times.

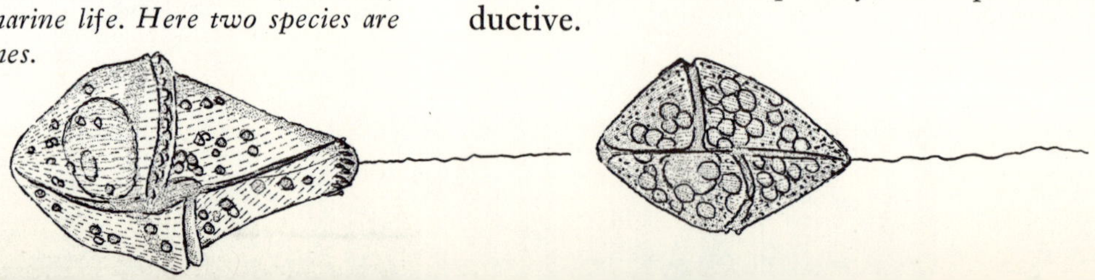

These fish were killed by a red tide said to have been triggered by man-made pollution. Photo, United States Naval Oceanographic Office, Washington, D.C.

A further ocean dumping cost is that of cleaning up or rehabilitating polluted beaches and shores. If projected increases in solid waste are dumped at sea, continuous and expensive clean-up operations will be required.

A report by the Council on Environmental Quality to the president of the United States about this vital matter of marine pollution stated that "knowledge of ocean pollution is rudimentary and generally it has not been possible to separate the effects of ocean dumping from the broader issue of ocean pollution. Yet one general conclusion is apparent: There is reason for significant concern. Dealing with ocean pollution requires that all sources be greatly reduced. If no action is taken and ocean dumping continues to increase, the long-term damage to the marine environment will be great."

Millions of dead fish mark an oncoming red tide. Photo, Saint Petersburg Times, Saint Petersburg, Florida.

The loss is well documented in San Francisco Bay. Prior to 1935, the annual harvest of soft-shell clams was between 100,000 and 300,000 pounds. Today clam digging in that area is virtually non-existent because of pollution. The annual commercial landings of shrimp prior to 1936 were as high as 6.5 million pounds; landings in 1965 were only 100,000 pounds.

Even where contamination levels are acceptable, the food may be discolored or tainted. Sludge decay can result in the production of hydrogen sulfide, which blackens the shells of clams and oysters and affects their taste and odor. Even in very small amounts, oil can taint the flesh of fish. The discharge residue from burning 2.6 gallons of a gasoline-oil mixture in an outboard motor has been proved to taint fish in 1 acre-foot of water.

Edible bivalves, like the ones on this Japanese oyster farm, are generally stationary and therefore susceptible to pollutants dumped into nearby waters. Photo, United States Naval Oceanographic Office, Washington, D.C.

22

Shellfish

When we speak of shellfish, we normally mean edible oysters, clams, mussels, and barnacles. These animals are known as mollusks and under most conditions are safe to eat. However, because such bivalves are generally stationary and feed on tiny plants and animals that float through the water, their condition is dependent upon the sanitation or wholesomeness of the ocean wherein they reside. These tiny edibles (dinoflagellates) on which shellfish live can be either a species of plant life, a form of animal matter or one that is part animal and part plant.

These dinoflagellates are an unusual group of single-celled microscopic organisms named protozoans. Dinoflagellates may possess both plant characteristics (such as chlorophyll and cellulose plates) and animal characteristics (such as the ingestion of food). Many marine forms are luminescent; when in great numbers, they are responsible for sheetlike luminescence seen in the water at night. In addition, great concentrations may create a red discoloration (called a red tide) that results in death to great masses of marine life. Other discolorations caused by these organisms may be yellow, green, or shades of brown; but the majority of these concentrations are nonpoisonous to marine life.

The red tide is the worst of the dinoflagellate-

There are many hundreds of species of conchs (top left) *and burrowing sand crabs, some of which are edible. Either could suddenly become toxic.* Photo, Department of the Interior, Washington, D.C.

Small shellfish such as these are food for other types of marine life. If they have been in contaminated waters, they might readily be toxic. Photo, Environmental Science Services Administration, Washington, D.C.

caused shellfish poisons. Outbreaks of shellfish poisoning in California have been caused by genus *Gonyaulax;* in Florida waters, the dinoflagellates causing trouble belong to the genus *Gymnodinium.* Sometimes one or another of these marine species will appear in concentrations of up to 50 million per quart of ocean water.

Late in 1946, the waters rimming the western shore of Florida took on a turbid reddish color. Fish began to die, indicating a definite toxicity in the water; an analysis showed a massive infestation of a one-celled marine organism, which of course was Dinoflagellata. By the time early spring arrived, the ocean water had literally turned thick and viscous with them, and oysters, clams, barnacles, turtles, fish, shrimp, and crabs were killed en masse. For a stretch of 60 miles, the beautiful Florida beaches were turned into a garbage heap. Spray from the water was so irritating to human eyes and flesh that shoreside schools and hotels had to close.

Seafood cannot be judged safe or poisonous by appearance alone. This stonecrab may or may not be a delectable morsel. Photo, National Oceanic and Atmospheric Administration, Washington, D.C.

A special frame of cultivated two-year-old flat oysters are readied for dissection in a Japanese contamination study. Photo, United States Naval Oceanographic Office, Suitland, Md.

In 1940, a serious outbreak of shellfish poisoning in California waters caused an intensive investigation to be launched. Dinoflagellates were the culprits.

Dinoflagellates are capricious. Sometimes they will sneak in to poison shellfish; at other times they will sweep through feeding grounds in suffocating waves. Frequently, one species will be proved definitely toxic, and a twin species, in the same area, will turn out to be harmless. It is not known for certain what causes them to grow in such abundance, nor what causes them to suddenly disappear. They come and go, mostly without warning.

Generally speaking, dinoflagellates are free-living and are widely distributed. Thus no sea is safe from their scourge and therefore no shellfish are free of the threat of toxicity. Skin divers, scuba divers, underwater swimmers, and people wading

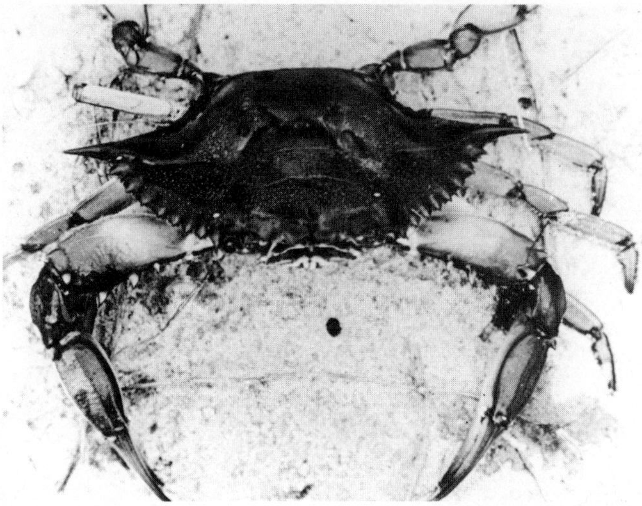

When eaten, this blue crab, if not fresh and properly cooked, could make a person sick. Photo, Department of the Interior, Washington, D.C.

the oceans of the world are discovering new shellfish to eat. Summer clams (*Schizothaerus nuttalli*), similar to those found along the northern shores of California, are also being discovered along the shores of northern Japan. Unfortunately, the Oriental variety suffer from a certain species of Dinoflagellata, which makes them uneatable during certain specific months. Microorganisms have been found in clams and oysters in one part of the world that are not known in another. Local fishermen can usually reveal any danger inherent in sectional shellfish. However, under normal conditions mollusks are generally considered edible. It is usually necessary to take extra care during the summer months, for it is during this time that numerous species of toxic Dinoflagellata abound.

There are three general types of shellfish poisoning that afflict man. First, there is the kind specifically caused by dinoflagellates, which has become more commonly known as clam, oyster, mussel, or gonyaulax poisoning. The most apparent symptom of this affliction is a tingling sensation, which the victim generally complains about in the area of the mouth and lips. This tingling spreads to other parts of the body and may result in paralysis. If the attack is sufficiently severe, death can follow. Second, there is gastrointestinal poisoning, which causes nausea, upset stomach, diarrhea, and gas pains. This is not known to be fatal. Third, there is an allergic type of shellfish poisoning, which causes rash, swelling, itching, headaches, and other symptoms common to allergy. This is not a common reaction and is probably suffered only by those individuals who have a specific allergy.

Unfortunately, it is impossible to provide any rule of thumb for judging the edibility of shellfish. Only local authorities are to be trusted. Pay attention to their announced quarantines.

Hermit crabs range in size from a few inches to over one foot in length. The larger species (Dardanus megistos) *are reputed to cause gastrointestinal poisoning during certain months of the year.* Photo, Naval Photographic Center, Washington, D.C.

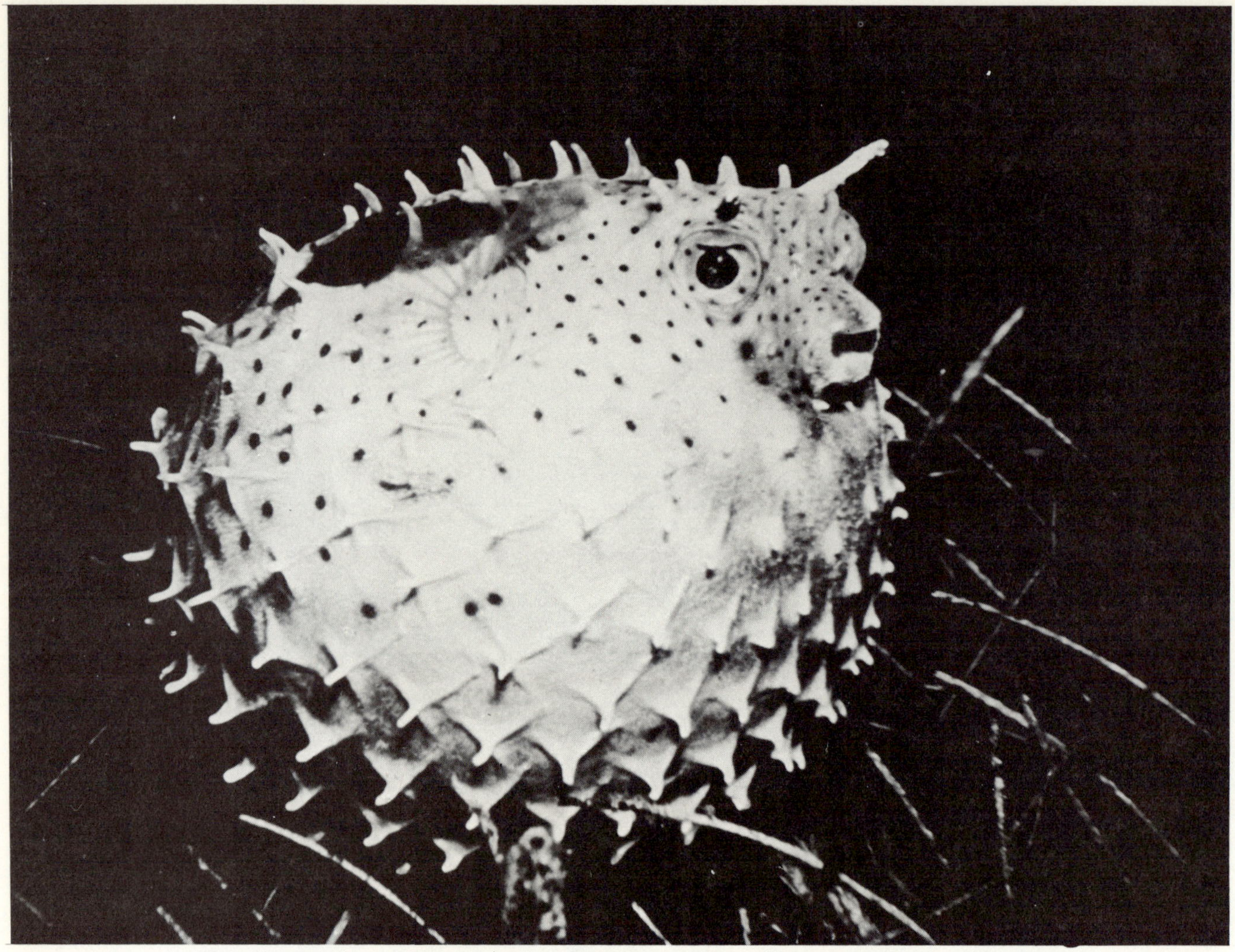

There are various species of porcupine puffer fishes. All are poisonous when eaten.
Photo, Department of the Interior, Washington, D.C.

23
Fish

The meat of some fish is known to be poisonous at all times of the year. The maki-maki of Hawaiian waters, numerous species of toadfish, filefish, and puffers, are all well established as killers when eaten by man. Ironically, although the puffer is rated as one of the most deadly, it is nevertheless valued as a food fish in Japan, where it is considered a delicacy.

Because the puffer, called fugu in Japan, holds a unique place in the history of poisonous fish, its background is worth discussing. Toward the end of the nineteenth century, concern over puff-fish poisoning developed in the Far East. Over a hundred Japanese were dying each year from eating fugu. Despite the fact that this fish (*Fugu rubripes rubripes*) was known to be toxic, people were willing to risk death for its unique taste.

One of the earliest Japanese reports on fugu poisoning was published in 1825, in the magazine *Toyen Shosetsu,* wherein the experimenter related how he fed poisonous puffers to cats and dogs. According to the detailed results, the cat vomited and lived, but the dog retained its food and died. From that time on, constant experiments were conducted and the results published. Regardless, humans still displayed a yearning for fugu.

There are nearly a hundred known species of puffers. All have the ability to inflate themselves by gulping large amounts of air or water. Photo, John G. Shedd Aquarium, Chicago, Illinois.

In most parts of the world, commercial fisheries regard the fugu species as a trash fish to be discarded. However, in Japan they are considered a lucrative byproduct of croaker fisheries. Although there is no single enterprise devoted exclusively to fugu fishing, the largest number are caught by trawlers working in the East China Sea. When netted, fugu are either iced or placed in saltwater tanks live for shipment to the larger cities. In recent months, it has been reported that in Tokyo alone, there are some two thousand dealers specializing in such palatable and uncommon species as puffers, porgies, and abalone.

Fugu restaurants are of a high order because only government-licensed restaurants are allowed to serve fugu. Cooks must display training certificates to prove that they have been properly schooled. A fugu enthusiast generally consumes about three pounds of meat at one sitting, and the aftereffects are notable because they leave a slight tingling in the extremities of the limbs.

In recent years the poison from the puff-fish has been widely used as a therapeutic agent in the treatment of asthma and numerous other disorders. Oriental physicians claim great successes for the poison as a pain reliever. They have found it to be effective in the treatment of migraine headaches.

Another edible fish, normally prized as seafood in the Orient, is the shark. Eating the flesh and liver of any shark, especially of large ones, should be questioned unless the fish is known to be safe. There are three steps that can be followed:

1. Feed a strip of the shark meat intended as food to a cat, dog, or other small meat-eating animal. A few hours of close observation might reveal if the meat is toxic or not. Obviously, if the animal becomes ill, the meat was tainted.

2. Shark meat should be cut into strips and hung to dry. However, this is not a simple procedure; great care must be taken to see that the pieces are allowed to drain off. In some instances, where the meat of shark has been allowed to dry too rapidly in a strong sun, a noticeable crust forms on the meat. This crust appears to seal the poisonous substance into the meat, thereby acting as a preservative. Therefore, in order to make certain that the poisonous juices are dried out of the meat, it should be rotated from time to time in a not-too-hot sun and in an area of strong winds. Reports say that Eskimos have found that shark meat cut in the fall, frozen in the winter, and hung to thaw in the spring is almost certain to be free of any toxicity. This could be because shark poisons are generally water-soluble and the ice formed within the meat assists in the draining action.

3. Because shark poison is water-soluble, frequent and energetic washings will help flush away any poisons. If the meat is boiled, the water should be changed at numerous intervals and the broth thrown away. Shark soup, made from uncertain meat, should be avoided under any circumstances. Nevertheless, in many lands, shark-fin soup is considered a delicacy.

Shark poisoning was first reported shortly before 1770 by a man named Sauvages. He related how an entire family had become ill after eating the liver of a shark. In 1780, a scientist named Fabricus publicly warned that if shark meat were to be eaten, it should be boiled in water that is changed frequently. In 1914, another scientist, named Jensen, wrote of Greenland sled dogs poisoned from eating local shark meat. Following World War I, a great deal more research was conducted on the toxicity of sharks, and both reports and warnings became quite common.

A porcupine fish, member of the puffer family, displays its dangerous barbs for an underwater cameraman. Photo, United States Naval Photographic Center, Washington, D.C.

The effects of eating poisonous shark meat are generally mild but, of course, depend on how much of the food is consumed. Partaking of toxic shark liver is more serious and can even result in death. Symptoms are quite obvious: vomiting, diarrhea, abdominal pain, and severe headaches. They generally develop about thirty minutes after eating. The recovery period can be a long one, up to two weeks.

Eskimos, shown here butchering shark meat for cooking, first leave it hanging for one entire winter. This seems to rid the meat of toxicity. Photo, Library of Congress, Washington, D.C.

In Japan, puff-fish, despite their known toxicity, are considered a delicacy. Photo, John G. Shedd Aquarium, Chicago, Illinois.

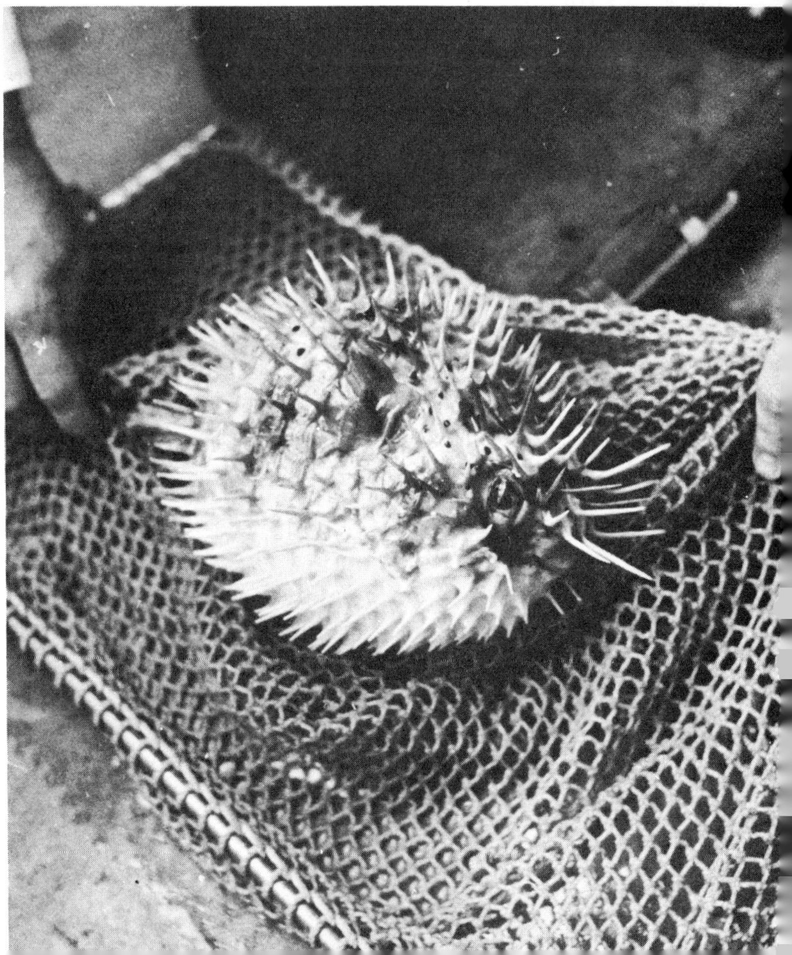

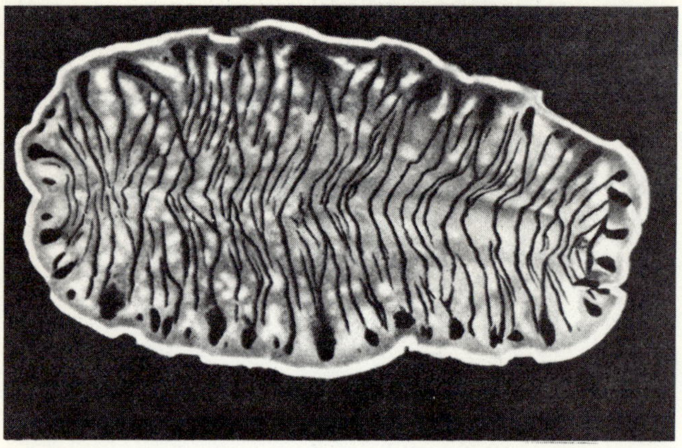

Parasitic marine flatworms, such as this 6-inch (family Pseudoceridae), which is light brown with dark stripes, come in thousands of shapes, colors, and sizes. When eaten by fish, they often cause the meat to become violently toxic to man.

There are many uncertainties concerning shark poisoning. For example, the liver of most sharks is warned against, but in one Greenland species of shark (*Somniosus microcephalus*), the poison is concentrated in the flesh, leaving the liver completely safe for eating.

From the few facts that are known about shark-meat poisoning, it can be concluded that safety rests in caution. Since this caution should be exercised with *all* shark meat, it is best that the casual observer be aware that the following more common sharks are indeed poisonous: blacktip reef shark, blue shark, soupfin shark, sharp-nosed shark, nurse shark, Greenland shark, seven-gilled shark, mud shark, and white shark; and there are many, many others. But names can be misleading. A name given to a harmless shark in one country may be given to a highly toxic shark in another country. The animal known as a white shark in Spain could be called a gray shark in Japan and a cow shark in England.

The ancient Greeks and Romans knew that skates and rays were poisonous. These animals were mentioned in the works of Rondelet and Oppian; in 1568, a man named Grevin wrote about them in his book *Deux Livres des Venins*. Following Grevin, the first of the more modern writers to deal with the subject, a number of books appeared over the years. In 1904 Gley wrote at some length about the poisonous effects of ray blood but failed to mention the meat.

Not much data are available on the lethal effects of eating skates or rays, as this type of bottom dweller is not particularly savored in most parts of the world. Additionally, those people who do eat rays and skates, are usually inhabitants of an economically depressed area. Educational levels are low; communications are nonexistent with outside people; and by the time word spreads about someone made ill from skate or ray meat, the true facts have long disappeared.

Eels that are poisonous to eat are thought to be mostly of the family Muraenidae, whose history extends into antiquity. Hippocrates, in the fifth century B.C., warned against the eating of moray eels. Also, moray eels had a unique place in the history of ancient man as a horror weapon. Pliny wrote about a wealthy Roman who possessed a man-made reservoir capable of supplying 6,000 morays for a single banquet. Because the meat of the moray was thought to grow more sweet through the consumption of human flesh, old slaves were frequently tossed alive to the hungry eels.

Little information is available on the feeding habits of morays and the causes that might turn their meat poisonous. The best advice is not to eat

In Spain, eels are trapped at night and rushed to the markets the next day. Once they are dead, the meat rapidly turns bad. Photo, Madrid Maritime Research Foundation, Madrid, Spain.

them. Many natives in various parts of the world do eat moray eels, but they are familiar with the nontoxic parts and know how to prepare the meat for safe eating.

"Ciguatera poisoning" is a much misused and misunderstood poisoning that has come to mean almost any kind of toxic reaction suffered through the eating of fish. The *Glossary of Oceanographic Terms*, published by the United States Naval Oceanographic Office, states that ciguatera is: "An intoxication in humans resulting from the ingestion [eating] of various tropical reef and inshore fishes and possibly certain invertebrates. The most common symptoms are tingling and numbing of the tips, tongue, hands, and feet, diarrhea, joint and muscular pain, burning urination, inability to coordinate voluntary muscular movements, and difficult breathing. The fatality rate is about 7 percent. The fishes involved usually are large carnivorous forms such as barracudas, snappers, groupers, and jacks. Many other common food fishes have been implicated in sporadic outbreaks of this type of poisoning, as well as certain marine snails and sea urchins."

In ancient times, around 300 B.C., Diphilus of Siphinium said that parts of the wrasses, if eaten by man, will cause "diarrhea and all the symptoms of cholera." Alexander the Great is reported to have forbidden his troops to eat certain toxic marine foods while in foreign lands. In the days of the *Odyssey*, ancient Greeks wrote of meat from certain fish that was known to inflict miseries on those who would partake. The early Greeks and Romans were not the only ones aware of toxicity in fish. A Chinese philosopher named Ch'en T'sang-ch'i, who lived during the ancient T'ang dynasty, wrote about a fish he called "yellow tail," which was known to cause sickness and death if eaten by humans.

An authority on ciguatera, Mr. John E. Randall, presented a review in the *Bulletin of Marine Sciences*, 1958. In this he outlined at some length certain observations that would reveal possible ciguatera in fish. Suspect fish are scattered over a wide area, having varied body shapes and characteristics. Their feeding and reproductive habits are remarkably diversified; as a result, it is almost impossible to generalize on what constitutes the "standard biology" of a ciguatoxic fish. However, based on Randall's work, the Naval Research Laboratory, has presented the following thirteen points that should be considered in a possible ciguatera fish:

1. Ciguatoxic fish are most often bottom dwellers, with some few being open-water species; all are generally found in depths of less than 300 feet.

2. Poisonous varieties are most often found in the vicinity of coral reefs. Fishes that are taken over grass, sand, or mud bottoms are less likely to be poisonous.

3. Most poisonous fish eaten by man are of moderate or large size. However, many ciguatoxic fish may be small.

4. Fishes that cause poisoning are generally carnivores or those that eat algae as opposed to plankton. However, of late, there are a significant number of plankton feeders that have also been incriminated.

5. There seems to be a correlation between the amount of fish in the diet and the degree of toxicity in poisonous fishes that feed primarily on fishes and crustaceans.

6. The food customs of surgeonfish and knowledge of their relative toxicity may shed light on the nature of the organism producing ciguatera toxin. Surgeonfish are an abundant reef fish, most often strongly toxic.

7. Some gastropods (those popularly called *shellfish* and such others as sea snails and sea slugs, and so forth) thought to be plant eaters have been reported poisonous.

8. The geographical distribution of poisonous fish is spotty. Fishes on one side of a small island may be dangerous to eat; whereas those on the other side of the island may be entirely safe.

9. This fact of area poisonings in fish indicates that they are not migratory.

10. A region once known to produce poisonous fish may turn safe, or exactly the reverse may occur.

11. Severe storms, hurricanes, earthquakes, or man-made catastrophic events (such as an underwater explosion) may produce a population of poisonous fish.

Neither this 8-foot green moray nor the reef fish shown with it are safe for man to eat. Both are considered toxic. Photo, Marine Studios, Marineland, Florida.

12. An area having freshwater drainage may support a poisonous fish population.

13. The majority, but not all, of poisonous fishes are found in the coral reef belt.

On the basis of these general observations, scientists have reached a number of conclusions. For example, it is obvious that some local food causes the meat of fish to be poisonous. Hence, it has been established that the basic poison is *benthic*—that is, stemming from fauna and flora of the sea bottom—on which count it would then probably be a fungus, a protozoan, or a bacterium. Taking this reasoning one step further, it can be assumed that the source of the poison is a blue green algae, members of the Cyanophyta, since this same group is involved in freshwater poisoning.

In *Poisonous and Venomous Marine Animals*, Halstead says that "ciguatera fish poisoning in its simplest uncomplicated form develops within 3 to

This brilliantly colored fish is a member of the genus Zebrasoma *and, like most reef fish, has meat that is deadly poisonous to man.* Drawing by the author.

5 hours after the fish is eaten. There is a sudden onset of abdominal pain followed by nausea, vomiting, and a watery diarrhea. The gastro-intestinal symptoms will occur in about 40 or 75 percent of the cases. The victim feels weak, generally ill, and may experience muscle aches throughout the back and thighs in about 10 percent or more of the cases. Soon after, the victim complains of numbness and tingling in and about the mouth which then extends to the extremities (present in about 50 percent or more of the cases). Fever, headache, and rash are generally absent, and the patient has no desire for food. The acute symptoms usually subside in about 8 to 10 hours, and within 24 hours after onset most of the patient's symptoms will have completely subsided except for a feeling of weakness. However, the numbness and tingling may continue to a lesser extent for a period of 4 to 7 days. . . . The foregoing résumé is typical of the majority of uncomplicated ciguatoxications that are generally encountered by the practicing physician in an endemic ciguatoxic locality such as the Marshall Islands."

In severe cases of poisoning, Halstead states, recovery is slow and may be prolonged. Convalescence is quite likely to be accompanied by extreme weakness, sensory disturbances and the loss of a considerable amount of weight. There are cases on record where patients required two or more years to make a complete physical recovery.

Regardless of the many statements current among knowledgeable fishermen, there is no dependable method of visually judging whether a fish is poisonous to eat. Some natives claim that poisonous fishes can be detected by observing the color of their throat, which is supposedly orange-colored and streaked with thin brown lines when the animal is toxic. Others say that the stomach should be opened, and if worms are present, it is safe to eat. Pacific Islanders will frequently place small pieces of the flesh from a fish in the vicinity of an ant hill and consider it toxic if the ants refuse to eat

Channel catfish and bluegills have been known to turn toxic after feeding on a lengthy diet of bottom-of-the-sea poisons. Drawing, Bureau of Sport Fisheries and Wildlife, Washington, D.C.

Beautifully colored reef fish called butterfly fish are toxic if eaten by man. Photo, John G. Shedd Aquarium, Chicago, Illinois.

it. Still others place a portion of the viscera out in the sun to dry, and if the flies refuse to settle on it, the meat is deemed poisonous. Black-colored teeth and a bitter-tasting liver are reputed to be evidences of toxicity in barracuda. If, when a fish is clean, the juices produce a burning or stinging sensation on the hands, the fish is said to be toxic. Some say to cook the fish and tear the flesh into bits; if small dark veins are present the fish is thought to be edible. Some people of the Marshall Islands place their live catch in brackish water heavy with organic matter for several days in order to let the fish detoxify. Okinawans are said to test for poisons by tasting the slime from a fish's eye, believing that a poisonous one will cause a tingling sensation. Red snappers whose pectoral fins extend

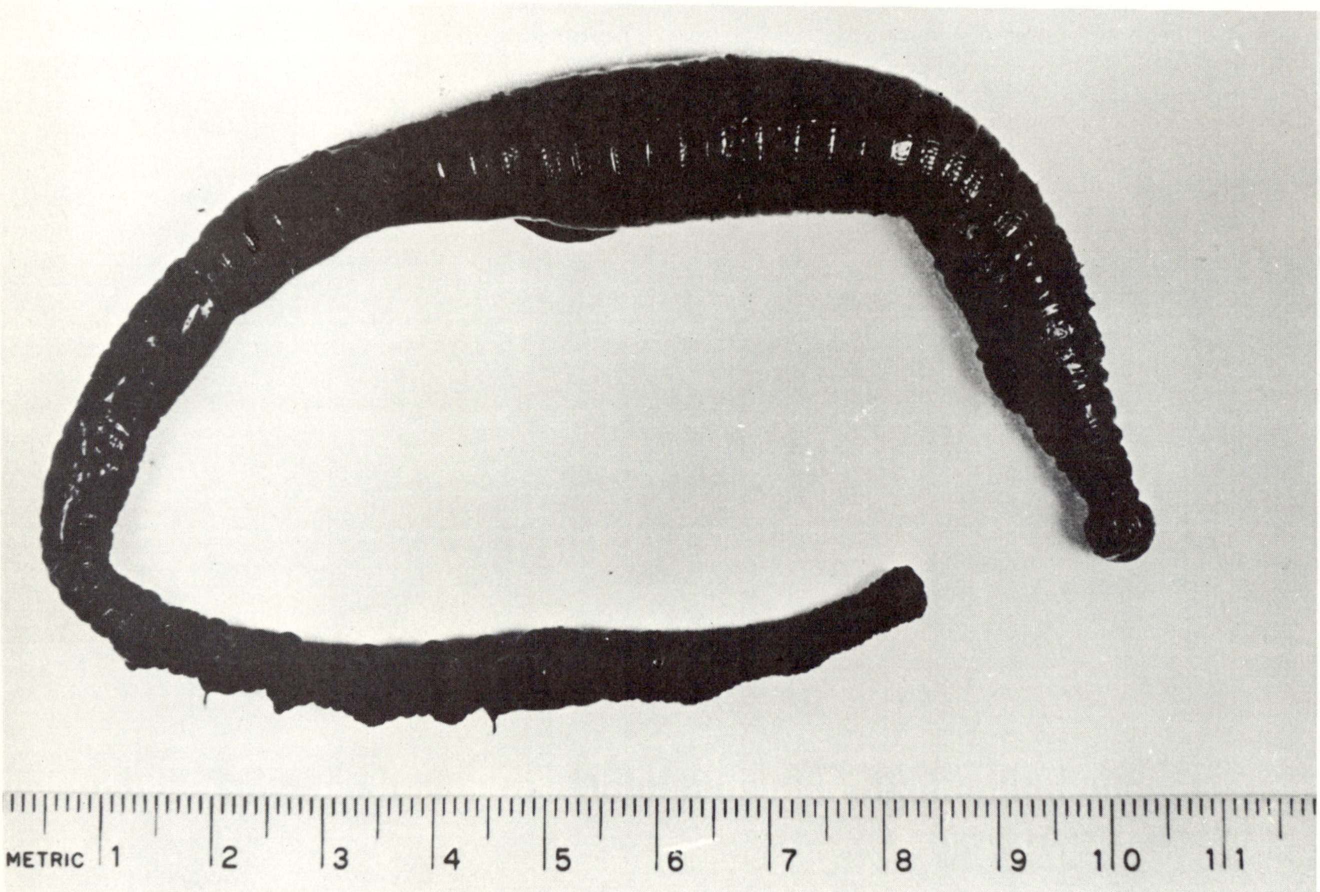

Man is not likely to consume marine worms, such as this lugworm, but their poisons can nevertheless be passed through several links in a food chain and thus be passed on to a human. Photo, Bureau of Commercial Fisheries, Washington, D.C.

Poisonous reef fish (bottom) *are seen here with a large edible tarpon.* Photo, Marine Studios, Marineland, Florida.

Queen-angle fish are beautifully colored, easy to catch, and deadly poisonous if eaten. Photo, Department of the Interior, Washington, D.C.

beyond the snout are reputedly poisonous. All fish from a specific location about an island are said to be toxic because of the appearance of the gills. It is believed by many in the West Indies and the Pacific Islands that if a silver coin is cooked with the fish, it will turn black if the fish is toxic. The Gilbertese say that if grated coconut is cooked inside the fish, it will turn green if the fish is toxic. Numerous other statements have appeared over the years among native groups, but most are based upon superstition and are not supported by any scientific evidence.

The U.S. Naval Research Laboratory, Washington, D.C., has suggested the five following steps in order that precautions may be taken against the eating of poisonous fish:

1. Never eat the viscera, liver, gonads, or intestines of tropical fishes.

2. The roe of any tropical fish is potentially dangerous.

3. Large reef fishes, such as snapper, barracuda, grouper, and jack should never be eaten.

4. Keep in mind that ordinary cooking, such as baking, frying, or drying, does not render a toxic fish safe to eat.

5. Tropical moray eels should never be eaten.

6. Eat only open-water fishes.

As a note of conclusion to the above, the Naval Research Laboratory adds that only about 500 species of marine fish are known to be toxic. Since there are an estimated 20,000 or more species of fish, the percentage of virulence is small.

The leatherback turtle, also called the leathery turtle, is the largest known species of sea turtle. It is somewhat rare and lives in tropical areas. Photo, Library of Congress, Washington, D.C.

24
Turtles

Members of the reptilian order Testudinata and Chelonia, commonly called *turtles*, but frequently referred to as *tortoises*, are especially numerous in warm waters. None are found in Arctic or Antarctic areas. All are seagoing, or pelagic, and therefore are found in many parts of the world. These marine animals display considerable diversity in structure, habits, and feeding patterns. The subclass to which turtles belong (*Chelonia*), may be divided into four suborders: 1. *Athecae*, in which the bony covering of the back is of a leathery consistency; 2. *Trionychoidea*, with a true back covering composed of dermal bones united at the ribs; 3. *Cryptodira*, in which the back is covered with horny plates; 4. *Pleurodira*, which is similar to *Cryptodira*, with slight differences in neck and pelvic structures. All of this suborder are freshwater types.

The huge leatherback, or leathery turtle (*Dermochelys coriacea*), is the largest of all contemporary turtles. Soft-shelled turtles (*Pelochelys bibroni*) belong to the family Trionychidae. They are freshwater turtles found in Africa, Asia, and North America. They have the plastron (breastplate) and carapace (back covering) incomplete and covered by a leathery skin. In this species, the neck is long and quite flexible, but it cannot be withdrawn into the shell for its entire length. This turtle has a small head, with a slender snout, at the tip of which is a pair of nostrils. The feet, in this type at least, are broadly webbed and have claws that are long and straight. Another type, *Amyda mutica*, has an upper jaw that is serrated and a carapace that is without tubercles (ridges). This particular variety is common in the Great Lakes region and portions of the Ohio River.

Aspidonectes spinifer is abundant in areas throughout the upper Mississippi River and the Great Lakes. They are aquatic, lay about fifty spherical, hard-shelled eggs, and spend much of their time buried in bottom mud. Carnivorous, they consume all kinds of fishes, frogs, mussels, snails, and insects. They are strong and durable swimmers.

Pleurodira, a turtle that inhabits the Amazon, is of great commercial value and is found in Australia and Africa as well as in South America. Unlike most turtles, it is strictly vegetarian.

It is to Cryptodira that most living tortoises and turtles belong. In this suborder, there are at least twelve families. The Cheloniidae includes the true sea turtles, which have flat limbs, inflexible paddles, and a head that cannot be retracted into the shell. All genera and species are widely distributed. The best-known species is the green turtle (*Chelonia mydas*), the flesh of which is renowned for soup. It is also recognized as the largest of all the sea turtles, for it frequently reaches 1,000 pounds. This green turtle is found in all parts of the world but is most common in the tropical archipelagos. They are hunted on moonlit nights, during summer months, when the females come ashore to lay their eggs. Hunters simply turn them on to their backs, where they remain helpless until they are taken away. Turtles of this species can survive for many weeks without water or food and are therefore easy to transport in the holds of small sailships. Prior to the advent of refrigeration, mariners often carried them in their ships as a reserve supply of fresh meat. Because of this feature, green turtles were hunted almost to extinction. For food, green turtles rely heavily on the roots of eelgrass during certain periods of each year; and because of this, they can be spotted by the action of the tops of the grass. Hunters go after them with both nets and spears. American markets generally receive their supplies of green turtle from the Florida reefs.

Still another sizable and valuable turtle is the one that supplies the world with tortoiseshell, namely the hawksbill. A third sea turtle is the loggerhead (*Caretta caretta*), which is more numerous along the coasts of the Middle Atlantic and New England states than are either of the others. It has no commercial value, frequently weighs as much as 500 pounds, and is sometimes passed off as a green turtle. However, unlike the green turtle, it has thin, nonimbricated plates of the carapace and has a central scale on the uppermost part of the head that is encircled by twelve small scales; whereas the green turtle has only seven. Furthermore, it has two claws on the feet, whereas the green turtle has just one.

There are numerous other species, such as snapping turtles, musk turtles, and diamondback terrapins, which are to be found in swamps, sluggish streams, and ponds. Our interest here remains with the true sea turtle and those most likely to be captured along coastal areas.

In this respect, the book *A Voyage Round the World*, written by British Admiral George Anson following his lengthy trip during the years of 1740 to 1744, told in some detail of two cases of poisoning resulting from the eating of marine turtles. This was the first reported incident of such poisoning. He said that in his travels within waters adjacent to Mexico, loggerhead turtles (*Caretta caretta gigas*), as well as trunk turtles (*Dermochelys coriacea*), were known to be both rank and unwholesome. The admiral further reported that Spanish explorers of the age deemed the green turtle, of the eastern Pacific region, also poisonous as food. Before they knew this, the admiral and his party ate green turtle and experienced no ill effects whatever.

Admiral Anson's report on poisonous turtle meat is thought to be the sole source of information prior to the middle of the eighteenth century.

Discoveries regarding poisonous turtles, at least during the 1700s, were mostly contributed by doctors, explorers, naturalists, historians, missionaries, and scientifically curious travelers making out-of-the-way voyages. No programed biological voyages were ever launched. Nevertheless, the geography and movements of poisonous turtles began to be compiled from case histories of turtle poisoning around the world. Poisoning from the eating of marine turtles, puff-fish, sharks, stingrays, and many other marine animals were also recorded. These reports came from New Caledonia, Vancouver Island, Japan, China, Latin America, and numerous other places.

In 1821, a French economist named Alexandre Moreau de Jonnes published a paper, "Sur les Poissons Toxicophores des Indes Occidentales." This treatise sought to determine the cause of turtle and fish poisoning by expostulating on his own previous experiments. Although he was neither a physician nor a biologist, he conducted many highly remarkable experiments within his capabilities. One of his more unusual tests had been to take coral thought to be poisonous, make it into a paste, feed it to fish, and then feed the flesh to animals to see if the meat of the fish still carried those same poisons.

According to different reports on the subject, turtle poisoning is usually felt a few hours after eating. Symptoms consist of nausea, diarrhea, vomiting, cramps in the upper abdominal area, dizziness, a burning sensation in the mouth, and a dryness of the throat. These seem to be the initial stages; the victim's condition may become progressively worse until death follows, after severe drowsiness.

Some sea turtles suspected of having a poisonous liver or other toxic parts are the leatherback turtle (*Dermochelys coriacea*), the green turtle (*Chelonia mydas*), and the hawksbill turtle (*Eretmochelys imbricata*).

Special Libraries and Aquariums

Amsterdam University Library, *Amsterdam, Netherlands*

Aquatarium, *Saint Petersburg Beach, Florida 33736*

Bermuda Aquarium, *Flatts, Bermuda*

Boston Public Library, *Boston, Massachusetts 02117*

Bounty Exhibit, The, *Saint Petersburg, Florida 33701*

British Guiana Aquarium, *Georgetown, British Guiana*

California Academy of Sciences, *San Francisco, California 94118*

Caribbean Aquarium, *Piscadera Bay, Curacao, Dutch West Indies*

Chicago Museum of Natural History, *Chicago, Illinois 60605*

Cincinnati, Zoological Society of, *Cincinnati, Ohio 45220*

Clearwater Sun Library, *Clearwater, Florida*

Comparative Zoology, Museum of, *Cambridge, Massachusetts*

Fish and Wildlife Service, *Washington, D.C.*

Fisheries Bureau Library, *Tokyo University, Japan*

Hawaii, University of, *Honolulu, Hawaii 96822*

Hope Zoo-Aquarium, *Jamaica, British West Indies*

International Oceanographic Foundation, *Key Biscayne, Florida 33149*

Jardin Zoologico "Bararida," *Barquisimeto, Venezuela*

Mairie de Saint-Malo Musées, *Saint-Malo, France*

Marineland, *Saint Augustine, Florida 32084*
Marineland of Pacific, *Palos Verdes Estates, California 90274*
Mariners Museum, *Newport News, Virginia 23606*
Mason County Historical Society, *Ludington, Michigan 49431*
Massachusetts Institute of Technology, *Cambridge, Massachusetts 02139*
Miami Seaquarium, *Miami, Florida 33149*
Mid-Pacific Sport Divers, *Wake Island 96930*
Mol, Thijs, *Dr Nolenslaan 5, Arnhem, Netherlands*
Montreal Aquarium, *Montreal, Canada*
Municipal Aquarium, *Mendoza, Argentina*
Museum of Fine Arts, *Boston, Massachusetts 02115*
Nasjonalgalleriet, *Oslo 1, Norway*
National Institutes of Health Library, *Bethesda, Maryland*
National Scheepvaartmuseum, *Antwerp, Belgium*
Naval Historical Foundation, *Washington, D.C. 20360*
Naval Oceanographic Office Library, *Suitland, Maryland 20390*
New England Aquarium, *Boston, Massachusetts 02110*
Niagara Falls Aquarium, *Niagara Falls, New York 14301*
Ocean World, *Fort Lauderdale, Florida 33316*
Ohio Historical Society, *Columbus, Ohio 43210*
Peabody Museum, *Salem, Massachusetts 01970*
Point Defiance Park Aquarium, *Tacoma, Washington*
Rhode Island University Library, *Narragansett, Rhode Island 02882*
Roosevelt Library, *Hyde Park, New York 12538*
Royal Scottish Museum, *Edinburgh, Scotland*
Sea-Arama Marineworld, *Galveston, Texas 77550*
Seamen's Church Institute Library, *New York, New York 10004*
Shedd Aquarium, *Chicago, Illinois 60605*
Steinhart Aquarium, *San Francisco, California 94118*
Taiwan University, *Taipei, Taiwan, China*
Tales of Cape Cod, Inc., *Hyannis, Massachusetts 02601*
Tokyo University of Mercantile Marine, *Fukagawa, Japan*
Vaughn Aquarium, Scripps Institute of Oceanography, *San Diego, California 92037*
Virginia Institute of Marine Science, *Gloucester Point, Virginia 23062*
West Indies Aquarium, *Kingston, West Indies*
Yale University Peabody Museum, *New Haven, Connecticut 06520*

Bibliography

ADAMS, W. H. *Mysteries of the Ocean.* London: T. Nelson, 1870.

ALPERS, A. *Dolphins: The Myth and the Mammal.* Boston: Houghton Mifflin, 1961.

ARNOV, B., JR. *Oceans of the World.* Indianapolis: Bobbs-Merrill, 1964.

BASCOM, W. *A Hole in the Bottom of the Sea.* New York: Doubleday, 1961.

BASSETT, F. S. *Legends and Superstitions of the Sea and Sailors.* London: Sampson-Low, 1885.

BATCHELDER, EUGENE. *A Romance of the Sea Serpent, or the Ichtyosaurus.* Cambridge: Cambridge University Press, 1850.

BENWELL, G. *The Sea Enchantress: The Tale of the Mermaid.* London: 1961.

BOUDET, JACQUES. *Man and Beast.* New York: Golden Press, 1964.

BREDER, C. M., JR. *Marine Fishes of the Atlantic Coast.* New York: Putnam's, 1948.

BUCHSBAUM, RALPH. *Animals Without Backbone.* Chicago: University of Chicago Press, 1938.

Buehr, Walter. *Sea Monsters.* New York: Norton, 1966.
Caras, Roger A. *Dangerous to Man.* Philadelphia: Chilton, 1964.
Carrington, R. *A Biography of the Sea.* New York: Basic Books, 1960.
———. *Mermaids and Mastodons.* New York, Rinehart, 1957.
Carson, R. *The Sea Around Us.* New York: Oxford University Press, 1961.
Cassel, Bo. *Havet Dykaren Fynden Doktryckeriet.* Stockholm, 1967.
Challanger Society. Sciences of the Sea. New York: Dutton, 1912.
Chapman, V. J. *Seaweeds and Their Uses.* London: Methuen Co. Ltd., 1950.
Clarke, W. D. *The Young Pathfinders Book of Oceans.* New York: Hart, 1964.
Cousteau, Jacques-Yves. *The Living Sea.* New York: Harper & Row, 1963.
D'Arcs, T. W. *Science of the Sea.* London, 1912.
Darling, Lois, and Darling, Louis. *Sea Serpents Around Us.* Boston: Little Brown, 1965.
Daugherty, C. M. *Searchers of the Sea.* New York: Viking, 1961.
Del Rey, L. *The Mysterious Sea.* New York: Chilton, 1961.
Dietrich, Gunter. *General Oceanography.* New York: Interscience, 1963.
Dinsdale, Tim. *Loch Ness Monster.* Philadelphia: Chilton, 1961.
Dugan, J. *Man Under the Sea.* New York: Collier, 1964.
Dunkel, Ulrich. *Abenteuer mit Seeschlangen.* Stuttgart: Kreuz-Verlag, 1961.
Ekman, Sven. *Zoogeography of the Sea.* London: Sidgwick and Jackson, 1953.
Forsee, A. *Beneath Land and Sea.* Philadelphia: Macrae Smith, 1962.
Gaskell, T. F. *World Beneath the Oceans.* New York: Natural History Press, 1964.
Gilbert, Perry W. *Sharks and Survival.* Boston: Heath, 1963.
Gould, R. T. *The Case of the Sea Serpent.* New York: Putnam's, 1934.
———. *Enigmas: Another Book of Unexplained Facts.* New Hyde Park, New York: University Books, 1965.
Halstead, B. W. *Poisonous and Venomous Marine Animals of the World.* Washington, D.C.: U.S. Government Printing Office, 1967.
Hanna, S. W. Bulletin No. 26 of U.S. Fish Commission. Washington, D.C., 1883.
Hardy, A. C. *The Open Sea: Its Natural History.* Boston: Houghton Mifflin, 1956.
Helm, Thomas. *Monsters of the Deep.* New York: Dodd, Mead, 1962.
———. *Shark.* New York: Dodd, Mead, 1961.
Heuvelmans, Bernard. *On the Track of Unknown Animals.* New York: Hill & Wang, 1962.
———. *In the Wake of the Sea Serpent.* New York: Hill & Wang, 1965.
Hull, S. *The Bountiful Sea.* Englewood Cliffs, N.J.: Prentice-Hall, 1964.
Idyll, C. P. *Abyss: The Deep Sea and the Creatures That Live in It.* New York: Crowell, 1964.
Jargerstein, Gosta, and Nelson, L. *Life in the Sea.* New York: Basic Books, 1964.
Knowlton, William. *Sea Monsters.* New York: Knopf, 1959.
Koch-Isenburg, Ludwig. *Der Jumgend-Brehm.* Minden, Germany: Mindener Tagblatt, n.d.
Lancaster, E. R. *A Treatise on Zoology.* London: Adam and Charles Black, 1906.
Lane, Frank W. *Kingdom of the Octopus.* London: Jarrolds, 1957.

LEE, HENRY. *Seamonsters Unmasked.* London: Clowes, 1883.

LEY, WILLY. *Lungfish and Unicorn.* New York: Viking, 1948.

LILLY, J. C. *Man and Dolphin.* Garden City, N.Y.: Doubleday, 1961.

LONG, JOHN E. *Ocean Sciences.* Annapolis, Md.: Naval Institute, 1964.

MACGINITIE, G. E., and MACGINITIE, NETTIE. *Natural History of Marine Animals.* New York: McGraw-Hill, 1959.

MARMER, H. A. *The Sea.* New York: Appleton, 1930.

———. *The Tide.* New York: Appleton, 1926.

MARSHALL, N. B. *Aspects of Deep Sea Biology.* New York: Philosophical Library, 1954.

MILLER, ROBERT L. *Papers in Marine Biology.* New York: Macmillan, 1964.

MINER, R. W. *Field Book of Seashore Life.* New York: Putnam's, 1950.

NICOL, JOSEPH A. C. *Biology of Marine Animals.* New York: Interscience, 1960.

NIKOLSKY, G. V. *Ecology of Fishes.* New York: Academic Press, 1962.

NORMAN, J. R. *A History of Fishes.* New York: A. A. Wyn, 1948.

OKADA, Y. *Fishes of Japan.* Tokyo: Maruzen Co., 1955.

OUDEMANS, A. C. *The Great Sea Serpent.* Published by the author, 1892.

PERLMUTTER, ALFRED. *Guide to Marine Fishes.* New York: University Press, 1961.

RAMANOVSKY, V. *La Mer.* Paris: Librairie Larousse, 1953.

RANG, P. C., and SOULAYET, L. *Histoire Naturelle des Mollusques Pteropodes.* Paris: J. B. Ballière, 1852.

ROUGHLY, T. C. *Wonders of the Great Barrier Reef.* New York: Scribner's, 1959.

RUSSELL, SIR F. S. *Advances in Marine Biology.* London: Academic Press, 1965.

SHEPARD, FRANCIS P. *Submarine Geology.* New York: Harper & Row, 1963.

SMITH, J. L. B. *The Search Beneath the Sea.* New York: Henry Holt, 1956.

SNOW, E. R. *Incredible Mysteries and Legends of the Sea.* New York: Dodd, Mead, 1967.

SOULE, GARDNER. *The Maybe Monsters.* New York: Putnam's, 1963.

SPILHAUS, A. *Turn to the Sea.* Washington, D.C.: National Research Council, 1965.

SWEENEY, JAMES B. *A Pictorial History of Oceanographic Submersibles.* New YORK: Crown, 1970.

VAN DERVOORT, J. W. *The Water World.* New York: Union Publishing House, 1886.

VILLIERS, A. *The Ocean: Man's Conquest of the Sea.* New York: Dutton, 1963.

VON BRONSART, HUBERTA. *Judend am Fernohr.* Frankfort, Germany: Neue Press, n.d.

WALFORD, L. A. *Living Resources of the Sea.* New York: Ronald Press, 1958.

WEYER, E. M., JR. *Strangest Creatures on Earth.* New York: Sheridan House, 1953.

WHYMPER, FRED. *Romance of the Sea: Its Fictions, Facts and Folklore.* Young & Co., 1896.

WILLISTON, SAMUEL WENDELL. *Water Reptiles of the Past and Present.* Chicago: University of Chicago Press, 1914.

WILSON, ANDREW. *Sea Serpents of Science.* John B. Alden, 1883.

YASSO, WARREN E. *Oceanography.* New York: Holt, Rinehart & Winston, 1965.

ZIM, H. S., and INGLE, L. *Seashores.* New York: Golden Press, 1955.

Index

Illustrations are indicated by italic numbers.